Communications in Computer and Information Science 1100

Commenced Publication in 2007
Founding and Former Series Editors:
Phoebe Chen, Alfredo Cuzzocrea, Xiaoyong Du, Orhun Kara, Ting Liu,
Krishna M. Sivalingam, Dominik Ślęzak, Takashi Washio, and Xiaokang Yang

More information about this series at http://www.springer.com/series/7899

Michael W. Berry · Bee Wah Yap ·
Azlinah Mohamed · Mario Köppen (Eds.)

Soft Computing
in Data Science

5th International Conference, SCDS 2019
Iizuka, Japan, August 28–29, 2019
Proceedings

 Springer

Editors
Michael W. Berry
University of Tennessee
Knoxville, TN, USA

Azlinah Mohamed
Universiti Teknologi MARA
Shah Alam, Selangor, Malaysia

Bee Wah Yap
Universiti Teknologi MARA
Shah Alam, Selangor, Malaysia

Mario Köppen
Kyushu Institute of Technology
Fukuoka, Japan

ISSN 1865-0929   ISSN 1865-0937   (electronic)
Communications in Computer and Information Science
ISBN 978-981-15-0398-6   ISBN 978-981-15-0399-3   (eBook)
https://doi.org/10.1007/978-981-15-0399-3

This Springer imprint is published by the registered company Springer Nature Singapore Pte Ltd.
The registered company address is: 152 Beach Road, #21-01/04 Gateway East, Singapore 189721, Singapore

Preface

We are pleased to present the proceeding of the Fifth International Conference on Soft Computing in Data Science 2019 (SCDS 2019). SCDS 2019 was a collaboration with Kyushu Institute of Technology and was held at the Centre of Iizuka Research and Development, in Iizuka, Japan, during August 28–29, 2019. The theme of the conference was "Science in Analytics: Harnessing Data and Simplifying Solutions." SCDS 2019 aimed to provide a platform for knowledge sharing on theory and applications of advanced technologies and techniques for Big Data Analytics (BDA). Universities play an important role in producing data professionals and data scientists who are in great demand by the industries which need to harness their data to gain competitive advantage and to improve their business process, efficiency, and productivity.

The papers in this proceeding cover the issues, challenges, theory, and innovative applications of data science and big data analytics including, but not limited to: data capture and storage; information and customer analytics; data visualization; image processing; data mining tools and techniques; big data technologies; machine learning and deep learning algorithms for big data. For the advancement of society in the 21st century, there is a need to transfer knowledge and technology to industrial applications to solve real-world problems that benefit the global community. Research collaborations between academia and industry can lead to the advancement of innovative analytics and computing applications to facilitate real-time insights and solutions.

We were delighted this year to have received paper submissions from a diverse group of national and international researchers. We received 75 paper submissions, among which 30 were accepted. SCDS 2019 utilized a double-blind review procedure. All accepted submissions were assigned to at least two independent reviewers in order to ensure a rigorous, thorough, and convincing evaluation process. A total of 74 reviewers were involved in the review process. The conference proceeding volume editors and Springer CCIS Editorial Board made the final decisions on acceptance, with 30 of the 75 submisssions (40%) being published in the conference proceedings.

We would like to thank the authors who submittted manuscripts to SCDS 2019. We thank the reviewers for voluntarily spending time to review the papers. We thank all conference committee members for their time, ideas, and efforts in ensuring the success of SCDS 2019. We also wish to thank the Springer CCIS Editorial Board, organizations, and sponsors for their continuous support. We sincerely hope that SCDS 2019 has provided a venue for knowledge sharing, publication of good research findings, and new research collaborations. Last but not least, we hope everyone gained from the keynote and parallel sessions, and had an enjoyable and memorable experience at SCDS 2019 and in Iizuka, Japan.

August 2019

Michael W. Berry
Bee Wah Yap
Azlinah Mohamed
Mario Köppen

Organization

Patron

Mohd Azraai Kassim   Universiti Teknologi MARA, Malaysia
  (Vice-chancellor)

Honorary Chairs

Haryani Haron   Universiti Teknologi MARA, Malaysia
Takeshi Yamakawa   Kyushu Institute of Technology, Japan
Azlinah Mohamed   Universiti Teknologi MARA, Malaysia
Michael W. Berry   University of Tennessee, USA
Surina Shukri   Malaysia Digital Economy Corporation, Malaysia

Conference Chairs

Bee Wah Yap   Universiti Teknologi MARA, Malaysia
Mario Köppen   Kyushu Institute of Technology, Japan

Secretary

Siti Shaliza Mohd Khairi   Universiti Teknologi MARA, Malaysia

Secretariat

Mohd Alias Jais   Universiti Teknologi MARA, Malaysia
Zuraini Dollah   Universiti Teknologi MARA, Malaysia
Norkhalidah Mohd Aini   Universiti Teknologi MARA, Malaysia
Natsuki Kouyama   Kyushu Institute of Technology, Japan

Finance Committee

Sharifah Aliman   Universiti Teknologi MARA, Malaysia
Nur Huda Nabihan Md Shaari   Universiti Teknologi MARA, Malaysia
Azira Mohamed Amin   Universiti Teknologi MARA, Malaysia
Muhd Azuzuddin Muhran   Universiti Teknologi MARA, Malaysia

Technical Program Committee

Azlin Ahmad	Universiti Teknologi MARA, Malaysia
Dhiya Al-Jumeily	Liverpool John Moores University, UK
Shamimi A. Halim	Universiti Teknologi MARA, Malaysia
Ezzatul Akmal Kamaru Zaman	Universiti Teknologi MARA, Malaysia

Sponsorship Committee

Nuru'l-'Izzah Othman	Universiti Teknologi MARA, Malaysia
Norhidayah A. Kadir	Universiti Teknologi MARA, Malaysia
Norhayati Shuja'	Jabatan Perangkaan Malaysia
Haliza Abdul Hamid	Universiti Teknologi MARA, Malaysia
Mathuri A/P Selvarajoo	Universiti Teknologi MARA, Malaysia

Publication Committee (Program Book)

Zainura Idrus	Universiti Teknologi MARA, Malaysia
Nur Atiqah Sia Abdullah	Universiti Teknologi MARA, Malaysia
Marshima Mohd Rosli	Universiti Teknologi MARA, Malaysia

Website Committee

Mohamad Asyraf Abdul Latif	Universiti Teknologi MARA, Malaysia

Publicity and Corporate Committee

Nur Aziean Mohd Idris	Universiti Teknologi MARA, Malaysia
Ezzatul Akmal Kamaru Zaman	Universiti Teknologi MARA, Malaysia
Kenichi Kourai	Kyushu Institute of Technology, Japan

Media/Photography/Montaj Committee

Marina Ismail	Universiti Teknologi MARA, Malaysia
Norizan Mat Diah	Universiti Teknologi MARA, Malaysia
Masahiro Shibata	Kyushu Institute of Technology, Japan

Logistics Committee

Masato Tsuru	Kyushu Institute of Technology, Japan
Kazuya Tsukamoto	Kyushu Institute of Technology, Japan
Hamdan Abdul Maad	Universiti Teknologi MARA, Malaysia
Abdul Jamal Mat Nasir	Universiti Teknologi MARA, Malaysia

International Scientific Committee

Dhiya Al-Jumeily	Liverpool John Moores University, UK
Adel Al-Jumaily	University of Technology Sydney, Australia
Chidchanok Lursinsap	Chulalongkorn University, Thailand
Agus Harjoko	Universitas Gadjah Mada, Indonesia
Sri Hartati	Universitas Gadjah Mada, Indonesia
Min Chen	Oxford University, UK
Simon Fong	University of Macau, China
Mohammed Bennamoun	University of Western Australia, Australia
Yasue Mitsukura	Keio University, Japan
Richard Weber	University of Chile, Santiago, Chile
Jose Maria Pena	Technical University of Madrid, Spain
Yusuke Nojima	Osaka Perfecture University, Japan
Siddhivinayak Kulkarni	University of Ballarat, Australia
Tahir Ahmad	Universiti Teknologi Malaysia, Malaysia
Daud Mohamad	Universiti Teknologi MARA, Malaysia
Mazani Manaf	Universiti Teknologi MARA, Malaysia
Jasni Mohamad Zain	Universiti Teknologi MARA, Malaysia
Suhartono	Insititut Teknologi Sepuluh Nopember, Indonesia
Wahyu Wibowo	Insititut Teknologi Sepuluh Nopember, Indonesia
Edi Winarko	Universitas Gadjah Mada, Indonesia
Retantyo Wardoyo	Universitas Gadjah Mada, Indonesia
Dittaya Wanvarie	Chulalongkorn University, Thailand
Jaruloj Chongstitvatana	Chulalongkorn University, Thailand
Pakawan Pugsee	Chulalongkorn University, Thailand
Krung Sinapiromsaran	Chulalongkorn University, Thailand
Saranya Maneeroj	Chulalongkorn University, Thailand
Saiful Akbar	Institut Teknologi Bandung, Indonesia

Additional Reviewers

Albert Guvenis	Bogazici University, Turkey
Ali Qusay Al-Faris	University of the People, USA
Aida Mustapha	Universiti Tun Hussein Onn Malaysia, Malaysia
Ahmad Farid Abidin	Universiti Teknologi Mara, Malaysia
Ahmad Nazim Aimran	Universiti Teknologi Mara, Malaysia
Azlan Iqbal	Universiti Tenaga Nasional, Malaysia
Azlan Ismail	Universiti Teknologi MARA, Malaysia
Azlin Ahmad	Universiti Teknologi MARA, Malaysia
Bee Wah Yap	Universiti Teknologi MARA, Malaysia
Bong Chih How	Universiti Malaysia Sarawak, Malaysia
Chew XinYing	Universiti Sains Malaysia, Malaysia
Chidchanok Lursinsap	Chulalongkorn, University, Thailand
Chin Kim On	Universiti Malaysia Sabah, Malaysia
Daud Mohamad	Universiti Teknologi MARA, Malaysia
Dedy Dwi Prastyo	Institut Teknologi Sepuluh Nopember, Indonesia

Deepti Prakash Theng	G. H. Raisoni College of Engineering and RTMNU, India
Dhiya Al-Jumeily	Liverpool John Moores University, UK
Dittaya Wanvarie	Chulalongkorn University, Thailand
Edi Winarko	Universitas Gadjah Mada, Indonesia
Ely Salwana	Universiti Kebangsaan Malaysia, Malaysia
Eng Harish Kumar	King Khalid University, Saudi Arabia
Fam Soo Fen	Universiti Teknikal Malaysia, Malaysia
Hanaa Ali	Zagazig University, Egypt
Hamidah Jantan	Universiti Teknologi MARA, Malaysia
Hamzah Abdul Hamid	Universiti Malaysia Perlis, Malaysia
Hasfazilah Ahmat	Universiti Teknologi MARA, Malaysia
Izzatdin Abdul Aziz	Universiti Teknologi PETRONAS, Malaysia
J. Vimala Jayakumar	Alagappa University and Karaikudi, India
Karim Hashim Al-Saedi	University of Mustansiriyah, Iraq
Layth Sliman	EFREI-Paris, France
Lee How Chinh	Universiti Tunku Abdul Rahman, Malaysia
Leong Siow Hoo	Universiti Teknologi MARA, Malaysia
Liong Choong-Yeun	Universiti Kebangsaan Malaysia, Malaysia
Mario Köppen	Kyushu Institute of Technology, Japan
Mas Rina Mustaffa	Universiti Putra Malaysia, Malaysia
Mashitoh Hashim	Universiti Pendidikan Sultan Idris, Malaysia
Mazani Manaf	Universiti Teknologi MARA, Malaysia
Michael Loong Peng Tan	Universiti Teknologi Malaysia, Malaysia
Mohamed Imran Mohamed Ariff	Universiti Teknologi MARA, Malaysia
Mohd Fadzil Hassan	Universiti Teknologi PETRONAS, Malaysia
Mohd Zaki Zakaria	Universiti Teknologi MARA, Malaysia
Neelanjan Bhowmik	Durham University, UK
Ng Kok Haur	University of Malaya, Malaysia
Nikisha Jariwala	VNSGU, India
Noryanti Muhammad	Universiti Malaysia Pahang, Malaysia
Noor Azilah Muda	Universiti Teknikal Malaysia, Malaysia
Nor Azizah Ali	Universiti Teknologi Malaysia, Malaysia
Nor Saradatul Akmar Zulkifli	Universiti Malaysia Pahang, Malaysia
Norshita Mat Nayan	Universiti Kebangsaan Malaysia, Malaysia
Nursuriati Jamil	Universiti Teknologi MARA, Malaysia
Ong Seng Huat	Universiti Malaya, Malaysia
Pakawan Pugsee	Chulalongkorn University, Thailand
Peraphon Sophatsathit	Chulalongkorn University, Thailand
Puteri Nor Ellyza Nohuddin	Universiti Kebangsaan Malaysia, Malaysia
Retantyo Wardoyo	Universitas Gajah Mada, Indonesia
Richard C. Millham	Durban University of Technology, South Africa
Rizauddin Saian	Universiti Teknologi MARA, Malaysia
Rodrigo Campos Bortoletto	São Paulo Federal Institute of Education, S&T, Brazil

Roselina Sallehuddin	Universiti Teknologi Malaysia, Malaysia
Ruhaila Maskat	Universiti Teknologi MARA, Malaysia
Saiful Akbar	Institut Teknologi Bandung, Indonesia
Saranya Maneeroj	Chulalongkorn University, Thailand
Sayang Mohd Deni	Universiti Teknologi MARA, Malaysia
Shamimi A. Halim	Universiti Teknologi MARA, Malaysia
Shuzlina Abdul-Rahman	Universiti Teknologi MARA, Malaysia
Siddhivinayak Kulkarni	Griffith University, Australia
Siti Sakira Kamaruddin	Universiti Teknologi MARA, Malaysia
Sofianita Mutalib	Universiti Teknologi MARA, Malaysia
Sri Hartati	Gadjah Mada University, Indonesia
Suhartono	Institut Teknologi Sepuluh Nopember, Indonesia
Teh SinYin	Universiti Sains Malaysia, Malaysia
Wahyu Wibowo	Institut Teknologi Sepuluh Nopember, Indonesia
Widhyakorn Asdornwised	Chulalongkorn University, Thailand
Zainura Idrus	Universiti Teknologi MARA, Malaysia

Organized by

Hosted by   Technical Co-sponsor

In Co-operation with

Supported by

Contents

Big Data Analytics

Computational and Artificial Intelligence

Social Network and Media Analytics

Information and Customer Analytics

Entropy Analysis of Questionable Text Sources by Example of the Voynich Manuscript

Natsuki Kouyama$^{(\boxtimes)}$ and Mario Köppen

Kyushu Institute of Technology, 680-4 Kawazu, Iizuka, Fukuoka 820-8502, Japan
kouyama.natsuki731@mail.kyutech.jp, mkoeppen@ieee.org

Abstract. The Voynich Manuscript (referred to as 'VMS') can be considered as one of the oldest puzzles which remained unsolved until now. While some say VMS is genuine, others say that it is just a hoax. In this research, we propose three methods to analyze VMS and verify the effectiveness of them. The reason for using this manuscript is that no one on this planet knows the meaning of the text of VMS, so we can tackle research without being biased. We analyze from the viewpoint of character frequency and entropy of VMS. Statistical analysis of word frequency already exists, in contrast, we adopt analysis based on a character unit. In that respect, there is diversity from other research. We stated the following three hypotheses as H1 through H3. H1: VMS is a Ciphertext. H2: VMS is close to a programming language. H3: VMS is close to a natural language. The experimental results demonstrate that our methods are valid and efficient. The methods can be applied to all text sources and different classes of them can be distinguished. The possibility that VMS is likely a Ciphertext can be rejected. This mysterious manuscript can be concluded as a meaningful human art, not a hoax. Moreover, the VMS is not encrypted. (However, we are not able to exclude being encoded.)

Keywords: Voynich manuscript · Character frequency analysis · Entropy analysis · Ciphertext analysis · Encrypted text analysis · Shannon entropy · Rényi entropy

1 Introduction

The Voynich Manuscript (referred to as 'VMS') is one of the most mysterious manuscripts in the world. Lots of people have tried to decipher it so far, nevertheless, even today no one has been successful. While some say VMS is genuine, others say that it is just a hoax.

Figure 1 refer to provide you more specific image of VMS. These figures are showing that VMS has a number of different genres in their pages [1]. In this research, we propose three types of methods to analyze VMS and verify

© Springer Nature Singapore Pte Ltd. 2019
M. W. Berry et al. (Eds.): SCDS 2019, CCIS 1100, pp. 3–13, 2019.
https://doi.org/10.1007/978-981-15-0399-3_1

(1) Naked ladies bathing in $f.84r$.

(2) Recipes-like illustration in $f.99r$.

Fig. 1. Example pages of VMS.

the effectiveness of them. The reason for using this manuscript for research is that no one on this planet knows the meaning of the text of VMS, so we can tackle research without being biased. We analyze from the viewpoint of character frequency and entropy of VMS. Statistical analysis of word frequency already exists, in contrast, we adopt analysis based on a character unit. Therefore, in that respect, we can say there is diversity from other research. Is VMS, which is the text with unknown writing system, meaningful or not? Besides, is it encrypted or not? Under these circumstances, we stated the following three hypothesis as H_1 though H_3.

Hypothesis 1 (H_1): VMS is a Ciphertext.
Hypothesis 2 (H_2): VMS is a programming language.
Hypothesis 3 (H_3): VMS is a natural language.

2   Related Work

As a related work, majority article adopted word unit based analysis. For example, [2] addresses analyzing VMS in his study as well. However, his analysis focused on the words unit. A matter of concern here is: what if the place of the space was wrong? For example below, in the text of type A, space positioned into the right place. However, type B, the same number of spaces positioned completely different places. Although it is surprisingly simple, we assume that it is enough to create Ciphertext.

A. entropy analysis of the example of the voynich manuscript
B. ent ropya nalysiso fthee xamp leofth evoyni chman uscript

For those reasons, we adopt to analyze VMS with the character unit in our project.

3 Materials and Methodology

3.1 Materials

Transcription. It is impossible to read the text of VMS. However, there is a transcription, which is a conversion of the handwritten text of VMS into a computer-readable format. A is the original, and B is the transcription of figure A. The transcription that has been adopted for this project is based on FSG because of its simplicity and available to download the file even today (Fig. 2).

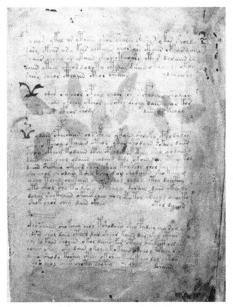

(1) First page of VMS   (2) FSG transcription of first page of VMS

Fig. 2. Manuscript vs transcript of first page of VMS.

Source Texts. To verify the hypothesis, we prepared 3 classes of the source text. As H1, an original encrypted text was created using a website that enables a plaintext encryption and decryption. And as an example of random numbers, we used 338 of IQOS [3] package code. And 5 Programming languages for H2, from Github and our own programs. And also, 5 natural languages for H3, which are randomly picked sentences from Wikipedia pages of each language (Table 1).

Table 1. Classes and its character count of each source text

Class	Source text	Character count
VMS	FSG transcription	152075
H_1: Ciphertext	Encrypted text	239552
	IQOS package code	4056
H_2: Programming language*	Python (object-oriented)	150697
	C (procedural)	207943
	Mathematica (procedural+functional)	280215
	FORTRAN (procedural)	211489
	Haskell (functional)	232292
H_3: Natural language	Italian	189290
	German	212767
	Turkish	197805
	Finnish	203432
	Russian	189384

*Most of the natural language comments in the source code have been removed

3.2 Methods

Algorithms used for text mining analyses in this project are as follows.

Step 1: Character Frequency-Based Source-Text Mining (CFST-M). The frequency analysis is defined as the calculation of how many times a symbol appears in some text. If the text has symbols approximately uniformly distributed, the text is likely a Ciphertext. If the distribution is not uniform, it means weak encryption, i.e. more likely a natural language [4].

Step 2: Shannon and Rényi Entropy-Based Source-Text Mining (SEST-M and REST-M). The concept of entropy in information theory describes how much information there is in a signal or an event. In information theory, the Rényi entropy(RE) generalizes the Shannon entropy (SE), and other kinds of entropy. Entropies quantify the diversity, uncertainty, or randomness of a system. RE of order α, where $\alpha \geq 0$ and $\alpha \neq 1$, is defined as the following formula. N is defined as a number of character. Following Shannon's convention, we used base 2.

$$H_\alpha(X) = \frac{1}{1-\alpha} \log(\sum_{i=1}^{N} p_i^\alpha) \tag{1}$$

As α approaches zero, RE increasingly weights all possible events more equally, regardless of their probabilities. In the limit for $\alpha \to 0$, RE is just the logarithm of the size of the support of X. The limit for $\alpha \to 1$ is SE. As α approaches infinity, RE is increasingly determined by the events of highest probability.

The limiting value of H_α as $\alpha \to 0$ is SE, is defined as

$$H_{\alpha=1}(X) \equiv \lim_{\alpha \to 1} H_\alpha(X) = -\sum_{i=1}^{N} p_i \log p_i. \tag{2}$$

As a second step, we name and define these algorithms as "Shannon Entropy-based Source-Text Mining (SEST-M)" and "Rényi Entropy-based Source-Text Mining (REST-M)". Applying formula above, we analyzed our source text. As a result, we verified that whether the method is valid and suitable.

4 Results and Discussions

As described in the Sect. 2, these are hypothesized as following;

Hypothesis 1 (H_1): VMS is a Ciphertext.
Hypothesis 2 (H_2): VMS is a programming language.
Hypothesis 3 (H_3): VMS is a natural language.

4.1 Step 1: Character Frequency-Based Source-Text Mining (CFST-M)

Regarding the CFST-M method, as described in Figs. 3, 4 and 5, trend of similar character frequencies to each other are able to be seen in the source text of each class.

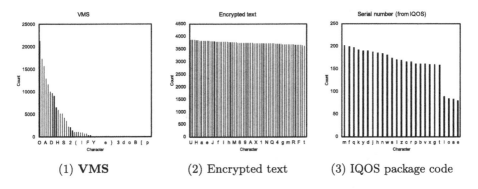

(1) **VMS**   (2) Encrypted text   (3) IQOS package code

Fig. 3. Results of CFST-M. VMS vs Ciphertext.

As figures are showing, H_1 has a lower correlation with VMS than H_2 and H_3. Moreover, as described in the third section, Murillo-Escobar (2014) [4] proposed that if the distribution of character frequency is not uniform, the text is likely a Ciphertext. Therefore, H_2 and H_3 are more reliable.

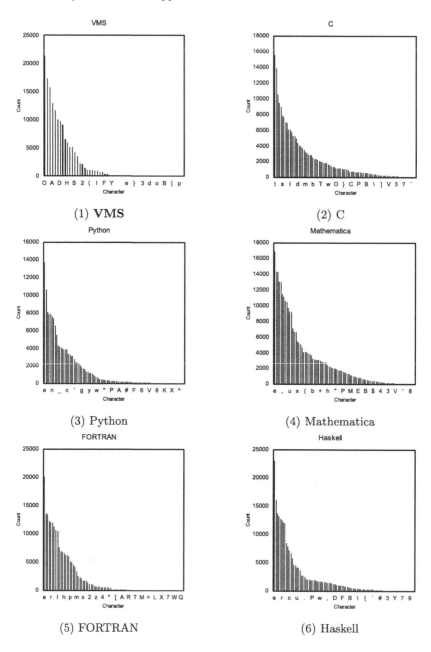

(1) **VMS**

(2) C

(3) Python

(4) Mathematica

(5) FORTRAN

(6) Haskell

Fig. 4. Results of CFST-M. VMS vs programming languages.

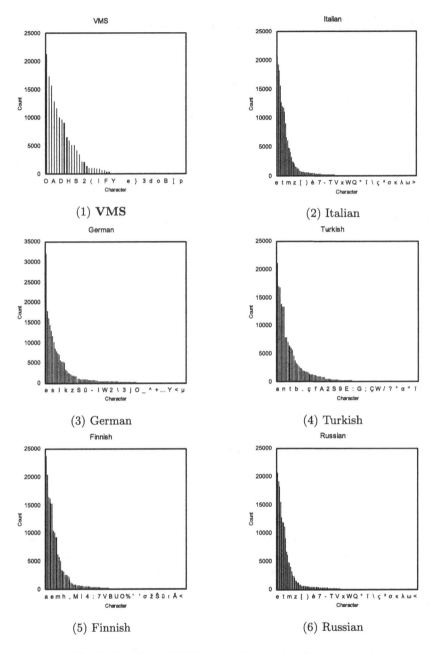

(1) **VMS**

(2) Italian

(3) German

(4) Turkish

(5) Finnish

(6) Russian

Fig. 5. Results of CFST-M. VMS vs natural languages.

4.2 Step 2-1: Shannon Entropy-Based Source-Text Mining (SEST-M)

Results are shown in Table 2. Entropy depends on the total number of possible characters. Those are different for languages, in addition, VMS has more different characters than other languages. (Note: in VMS there are no numerical characters and no upper and lower case). In order to compare the values, we divided our results by $\log |X|$ ($|X|$ means the size of set X. The number of different characters in each case) for normalization. $\log |X|$ is actually the largest possible entropy value when each character appears with the same probability of $1/|X|$.

Table 2. Results of SEST-M. Values of $H/\log |X|$ in each source text are shown.

Class	Source text	$|X|$	$H/\log |X|$
VMS	FSG transcription	56	0.696
H_1: Ciphertext	Encrypted text	65	0.996
	IQOS package code	25	0.991
H_2: Programming language	Python	94	0.789
	C	94	0.884
	Mathematica	93	0.845
	FORTRAN	86	0.759
	Haskell	94	0.794
H_3: Natural language	Italian	140	0.645
	German	108	0.712
	Turkish	113	0.718
	Finnish	120	0.663
	Russian	141	0.644

H_1 can be rejected, H_2 and H_3 cannot be decided for sure. The problem with this method is that a single value does not allow for strong conclusions, which is a disadvantage of the SE method. As described in Table 2 and Fig. 6, each text class (Ciphertext, programming and natural language) has a similar value of $H/\log N$. Moreover, its value of VMS is more close to H_2 and H_3. The values of SE of Ciphertext are higher compared to other source texts. It is defined that the entropy is high as unpredictability is high as to what will happen in the future. Since VMS has a lower value of SE than Ciphertext, it was verified that the next character of VMS is better predictable than Ciphertext.

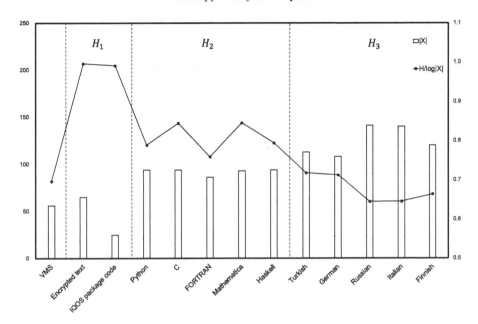

Fig. 6. Results of SEST-M. Number of character as $|X|$ and SE divided by logarithm $|X|$ as $H/\log|X|$.

4.3   Step 2-2: Rényi Entropy-Based Source-Text Mining (REST-M)

Exercising REST-M in the second step, the results of our experiment are described. Results are corresponding to H_1 through H_3 as well. In addition, since it is not possible to calculate SE with RE formula, Fig. 8 shows the results of SEST-M + REST-M at the same time.

The RE is a monotonic function of the information. However, as Kendall (1946)[5] states, these measures are scale-dependent when applied to continuous distributions, and so their absolute values are meaningless, it is needed to see the whole chart as we exercised. Therefore, they can generally only be used in comparative or differential processes. According to our experiment, it is obviously confirmed. Even for different source texts, there is a trend among each class. As Fig. 7 shows, we can say VMS is more likely a natural language as H_3.

Regarding H_2, the programming languages, we predicted that the curves would be lower than VMS and natural languages since we thought that the words used in the programming languages were limited. Surprisingly, however, the result was contrary to expectation.

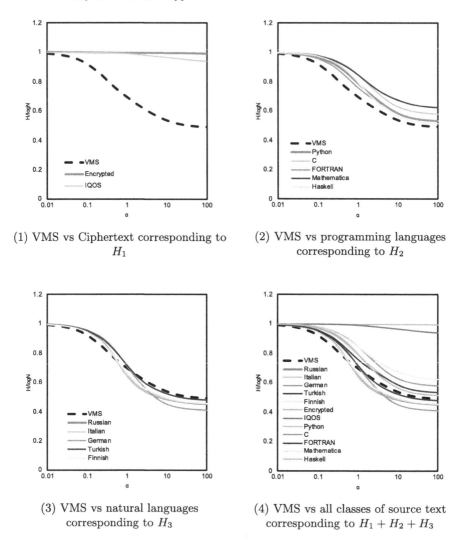

(1) VMS vs Ciphertext corresponding to H_1

(2) VMS vs programming languages corresponding to H_2

(3) VMS vs natural languages corresponding to H_3

(4) VMS vs all classes of source text corresponding to $H_1 + H_2 + H_3$

Fig. 7. Results of REST-M. Comparison between VMS vs each class of source text.

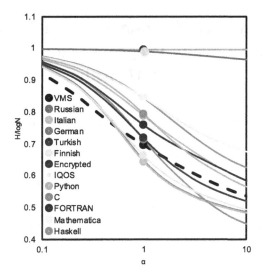

Fig. 8. Results of SEST-M, which is overlapped on the results of REST-M.

5 Conclusion

Our methods are valid and efficient. It is concluded that the methods can be applied to all text sources. Also, different classes of text sources can be distinguished. According to our experiment, the possibility that VMS is likely a Ciphertext can be rejected. Also, as the majority article said, we would like to conclude this mysterious manuscript as a meaningful human art, not a hoax. We have shown that VMS is not encrypted. (However, we are not able to exclude being encoded.) We can also see that VMS mostly resembles natural language, however, comes closer to a programming language. Therefore, it appears as a unique piece of text, nevertheless.

References

1. Skinner, S., Prinke, R.T., Zandbergen, R.: The Voynich Manuscript: The Worlds Most Mysterious and Esoteric Codex. Watkins Publishing, London (2017)
2. Rajkarnikar, M.: Analyzing the Voynich manuscript (2017). Unpublished
3. Philip Morris International. Frequently Asked Questions. IQOS Explained. Accessed 6 Feb 2019
4. Murillo-Escobar, M.A., Abundiz-Pérez, F., Cruz-Hernández, C., López-Gutiérrez, R.M.: A novel symmetric text encryption algorithm based on logistic map. In: Proceedings of the International Conference on Communications, Signal Processing and Computers, vol. 4953 (2014)
5. Kendall, M.G.: The Advanced Theory of Statistics, vol. 1, 4th edn. Charles Griffin and Co., Ltd., London (1948)

Decision Tree: Customer Churn Analysis for a Loyalty Program Using Data Mining Algorithm

Angela Siew Hoong Lee[1(✉)] (iD), Ng Claudia[1], Zuraini Zainol[2] (iD), and Khin-Whai Chan[1]

[1] School of Science and Technology, Department of Computing and Information Systems, Sunway University, 47500 Selangor, Malaysia
angela1@sunway.edu.my
[2] Department of Computer Science, Faculty of Defence Science and Technology, National Defence University of Malaysia, Sungai Besi Camp, 57000 Kuala Lumpur, Malaysia

Abstract. In the world of retailer, customers typically patronize multiple shops thus making loyalty programs a favorite among retailer to retain their customers. Loyalty programs are utilized across many different businesses as a marketing strategy to encourage customers to continuously shop or patronize the services provided by a certain organization. However, one of the biggest problem faced by these businesses is customer churn. The purpose of this research was to build a predictive model, which could predict customer churn, where visualization of data was generated to better understand the existing members and see the patterns and behavior demonstrated by members of the loyalty program. Through these, meaningful insights about the businesses' analysis on customers could be gathered and utilized for better actions which could be taken to address the issues which the company faces. At the end, based on the issues found, strategies were proposed to address the issues found. For this research, SAS Enterprise Miner was used to perform predictive analysis while Tableau was used to perform descriptive analysis.

Keywords: Decision tree · Customer analytics · Churn analysis · Loyalty program

1 Introduction

Loyalty program is also known as a reward program; it is often offered by a company to their customers. Through a loyalty program, customer may have early access to latest products, special promotions, or discounted items, which will motivate the customers to join the loyalty program. For a customer to be part of the loyalty program, they would typically register with their personal information and will be given a unique identifier like a member ID or membership card, which they would use when making a purchase. A customer exhibits customer loyalty when they consistently use your services over an extended period. Customer loyalty is crucial to a company because it can positively impact long-term profitability.

© Springer Nature Singapore Pte Ltd. 2019
M. W. Berry et al. (Eds.): SCDS 2019, CCIS 1100, pp. 14–27, 2019.
https://doi.org/10.1007/978-981-15-0399-3_2

Some customers who became a member might sometimes cease usage of the loyalty card after a period, which indicates that the customer may have stopped using the services provided by the loyalty program, these customers are known as churners. Losing customers or an increasing customer churn rate is one of the major problems companies may face because churn of good customers imposes irrecoverable damages for the company [1]. This is one of the reason customer loyalty is important, because when customer loyalty is present, customer churn rate will also reduce which in turn, retains customers. Many companies these days try to focus on customer acquisition to gain more new customers. However, it costs five times more to acquire new customers than to retain existing customers because companies often spend fortune on advertisements [2]. Therefore, the focus should shift from customer acquisition to customer retention and preventing churn.

1.1 Research Objectives and Problem Statement

There have been many researches done on identifying factors which can affect customer churn rate in loyalty cards on multiple different industries. However, there is less attention paid to improving loyalty programs and reducing customer churn rate, which in turn increase customer retention rate [3]. When customer churn rate is prevented, more customers will continue to patronize retailer services. Because retaining a customer imposes significant less cost than acquiring a new customer, it is better to focus on customer retention instead of customer acquisition [3].

Many malls have started providing loyalty programs with a main goal, to retain customers and keep them coming to the mall often. Retaining customers not as easy as it seems, due to all customers having different liking and preferences. When customers decide if they should patronize your services and use the loyalty card, they would always want rewards that can benefit them in return. Rewards can come in several different forms such as discounts, special promotions, vouchers, freebies, etc. Companies often face problems when they try to predict customer's churn behavior to understand factors which might retain or drive a customer away from using their services. This is a tedious task because it would require a lot of effort to predict customer's behavior based on the demographic data.

The objective of this research is to build a predictive model to predict customer's churn behavior and to propose potential strategies. This way, the company would be able to make better operational decisions and provide better rewards to the customer's liking hence potentially improve their overall experience on the loyalty program.

2   Literature Review

2.1   Loyalty Program

Loyalty programs are so common these days that they seem to be everywhere wherever we go. Whether be it in supermarkets, cinemas, or even restaurants all seem to provide membership benefits or rewards programs. A loyalty program is designed to improve customer's satisfaction and commitment [4], it is defined as a system that offers rewards to customers, in aim and goals to increase customer loyalty [5]. Loyalty programs utilize psychological principles by rewarding customers to achieve desired behavior of the customer, such as increased transactions or constant repurchases of the company's product [5].

Way back in the late 18th century, customer loyalty program seemingly started as "premium marketing" [6]. Green Shield stamps was one of the very first retail loyalty programs, they awarded stamps when a purchase is made at selected retailers, the stamps could later be redeemed for products, the aim was to encourage customers to make purchases at participating stores [7]. The strategy of using stamps as customer loyalty programs continued to lead on until the early 20th century when retailers began to introduce new ways to attract customers [6]. In year 1929, Betty Crocker introduced the box top program where customers collect box top clippings that can be accumulated and exchanged for items [7]. In year 1981, the modern-day loyalty program was launched by American Airlines known as "Frequent Flier" which was regarded the first full scale customer loyalty program in the era [6]. After that, card-based loyalty programs began to gain popularity in the 1990s because of the convenience it provides rather than collecting stamps or currency and it is still popular up till today.

Loyalty programs today come in many different forms such as collecting points which can be redeemed by the company itself or third-party companies, cash back rewards, member exclusive sales, extra discounts, or tier levels of benefit depending on the different level of membership. Benefits provided by customer loyalty programs can be divided into two types; tangible benefits and intangible benefits. Tangible benefits are benefits that are monetary, it comes in form of rebates, exclusive discounts or coupons while intangible benefits are non-monetary benefits [8].

2.2 Customer Loyalty and Satisfaction

Customer loyalty is the relationship between a customer and a seller after the primary transaction, it is known as "a commitment to repeatedly buy or patronize a particular product or service consistently, which causes repetitive same-brand or same-family brand purchases despite any influences that might cause switching behavior. Customer satisfaction happens when a person feels like his goals were achieved or attained [9] it can be measured by the customer's attitude towards the company's products or services [10]. Customer loyalty can also be known as an outcome measured by the frequency of repurchases made over a period. Traditionally, by just doing your best to satisfy a customer's needs is a way to achieve customer loyalty, building customer loyalty means to develop a positive relationship which mutually rewards each both the organization and the customer, this can be done by ensuring that the customers feel certain about the company's commitment to them in a way that the customers themselves have to know that they are important to the company and that they are always placed first. Countless companies and organizations have been finding ways and solutions to retain their existing customers and at the same time, attain and attract new customers [9]. Therefore, customer loyalty is prioritized in many organizations as it is a very important aspect as it would impact long-term profitability in a positive way, succeeding in achieving customer loyalty directly leads to successful customer retention which is very beneficial to the long-term growth of the company. Customer loyalty can only grow successfully when a company is aware of how their customers evolve and change each day, and to ensure that the customers knows exactly what benefits they are receiving from the company, positive connection between customers and the company makes the relationship between the two more stable and weighted, preventing other companies from winning the customers over.

3 Research Methodology

Figure 1 shows the data mining process that consists of seven main processes: (i) literature review, (ii) business understanding, (iii) data collection, (iv) data understanding, (v) data cleaning, (vi) data analysis, (vii) prediction model and, (viii) proposed strategies. The explanation of each stages will be explained in the next paragraph.

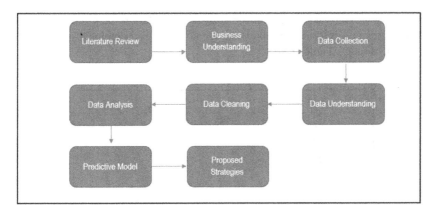

Fig. 1. Data mining process

A literature review was done to evaluate selected documents based on this topic to understand and identify the types of problems faced by other researchers while conducting the research, as well as to perceive the methods used by them to solve the problems. Based on the research done, it enabled us to broaden our understanding to perform this project. Business understanding allows us to identified and understand business problems. Data was collected from a loyalty card company. A total of four different datasets (CubeData, TransactionListing, MerchantListing & MemberListing) were given by the company. The duration of CubeData ranges from June 2016 – June 2017, TransactionListing has information for the entire year 2016, while MerchantListing and MemberListing contains the full data ranging from the beginning of the membership program until June 2017. Data understanding is to get familiar with the data to identify data quality problems and discover interesting insights about the data, and to reveal relationships between subsets that can form hypothesis for hidden information. Data understanding is a crucial process to understand the requirements and the business. Because data provided is raw, it contained noisy data like irrelevant variables and outliers. Therefore, data cleaning is crucial and necessary. For this step, SAS Enterprise Guide was used to perform data cleaning. Data analysis was the next step which was performed using SAS Enterprise Miner. Data Analysis was divided into two parts; Descriptive Analysis and Predictive Analysis. Descriptive Analysis is where the data was understood and explain, graphical representation of data was generated and described. Predictive Analysis will be explained in the latter stages. In SAS Enterprise Miner, a predictive model was built to predict customer churn rate and their behavior. Decision tree was built based on the target variable set. Through the analysis and findings, proposed strategies were presented to the loyalvty company with valid explanation and reasons to justify each proposition, which could potentially impact the business.

4 Descriptive Analysis

The following models are based on the dataset CubeData which show the general overall statistics and information about the members. These graphical representations were generated using Tableau.

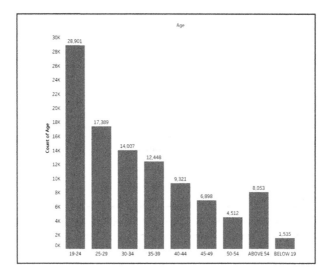

Fig. 2. Age group of customer

The above bar chart shows the overall age of customers (see Fig. 2). It is observed that the number of customers is the highest with an amount of 28,901 at age ranged 19–24 which is the younger crowd, possibly students. From age range of 25–29 onwards the frequency of customers reduces steadily as the age increases until age ABOVE 54 where a significantly higher frequency of customers was observed. Customers aged below 19 is the lowest at 1,535, this is because the company requires the customers to be at least 18 years old to register as a member. Hence, it explains the low number of customers in the age range as it only includes customers who are 18–19 years old.

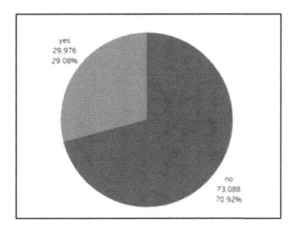

Fig. 3. Population of churners and non-churners

Next, shows the population of churners and non-churners (see Fig. 3). 'yes' indicates that they are churners while 'no' indicates that they are not churners. The members of this loyalty program are loyal members as there were only slightly more than a quarter of the population who have churned.

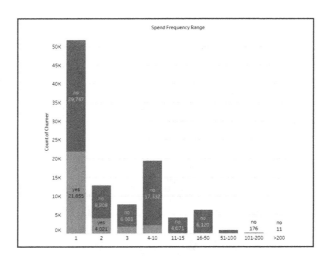

Fig. 4. Comparison of churners vs spend frequency range

When comparing Churner vs SpendFrequencyRange (see Fig. 4), it is noticeable that most of the churners (yes) SpendFrequencyRange of 1, 2, 3, and 4–10 which means that they have only spent less than 10 times. Looking at SpendFrequencyRange of 1, it is almost a half-half distribution where there is a close number of churners and non-churners. This means that the members under the 'no' category could be new members while the members in the 'yes' category never continued using the card after once.

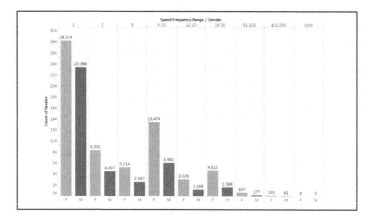

Fig. 5. Spend frequency range of customer vs gender

Above shows the spend frequency range of customers against gender (see Fig. 5). No matter the age range, female always has higher spend frequency range compared to male.

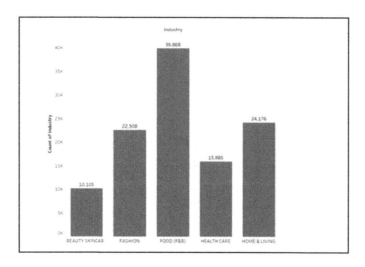

Fig. 6. Top 5 merchant industry

Lastly, this bar chart represents the Top 5 industries in this loyalty program (see Fig. 6). According to the model above, it can be immediately concluded that the most popular industry is the **FOOD (F&B)** industry. Followed by **HOME & LIVING, FASHION, HEALTHCARE**, and lastly, **BEAUTY SKINCARE/PERFUMERY**.

5 Predictive Analysis

The predictive model which was constructed using SAS Enterprise Miner (see Fig. 7). Starting from the left, the first node is CUBEDATA, which is the dataset 'CubeData' used for this model. The second node is Data Partition, this node was incorporated to separate the dataset into two distinct parts; 60% Training, 40% Validation. (Training is utilizing data containing target and input values to build and train predictive models, Validation is to validate the suitability of the data model created in Training.) After Data Partitioning, three different models were created - Default Tree, Depth & Branch Tree, and Maximum Tree. Decision trees were utilized because the target value 'CustomerActivity7to12Month' is binary, and three different decision trees were created to see if there are any notable differences between the three. The last node is Model Comparison, to compare and see which of the three models connected to it performs better.

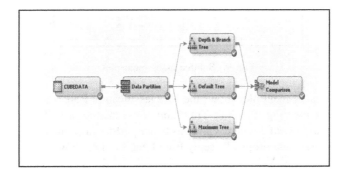

Fig. 7. Default tree

5.1 Default Tree

There are 61839 observations used for Train, and 41225 for Validation (see Fig. 8). The misclassification rate is 0.2625 for Train and 0.2615 for Validation. The average squared error for train and validation is 0.1740 and 0.1733 respectively which means that the both the misclassification rate and average squared error across the three partitions are not very significant.

Fit Statistics						
Target	Target Label	Fit Statistics	Statistics Label	Train	Validation	Test
Churner		NOBS	Sum of Frequencies	61839	41225	
Churner		MISC	Misclassification ..	0.262488	0.261589	
Churner		MAX	Maximum Absolut..	0.97272	1	
Churner		SSE	Sum of Squared E..	21514.64	14286.02	
Churner		ASE	Average Squared ..	0.173957	0.173269	
Churner		RASE	Root Average Squ..	0.417081	0.416256	
Churner		DIV	Divisor for ASE	123678	82450	
Churner		DFT	Total Degrees of ..	61839		

Fig. 8. Default tree

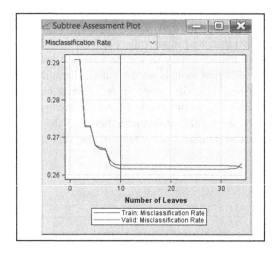

Fig. 9. Subtree assessment plot

The subtree assessment plot shows the Misclassification Rate against each sub-tree as the data splits (see Fig. 9). Both Train and Valid started at 0.291 on Leaf 2 and dropped steeply until Leaf 3 where Valid has better performance than Train. Both Train and Valid meet again and dropped steeply from Leaf 4 to Leaf 5 where Train had better performance. Train and Valid met again at Leaf 7 and dropped gradually with Valid having the best performance.

Variable Importance					
Variable Name	Label	Number of Splitting Rules	Importance	Validation Importance	Ratio of Validation to Training Importance
SpendFrequencyRange		4	1.0000	1.0000	1.0000
TotalSpentRMRange		1	0.4779	0.4518	0.9454
TotalPointsRange		1	0.4250	0.3876	0.9121
Age		2	0.2636	0.2509	0.9517
PointsExpiringIn4Month		1	0.1887	0.1837	0.9735
TotalRedeemedRMR...		0	0.0000	0.0000	
PointBalanceRange		0	0.0000	0.0000	
RedeemFrequencyRa...		0	0.0000	0.0000	
TotalRedeemedPoint...		0	0.0000	0.0000	
PointsExpiringIn2Month		0	0.0000	0.0000	

Fig. 10. Variable importance

This shows us the importance of each variable towards the decision tree (see Fig. 10). The most important variable starting from the top is SpendFrequencyRange, followed by

TotalSpentRMRange, TotalPointsRange, and PointsExpir-ingIn4Month. Which means that in Default Tree, these variables have highest influence towards the model.

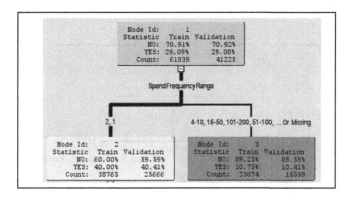

Fig. 11. First-level split

Firstly, for the chosen path the first-level split is at Node 1 to Node 2 and Node 3 (see Fig. 11). The topmost node (Node 1) represents the whole sample of the loyalty program members. 'NO' represents non-churners while 'YES' represents churners, both Train and Validation data subsets show that there are 71% non-churners and 29% churners.

The thicker connecting lines represent the volume of record going to each node, this can be justified by the total Count for each node – Node 2 has total Count of 64,431 (38765 + 25666) while Node 3 has total Count of 38,633 (23074 + 15559).

In this case, there are more records going to Node 2 which means that more members have a SpendFrequencyRange of 1 or 2 compared to the other SpendFre-quencyRange for Node 3. The lighter shading of the node indicates that the percentage of churners in Node 2 is higher compared to Node 3. In Node 2, 59.6% of members are non-churners, while 40.4% are churners.

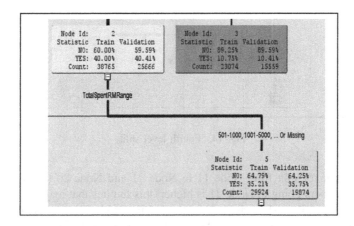

Fig. 12. Second level split

Figure 12 shows the second-level split. From Node 2, it splits to Node 5. Higher volume of members in Node 2 goes to Node 5, these members has Total-SpentRMRange of 501–1000, 1001–5000, etc. In Node 5, there are 64.25% non-churners and 35.75% churners.

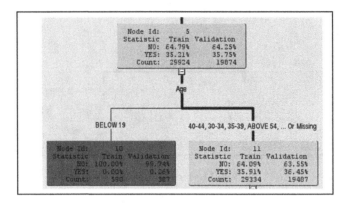

Fig. 13. Third level split

Next, the third-level split (see Fig. 13). There are higher volume of members from Node 5 going to Node 11. This means that majority of members from Node 5 are Age of NOT BELOW 19. For Node 11, there are 63.55% non-churners and 36.45% non-churners.

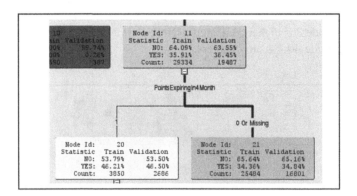

Fig. 14. Fourth level split

The fourth-level split from Node 11 to Node 21 and Node 20 (see Fig. 14). The volume of members going to Node 21 is higher, this implies that majority of members from Node 11 have PointsExpiringIn4Month of 0. For Node 21, there are 65.16% non-churners and 34.84% churners. The chosen path ends on Node 21.

6   Conclusion

The selected loyalty program for this research has been avidly promoting their loyalty program for the past few years, but in the beginning, there were not many participating merchants as compared to now where there are approximately 1,000 participating merchants in this program.

Based on the descriptive analysis, it allows understanding of the customer's background, characteristics, information and their spending pattern of participating merchants. These findings would be useful in identifying and targeting customers who are likely to churn and increase the possibility of customer retention. The descriptive analysis showed that most of the members of this loyalty program were loyal customers as there were only less than 30% who have churned. These churners consist mostly a of member who has spent less than 10 times. When further analyzed, it was found that the member does not feel that the loyalty program benefits them in any way after they have used it several times, and therefore ceased their usage.

It is understood that the younger the customers are, the more likely they would become a member of this program which the age group range from 19–34 represents almost 60% of the total members. The highest frequency of customers lies in the age group of 19–24 years old which are most probably students. The manager has also stated that generally there are higher number of female customers – as justified in the descriptive analytics previously, more than 60% of the members are females. As stated by him, females are more drawn to membership programs as compared to males. According to a survey done by CrowdTwist in 2016 [11] women are indeed more loyal customers and are stronger advocates of the loyalty program that they are using. In another research done by JakPat in 2016 [12], we have also observed that no matter which industry it is, females are always more inclined to participate in loyalty programs.

For the predictive analysis, Decision Trees were utilized for the model because the target 'Churner' is binary. Three decision trees with different settings were created to see the differences, but the Default Tree remains the best. Looking at the Variable Importance window for no matter which tree it is, the SpendFrequencyRange and TotalSpentRMRange which both represents member's spending behavior, are always the top two variables that shows most importance. This means that the number of times the member has spent and the total RM they have used while utilizing the card plays a massive impact on the prediction. It might be bias to just look at either one of these two variables because even though the member's SpendFrequencyRange is one of the highest, their TotalSpentRMRange might not be a lot. To add on, for some members who has very high TotalSpentRMRange could be property buyers who just signed up as a member for one-off benefit, it might not guarantee that they would continue patronizing or using the loyalty program. However, SpendFrequencyRange could tell more accurately whether a member is loyalty or not because if the member continuously uses it often, it means that they are potentially loyal members. At the end of the day, the objective of the model was to identify customer churn, hence it justifies the emphasis on SpendFrequencyRange over TotalSpentRMRange.

Moving on to the next part of the analysis, clustering was performed to group members of similar characteristics or behavior together and to see which the most common behavior which member exhibit. According to the results gained from clustering, it seems that majority of female members, aged mostly between 19-39 spent the most in the FOOD (F&B) industry, this is considering both churners and non-churners. People love getting free stuff, and they would share it with their friends and families therefore by offering a sign-up incentive it would encourage customers to join.

Firstly, based on the finding from descriptive analysis, that members who spent less than 10 times is very possibly a churner, one of the possible strategy to prevent customers with this behavior from churning is to provide newly joined/ signed up members with incentives dedicated specially to them. A few actions the company could take based on this strategy could be to provide the new members with rebates on the initial first few transactions. This could encourage the new members to get started on using the card more as they would want to use the rebate provided, while for the program, this can increase the likelihood of members to continuously utilize the membership program. Getting members engaged with the program is always the first step, once a member realizes the benefits, they could attain by using the card, they are more likely to continue patronizing.

Secondly, another strategy would be event-driven alerts. To send notifications or alerts to members of through the app to remind them of their point balance and the equivalent RM monetary value. The intention behind this idea was thought of because some members might have used the card previously but forgot about the points accumulated that can be used for redemption, while some members might not be aware of how much points they have collected thus far. This would benefit the members as they would be reminded to utilize the points they have collected, while for Sunway Pals it could encourage an increase of member's usage and increasing spend frequency.

Thirdly, this strategy is to encourage members engagement with the loyalty program. Increasing members spend frequency would increase members engagement. A way to increase the members spend frequency is to reward redeemable points to them after they have spent a certain amount. For example, to award RM10 worth of redeemable points to members for every RM200 they spent – in an accumulative manner. For members, they would feel that the loyalty program is worth their effort and participation. Through this research, we are able to identify that using these variables such as "spending frequency range, total spent range, total points collected and points expiring" are able to do an accurate and good prediction on customer churns especially in retail industry that deals with loyalty programs. We hope that with this analysis it will give practitioners and researcher in this field a better idea on targeting their customers and understand their spending behavior which may impact their loyalty program.

References

1. Soeini, R.A., Rodpysh, K.V.: Applying data mining to insurance customer churn management, p. 11 (2012)
2. Oyeniyi, A.O., Adeyemo, A.B.: Customer churn analysis in banking sector using data mining, p. 10 (2015)

3. Oentaryo, R.J., Lim, E.P., Lo, D., Zhu, F., Prasetyo, P.K.: Collective churn prediction in social network, p. 5 (2012)
4. Zakaria, I., Rahman, B.A., Othman, A.K., Yunus, N.A., Dzulkipli, M.R., Osman, M.A.: Loyalty Program. The Relationship between Loyalty Program, Customer Satisfaction and Customer Loyalty in Retail Industry: A Case Study, p. 8 (2013)
5. Toporek, A.: What Is a Loyalty Program? (And Will It Work For You), CTS Service Solutions, 25 June 2012. http://customersthatstick.com/blog/what-is/what-is-a-loyalty-program/. Accessed 31 May 2019
6. McEachern, A.: A History of Loyalty Programs, and How They Have Changed. smile.io. https://blog.smile.io/a-history-of-loyalty-programs. Accessed 31 May 2019
7. Loyalty Passport. Evolution of Customer Loyalty Programs! – Loyalty Passport. Loyalty Passport, 27 March 2016. https://loyaltypassport.wordpress.com/2016/05/27/evolution-of-customer-loyalty-programs/. Accessed 31 May 2019
8. Mulhern, T., Duffy, D.: Building loyalty at things remembered. J. Consum. Mark. **21**, 62–66 (2004)
9. Dehghan, A., Trafalis, T.B.: Examining churn and loyalty using support vector machine. Bus. Manag. Res. **1**(4), 153–161 (2012)
10. Wyse, S.E.: Customer Satisfaction vs. Customer Loyalty. Snap Surveys, 26 June 2012. https://www.snapsurveys.com/blog/customer-satisfaction-customer-loyalty/. Accessed 11 Oct 2018
11. CrowdTwist. Battle for the Sexes: Women Are More Brand Loyal Than Men, and Are Stronger Brand Advocates—Bulldog Reporter, 24 October 2016. https://www.bulldogreporter.com/battle-for-the-sexes-women-are-more-brand-loyal-than-men-and-are-stronger-brand-advocates/. Accessed 25 Jan 2019
12. JakPat. Membership and Loyalty Programs - Survey Report – JAKPAT, 10 May 2016. https://blog.jakpat.net/membership-and-loyalty-programs-survey-report/. Accessed 30 May 2019

Improving e-Commerce Severity Rating Measurement Using Consistent Fuzzy Preference Relation

Tenia Wahyuningrum[1,2(✉)], Azhari Azhari[1], and Suprapto[1]

[1] Universitas Gadjah Mada, Jl. Sekip Utara, Yogyakarta, Indonesia
[2] Institut Teknologi Telkom Purwokerto, Jl. DI Panjaitan 128,
Purwokerto, Indonesia
tenia@ittelkom-pwt.ac.id

Abstract. Usability is a critical success factor in business. However, many usability issues on e-commerce websites cannot be adequately handled because there is no priority scale. The severity rating helps focus on the significant problem. The severity triggered by the data is considered better than the severity triggered by the evaluator. Data obtained quantitatively used to determine essential values and usability scores. Previous research has applied the Fuzzy Analytical Hierarchy Process (FAHP) method with Extent Analysis (EA), and Fuzzy Preference Programming (FPP) approaches. We were assessing the weight of criteria to evaluate usability and determine severity rating. However, EA and FPP approaches have disadvantages. Among them, when determining the number of paired comparisons at the level of importance between criteria. The number of comparisons that must be assessed by Decision-Maker (DM) causes the fuzzy pairing matrix to be inconsistent. This inconsistency causes the weight between rules to be invalid. The Consistent Fuzzy Preference Relation (CFPR) method is present to overcome the problem of the number of paired comparisons. The CFPR method summarizes the comparison steps to facilitate DM in assessing the level of importance between criteria. The results show the results of rank similarity testing; the EA and CFPR methods have close relationships. The FPP and CFPR methods have a weak correlation in generating usability ranking and severity ratings.

Keywords: Consistent fuzzy preference relation · e-Commerce · Usability · Severity rating

1 Introduction

The rapid development of the internet has become an essential part of everyday life for most people. A user of the internet (called netizen) using that connectivity in almost every aspect of their lives, whether it is communicating with friends, playing games, or

This research was supported by the Ph.D grant of the Indonesia endowment fund for education (LPDP) ministry of Finance Republik Indonesia for Tenia Wahyuningrum (No. PRJ-4811/LPDP.3/2016).

M. W. Berry et al. (Eds.): SCDS 2019, CCIS 1100, pp. 28–39, 2019.
https://doi.org/10.1007/978-981-15-0399-3_3

buying products. The internet has a big contributed to the evolution of e-commerce in the world [1].

The first generation of e-commerce website (called e-commerce 1.0) was very static. It imitates the offline store, provides virtual shelves of product, little information, few pictures, challenging to navigate, and difficult to use. Along with the development of technology, version 2.0 of e-commerce is more easily accessible, readily usable, and involves users with social shopping environment [2]. E-commerce 3.0 engages more users in their online stores; they use new strategies to make consumers feel at home. Their efforts include using quizzes to give appropriate product curation, using Augmented Reality (AR), and using smart systems to find out user preferences for a product. User involvement in expressing comfort and satisfaction is an important thing that business people need to consider. Ease of learning, ease of use, and pleasure from the user's perspective in the Human-Computer Interaction area called usability [3, 4]. E-commerce developers and owners need to evaluate usability problem for improving the quality of the website regularly.

There are many usability issues on a website; however, not all usability problems are the same. Some are more serious than others. Therefore, there needs to be a degree of severity to help focus on the significant problem [5]. The better the user experience, the smaller the value of critical usability should be. Conversely, if the value of the significant usability level is high, it is necessary to give recommendations for improvements in design [5]. Evaluators and data assessments can trigger a critical level of usability. Critical driven data ratings are considered more reliable than evaluators evaluations because there are possible inconsistencies between and within evaluators [6]. Besides, evaluators do not have systematic data-based indicators, such as the level of completion of tasks and the proportion of products affected when setting critical standards. Data obtained quantitatively is then used to determine essential values and usability scores.

In previous studies, we used a quantitative method of determining severity ratings after measuring usability scores [7]. The use of the Extent Analysis (EA) and Fuzzy Preference Programming (FPP) techniques have been applied in determining the weight of usability criteria but has several weaknesses. The EA approach is considered to produce a lot of 0, so the rules deemed necessary in the assessment are ignored. Whereas the FPP method often results in negative membership degrees and generates multiple optimal solutions that make inconsistencies between fuzzy assessments. Another weakness in both EA and FPP methods is a large number of assessments that Decision-Maker (DM) makes in building pairwise comparison matrices. In both ways, it takes $n \times [(n-1)/2]$ the number of judgments where n is the number of elements compared. Too much analysis will allow for DM errors when giving experience [8, 9]. Herrera-Viedma makes a new method called Consistent Fuzzy Preference Relation (CFPR) to reduce the amount of DM assessment steps. It is only as much as $n-1$ to ensure consistency at the level with n criteria [10]. By using the CFPR methods, the weight for each usability criteria can be better for measuring usability score and adjusting severity ratings. This paper is organized as follows. Section 2 reviews the CFPR method. Section 3 explains the proposed method. Section 4 presents a result and discussion, and Sect. 5 draws the conclusions.

2 Consistent Fuzzy Preference Relations

Preference relations are usually in the form of matrix is the interest rate of the first criteria for eligibility for a second. The relationship of this assessment can be multiplicative preference relations of fuzzy preference relations. Multiplicative preference relations can be formulated as

$$R \subseteq A \times A, R = (r_{ij}), \forall i, j \in \{1, 2, \ldots, n\}, \tag{1}$$

where A is the set of criteria or alternatives, r_{ij} is the preference ratio of criteria or alternative a_i to a_j, $a_{ij} \cdot a_{ji} = 1, \forall i, j \in \{1, 2, \ldots, n\}$. Furthermore, the relationship will be represented by a pairwise matrix comparison $P = p_{ij}$, where the size is $n \times n$, $p_{ij} = \mu_p(a_i, a_j), \forall i, j \in \{1, 2, \ldots, n\}$ and the value of the membership function of fuzzy logic. Elements in the pairwise comparison matrix are calculated using several propositions [10].

Proposition 1. *Consider a set criteria or alternatives, $X = \{x_1, x_2, \ldots, x_n\}$ associated with a reciprocal multiplicative preference relation $A = (a_{ij})$ for $a_{ij} \in [1/9, 9]$. Then, the corresponding reciprocal fuzzy preference relation, $P = p_{ij}$ with $p_{ij} \in [0, 1]$ associated with A is given as*

$$p_{ij} = g(a_{ij}) = \frac{1}{2}(1 + \log_9 a_{ij}) \tag{2}$$

Proposition 2. *For each $P = g(A)$, where $P = (p_{ij})$, booth of Eqs. 3 and 4 are equivalent.*

$$p_{ij} + p_{jk} + p_{ki} = \frac{3}{2}, \forall i, j, k \tag{3}$$

$$p_{ij} + p_{jk} + p_{ki} = \frac{3}{2}, \forall i < j < k \tag{4}$$

Proposition 3. For each $P = (p_{ij})$, booth of Eqs. 5 and 6 are equivalent.

$$p_{ij} + p_{jk} + p_{ki} = \frac{3}{2}, \forall i < j < k \tag{5}$$

$$p_{i(i+1)} + p_{(i+1)(i+2)} + \cdots + p_{(i+k-1)(i+k)} + p_{(i+k)i} = \frac{k+1}{2}, \forall i < j. \tag{6}$$

Proposition 3 is used to construct a consistent fuzzy preference relation from the set of $n - 1$ values $\{p_{12}, p_{23}, \ldots, p_{n-1n}\}$. A decision matrix with entries that are not in the interval [0, 1], but in an interval $[-k, 1 + k]$, $k > 0$, can obtained by transforming the obtained values using a transformation function that preserves reciprocity and additive consistency. It is given by the function f: $[-k, 1 + k]$ to $[0, 1]$, $f(x) = (x+k)/(1+2k)$.

3 Proposed Method

CFPR methods are widely applied to various issues of decision-making [8, 9]. But have not found the use of the CFPR method to determine the usability evaluation and severity of ratings from e-commerce websites. We used the CFPR method to evaluate e-commerce usability. Figure 1 illustrates a detailed flowchart of the proposed method that consists of six steps. Determining usability criteria is the first step of evaluation. Developers and usability experts define essential rules in the assessment. The literature study activity often used to collect several papers relating to usability e-commerce, then look for the right criteria in measurement.

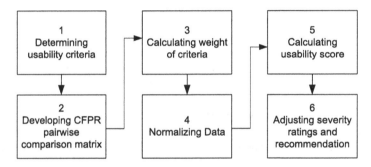

Fig. 1. Proposed method

Research conducted by [7], the e-commerce website usability performance can be presented by loading time (C_1), response time (C_2), page rank (C_3), frequency of update (C_4), traffic (C_5), site map(C_6), design optimization (C_7), page size (C_8), backlink (C_9), number of items (C_{10}), accessibility errors (C_{11}), markup validations (C_{12}), broken link (C_{13}) and ease of use (C_{14}). After determining criterias, the next step is developing CFPR pairwise comparison matrix. The third stage is calculating the weight of criteria. The average A_i and W_i was calculated using Eq. (7).

$$A_i = \frac{1}{n}\left(\sum_{j=1}^{n} p_{ij}\right), W_i = A_i/\sum_{j=1}^{n} A_i \tag{7}$$

where n is the number of criteria, $i = 1, 2, \ldots, n$. The normalized of the data result can be expressed using the Linear Weightage Model (LWM) [11] such as the Eq. (8).

$$r_{ij} = \begin{cases} \frac{\max - c_{ij}}{\max - \min}, \text{ for maximum threshold,} \\ \frac{c_{ij} - \min}{\max - \min}, \text{ for minimum threshold.} \end{cases} \tag{8}$$

where r_{ij} is normalized value and c_{ij} is criteria value. Max is the maximum value of the particular criteria among all websites, min is minimum value of the same criteria among the whole websites. The threshold for the "smaller is better" criteria must be maximum, and the threshold for the "larger is better" criteria must be minimum. The final step of this method is to determine severity ratings based on usability score obtained. Equation (9) used to calculating usability score, where l is the number of alternatives, n is the number of criteria, r_{ij} is normalized value, and w_j is weight of criteria.

$$\text{usability score} = \sum_{i=1}^{l} \sum_{j=1}^{n} w_j \times r_{ij}; i = 1, 2, \ldots, l, j = 1, 2, \ldots, n \qquad (9)$$

The normalized data are then multiplied by each criteria weight and summed to obtain usability score (9). Severity ratings was calculated by Eq. (10).

$$\text{severity ratings} = 100\% - \text{usability score} \qquad (10)$$

We assigns a severity rating on a 4-point scale (1 = irritant, 2 = moderate, 3 = severe, 4 = unusable). Next, we assigns a percentage of severity, also on a 4-point scale (1 = less than 10%; 2 = 11 to 50%; 3 = 51 to 89%; 4 = more than 90%). Recommendations are given on the website by looking at criteria that have poor values. For example, if the loading speed of a website exceeds 30 s, it can be said that the site is slow, so there needs to be a recommendation for improvement. Otherwise, for the backlink criteria, recommendations are given if the links to the website are few.

4   Result and Discussion

4.1   Determining the Weight of Criteria

Before determining the weight of criteria, the previous step is developing the CFPR pair-wise matrix comparison. Based on [7], we provided the fuzzy comparison matrix, as shown in Table 1. Table 1 represents the fuzzy paired comparison matrix of the case criteria for the usability of e-commerce websites in 14 × 14 arrays. By using the CFPR method, the number of comparisons to be used is only n − 1(=14 − 1) values. The usability expert as DM gives fuzzy judgments according to the level of importance between criteria [7]. For the requirements compared to the same criteria, it will automatically be worth (1, 1, 1) or equally important. The important level (2, 3, 4) means moderate importance, (4, 5, 6) is strong importance, (6, 7, 8) is very strong importance, (8, 9, 9) is extreme importance. The value (1, 2, 3), (3, 4, 5), (5, 6, 7), and (7, 8, 9) are represent the state between adjacent values.

Table 1. Fuzzy pairwise comparison matrix of criteria

C	C_1	C_2	C_3	C_4	C_5	C_6	C_7	C_8	C_9	C_{10}	C_{11}	C_{12}	C_{13}	C_{14}
C_1	(1, 1, 1)	(1, 2, 3)	z	z	z	z	z	z	z	z	z	z	z	z
C_2	z	(1, 1, 1)	(1, 2, 3)	z	z	z	z	z	z	z	z	z	z	z
C_3	z	z	(1, 1, 1)	(1, 2, 3)	z	z	z	z	z	z	z	z	z	z
C_4	z	z	z	(1, 1, 1)	(1, 1, 1)	z	z	z	z	z	z	z	z	z
C_5	z	z	z	z	(1, 1, 1)	(1, 1, 1)	z	z	z	z	z	z	z	z
C_6	z	z	z	z	z	(1, 1, 1)	(1, 1, 1)	z	z	z	z	z	z	z
C_7	z	z	z	z	z	z	(1, 1, 1)	(1, 2, 3)	z	z	z	z	z	z
C_8	z	z	z	z	z	z	z	(1, 1, 1)	(1, 1, 1)	z	z	z	z	z
C_9	z	z	z	z	z	z	z	z	(1, 1, 1)	(1, 2, 3)	z	z	z	z
C_{10}	z	z	z	z	z	z	z	z	z	(1, 1, 1)	(1, 2, 3)	z	z	z
C_{11}	z	z	z	z	z	z	z	z	z	z	(1, 1, 1)	(1, 2, 3)	z	z
C_{12}	z	z	z	z	z	z	z	z	z	z	z	(1, 1, 1)	(1, 2, 3)	z
C_{13}	z	z	z	z	z	z	z	z	z	z	z	z	(1, 1, 1)	(1, 1, 1)
C_{14}	z	z	z	z	z	z	z	z	z	z	z	z	z	(1, 1, 1)

Based on Table 1, the value of z is a matrix element entry (p_{ij}) that is not filled by DM. The Proposition 1 will calculate the value of p_{ij} when $i = j$ or $j = i + 1$. The elements on main diagonal ($i = j$) can be shown as bellow.

$$p_{11} = 1/2(1 + \log_9(1)) = 0.5;$$
$$p_{22} = 1/2(1 + \log_9(1)) = 0.5;$$
$$\vdots$$
$$p_{1414} = 1/2(1 + \log_9(1)) = 0.5.$$

The original data yield an average interval value. For example, the $C_1 \rightarrow C_2$ comparison is (1, 2, 3), and the average is $(1/2)(1 + 3) = 2$.

$$p_{12} = 1/2(1 + \log_9(2)) = 0.66;$$
$$p_{23} = 1/2(1 + \log_9(2)) = 0.66;$$
$$\vdots$$
$$p_{1314} = 1/2(1 + \log_9(1)) = 0.5.$$

The values below the main diagonal are calculated using Proposition 3 for each element as follows;

$p_{21} = 1 - p_{12} = 1 - 0.66 = 0.35; p_{21} = 1 - p_{12} = 1 - 0.66 = 0.34;$
$p_{31} = 1.5 - p_{12} - p_{23} = 1 - 0.66 - 0.66 = 0.34;$
$\vdots$

$p_{141} = 7 - p_{12} - p_{23} - p_{34} - p_{45} - p_{56} - p_{67} - p_{78} - p_{89} - p_{910} - p_{1011} - p_{1112} - p_{1213} - p_{1314}$
$p_{141} = 7 - 0.66 - 0.66 - 0.66 - 0.5 - 0.5 - 0.5 - 0.66 - 0.5 - 0.66 - 0.66 - 0.66 - 0.66 - 0.5 = -0.7.$

Table 2. CFPR pairwise comparison matrix of criteria

C	C_1	C_2	C_3	C_4	C_5	C_6	C_7	C_8	C_9	C_{10}	C_{11}	C_{12}	C_{13}	C_{14}
C_1	0.50	0.66	0.82	0.97	0.97	0.97	0.97	1.13	1.13	0.63	1.45	1.60	1.76	1.76
C_2	0.34	0.50	0.66	0.82	0.82	0.82	0.82	0.97	0.97	1.13	1.29	1.45	1.76	1.60
C_3	0.18	0.34	0.50	0.66	0.66	0.66	0.66	0.82	0.82	0.97	1.13	1.29	1.45	1.45
C_4	0.03	0.18	0.34	0.50	0.50	0.50	0.50	0.66	0.66	0.82	0.97	1.13	1.29	1.29
C_5	0.03	0.18	0.34	0.50	0.50	0.50	0.50	0.66	0.66	0.82	0.97	1.13	1.29	1.29
C_6	0.03	0.18	0.34	0.50	0.50	0.50	0.50	0.66	0.66	0.82	0.97	1.13	1.29	1.29
C_7	0.03	0.18	0.34	0.50	0.50	0.50	0.50	0.66	0.66	0.82	0.97	1.13	1.29	1.29
C_8	−0.13	0.03	0.18	0.34	0.34	0.34	0.34	0.50	0.50	0.66	0.82	0.97	1.13	1.13
C_9	−0.13	0.03	0.18	0.34	0.34	0.34	0.34	0.50	0.50	0.66	0.82	0.97	1.13	1.13
C_{10}	0.37	−0.13	0.03	0.18	0.18	0.18	0.18	0.34	0.34	0.50	0.66	0.82	0.97	0.97
C_{11}	−0.45	−0.29	−0.13	0.03	0.03	0.03	0.03	0.18	0.18	0.34	0.50	0.66	0.82	0.82
C_{12}	−0.60	−0.45	−0.29	−0.13	−0.13	−0.13	−0.13	0.03	0.03	0.18	0.34	0.50	0.66	0.66
C_{13}	−0.76	−0.60	−0.45	−0.29	−0.29	−0.29	−0.29	−0.13	−0.13	0.03	0.18	0.34	0.50	0.50
C_{14}	−0.76	−0.60	−0.45	−0.29	−0.29	−0.29	−0.29	−0.13	−0.13	0.03	0.18	0.34	0.50	0.50

Based on Table 2, it can be seen that the element of p_{141} is −0.76 (not in the interval [0, 1]. We applied the transforming function $f(x) = (x + k)/(1 + 2 k)$ for the elements on the upper main diagonal as follows;

$$p_{12} = f(0.66) = (0.66 + 0.76)/(1 + (2 \times 0.76)) = 0.56;$$
$$p_{23} = f(0.66) = (0.66 + 0.76)/(1 + (2 \times 0.76)) = 0.56;$$
$$\vdots$$
$$p_{1314} = f(0.5) = (0.5 + 0.76)/(1 + (2 \times 0.76)) = 0.5.$$

Table 3 shows the transformation of the pairwise comparison matrix. The average of each criteria A_i was calculated using Eq. (7).

$$A_1 = 1/14(0.5 + 0.56 + \ldots + 1) = 0.71;$$
$$A_2 = 1/14(0.44 + 0.5 + \ldots + 0.94) = 0.7;$$
$$\vdots$$
$$A_{14} = 1/14(0 + 0.06 + \ldots + 0.5) = 0.7;$$

$$\sum_{i=1}^{n} A_i = 0.72 + 0.7 + \ldots + 0.25 = 7.$$

The weight W_i was determined according to

$W_1 = 0.72/7 = 0.1;$ $W_2 = 0.72/7 = 0.1;$ $W_3 = 0.63/7 = 0.09;$
$W_4 = 0.57/7 = 0.08; W_5 = 0.57/7 = 0.08; W_6 = 0.57/7 = 0.08; W_7 = 0.57/7 = 0.08;$
$W_8 = 0.5/7 = 0.07; W_9 = 0.5/7 = 0.07;$ $W_{10} = 0.48/7 = 0.07;$
$W_{11} = 0.38/7 = 0.05;$ $W_{12} = 0.32/7 = 0.05; W_{13} = 0.25/7 = 0.04;$
$W_{14} = 0.25/7 = 0.04.$

Table 3. After transforming the matrix of criteria

C	C_1	C_2	C_3	C_4	C_5	C_6	C_7	C_8	C_9	C_{10}	C_{11}	C_{12}	C_{13}	C_{14}
C_1	0.5	0.56	0.63	0.69	0.69	0.69	0.69	0.75	0.75	0.25	0.88	0.94	1	1
C_2	0.44	0.5	0.56	0.63	0.63	0.63	0.63	0.69	0.69	0.75	0.81	0.88	1	0.94
C_3	0.38	0.44	0.5	0.56	0.56	0.56	0.56	0.63	0.63	0.69	0.75	0.81	0.88	0.88
C_4	0.31	0.38	0.44	0.5	0.5	0.5	0.5	0.56	0.56	0.63	0.69	0.75	0.81	0.81
C_5	0.31	0.38	0.44	0.5	0.5	0.5	0.5	0.56	0.56	0.63	0.69	0.75	0.81	0.81
C_6	0.31	0.38	0.44	0.5	0.5	0.5	0.5	0.56	0.56	0.63	0.69	0.75	0.81	0.81
C_7	0.31	0.38	0.44	0.5	0.5	0.5	0.5	0.56	0.56	0.63	0.69	0.75	0.81	0.81
C_8	0.25	0.31	0.38	0.44	0.44	0.44	0.44	0.5	0.5	0.56	0.63	0.69	0.75	0.75
C_9	0.25	0.31	0.38	0.44	0.44	0.44	0.44	0.5	0.5	0.56	0.63	0.69	0.75	0.75
C_{10}	0.75	0.25	0.31	0.38	0.38	0.38	0.38	0.44	0.44	0.5	0.56	0.63	0.69	0.69
C_{11}	0.13	0.19	0.25	0.31	0.31	0.31	0.31	0.38	0.38	0.44	0.5	0.56	0.63	0.63
C_{12}	0.06	0.13	0.19	0.25	0.25	0.25	0.25	0.31	0.31	0.38	0.44	0.5	0.56	0.56
C_{13}	0	0.06	0.13	0.19	0.19	0.19	0.19	0.25	0.25	0.31	0.38	0.44	0.5	0.5
C_{14}	0	0.06	0.13	0.19	0.19	0.19	0.19	0.25	0.25	0.31	0.38	0.44	0.5	0.5

4.2 Determining the Severity Ratings

The proposed method was tested using five popular e-commerce websites in Indonesia, i.e., Lazada, Blibli, Shopee, Jdid and Mataharimall (based on B2C e-commerce sites rank in Indonesia by eMarketer chart, and Indonesian e-commerce map by iprice.com). We assigned the normalized values, the weight of the EA method, and the weight of the FPP method as in Table 4 [7]. The weight of CFPR was calculated on the Subsect. 4.1.

Table 4. Data normalization

Criteria	Lazada	Blibli	Shopee	JD id	Matahari mall	Weight (W)		
						EA	FPP	CFPR
C_1	0.41	0.84	0.33	0.00	1.00	0.33	0.21	0.10
C_2	0.55	0.00	0.95	0.81	1.00	0.25	0.13	0.10
C_3	1.00	0.99	0.93	0.82	0.00	0.23	0.36	0.09
C_4	1.00	0.83	0.94	0.44	0.00	0.05	0.01	0.08
C_5	0.30	0.00	0.64	0.14	1.00	0.04	0.01	0.08
C_6	0.58	1.00	0.00	1.00	0.00	0.04	0.01	0.08
C_7	0.73	0.00	1.00	0.93	0.70	0.04	0.01	0.08
C_8	1.00	0.00	0.79	0.59	0.95	0.00	0.03	0.07
C_9	0.41	1.00	0.03	0.00	0.01	0.00	0.03	0.07
C_{10}	1.00	0.00	0.79	0.58	0.95	0.00	0.01	0.07
C_{11}	0.00	0.25	0.75	0.75	1.00	0.00	0.05	0.05
C_{12}	0.25	0.00	0.75	1.00	0.98	0.00	0.02	0.05
C_{13}	0.00	0.87	1.00	0.95	1.00	0.00	0.05	0.04
C_{14}	1.00	0.33	0.33	0.78	0.00	0.00	0.05	0.04

The CFPR weight of usability e-commerce website can be ranked as follows $C_1 > C_2 > C_3 > C_4 > C_5 > C_6 > C_7 > C_8 > C_9 > C_{10} > C_{11} > C_{12} > C_{13} > C_{14}$ as shown in the Table 4. Based on the result, load time is the most important criteria. Loading time is time median required to load a web page on the browser of the users concretely. Pingdom measures the load time criteria on each website. The last rank of weight criteria is ease of use. The ease of use value calculated by the System Usability Scale (SUS) questionnaire [7]. Table 4 shows that the CFPR and FPP method are responding to the weaknesses of the EA method, which often produces 0 weights. The improvement in the weight loss method using the CFPR method and the FPP method gives weights more significant than 0. However, the advantages of the CFPR method compared to the FPP method are in the number of comparisons between fewer criteria.

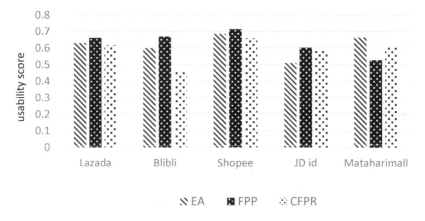

Fig. 2. The final result of usability performance

Figure 2 shows the final result of the usability calculation. Based on Fig. 2, the EA, FPP, and CFPR method gives the first rank is Shopee but provides the different result on the last position. The best-ranking law in the CFPR method are Shopee > Lazada > Mataharimall > JDid > Blibli. Shopee, Lazada, JD, and Mataharimall are in the moderate category. Blibli is in a severe type. As the last ranking, Blibli has some shortcoming in their performance. Criteria for response time, traffic, design optimization, size, the number of items have a poor value. Recommendations for improvement can be seen from the problem. Then traced back to crucial issues according to these criteria.

Recommendations, in this case, are given to improve system performance based on usability evaluation results (Table 5). On the Blibli site, it is necessary to look at the criteria for response time C_2, traffic C_5, design optimization C_7, size C_8, number of items C_{10} and markup validation C_{12}. How to improve website performance on these criteria includes reducing the size of files sent from the server to increase the speed released to the browser and optimize images to decrease the page size. Images are often expressed as a large portion of bytes downloaded on a web page and also often crawl the visual space significantly.

Optimizing images is often considered to save the most significant byte and increase performance for websites: the fewer bytes the browser has to download, the less competition for client bandwidth and the faster the browser can download and make it useful on the content screen. How to increase visitor traffic is closely related to the content of a website. Therefore, some activities are needed to make an attractive headline, good content, and update content regularly. These activities will not only increase the amount of traffic but also increase the number of backlinks on the website.

4.3 Comparative Analysis

We compared the CFPR method with the EA and FPP methods in previous studies in determining website ranking based on usability score and severity ratings [7]. This comparison is made to see whether there are differences in the three ways of classification. Rank similarity testing is done using the help of Statistical Packages for the Social Science (SPSS) software by calculating the Spearman's Rank correlation coefficient. Spearman's rank correlation coefficient ρ given as (11)

$$\rho = 1 - \frac{6 \sum_{i=1}^{n} d_i^2}{n(n^2 - 1)} \tag{11}$$

where d_i is the difference in rankings for each object i, $i \in \{1, 2, \ldots, n\}$.

Table 5. Spearman's rank correlations test for EA, FPP and CFPR method

Alternatives	Ranks			$d_{(EA\_CFPR)}$	$d^2_{(EA\_CFPR)}$	$d_{(EA\_CFPR)}$	$d^2_{(EA\_CFPR)}$
	EA	FPP	CFPR				
Lazada	3	3	2	1	1	1	1
Blibli	4	2	5	−1	1	−3	9
Shopee	1	1	1	0	0	0	0
JDid	5	4	4	1	1	0	0
Mataharimall	2	5	3	−1	1	2	4
Total				$\sum_{i=1}^{n} d_i^2$	4	$\sum_{i=1}^{n} d_i^2$	14
Sig. (2 tailed)				EA-CFPR	0.624	FPP-CFPR	0.104

The similarity rank between EA and CFPR is 0.8 means close to the strong relationship. Correlation is positive, meaning that the increase in usability value using the EA technique is directly proportional to CFPR. The similarity rank between FPP and CFPR is 0.3 means weak. The levels of EA and CFPR correlations are carefully suspected because the ranking of weights between criteria has the same pattern. It can be seen that the highest weight is in the first criterion (load time), so on until the lowest pressure is the criteria of fourteen (ease of use). In contrast to FPP's performance, the highest importance is the third criterion (page rank), while the lower-ranking criteria are frequency of update, traffic, site map design optimization, and number of items. The difference in the pattern on the weighting ranking causes a difference in the ranking pattern of severity ratings.

Correlation is positive, meaning that the increase in usability value using the FPP technique is directly proportional to CFPR. The EA-CFPR value of sig. (2 tailed) is 0.104. The FPP-CFPR value of sig. (2 tailed) is 0.624. Both of the sig. (2 tailed) value gives a significance level of greater than 5%. That means that the correlation between EA, FPP, and CFPR rank is not significantly.

5   Conclusion

The CFPR method has been implemented to improve the quality of e-commerce usability score and severity rating measurement. The CPFR method is proven to overcome the shortcomings of the EA and FPP methods in determining the number of comparisons of expert evaluations. The CFPR used to reduce errors when determining the level of importance between usability criteria. The CFPR method can overcome the shortcomings of EA in weighting. The weight generated by the CFPR method is always more than 0. This fact shows that the usability evaluation model has considered all the criteria weights in the calculation. Based on the results of rank similarity testing, the EA and CFPR methods have a close relationship. The FPP and CFPR methods have a weak correlation in generating usability ranking and severity ratings. However, the fact that two variables correlate cannot prove anything - only further research can confirm that one thing affects the other. Data reliability is related to the size of the sample: more collected data, the more reliable the result. So, for further research, we need to add the alternative to the profound about the differences of the rank.

References

1. Salwani, I., et al.: E-commerce usage and business performance in the malaysian tourism sector: empirical analysis. Inf. Manag. Comput. Secur. 17(2), 166–185 (2009)
2. The Evolution of Ecommerce Business Models in Southeast Asia. https://ecommerceiq.asia/evolution-ecommerce-business-models-southeast-asia/. Accessed 13 Sept 2018
3. Yang, H.: Measuring software product quality with ISO standards base on fuzzy logic technique. In: Luo, J. (ed.) Affective Computing and Intelligent Interaction. AISC, vol. 137, pp. 59–67. Springer, Heidelberg (2012). https://doi.org/10.1007/978-3-642-27866-2_8
4. Abdallah, S., Jaleel, B.: Website appeal: development of an assessment tool and evaluation framework of E-marketing. J. Theor. Appl. Electron. Commer. Res. 10(3), 45–62 (2015)
5. Tullis, T., Albert, B.: Measuring The User Experience. Morgan Kaufmann, Burlington (2013)
6. Sauro, J.: The relationship between problem frequency and problem severity in usability evaluations. J. Usability Stud. 10(1), 17–25 (2014)
7. Wahyuningrum, T., et al.: A comparison of extent analysis and fuzzy preference programming for evaluating B2C website usability. In: 8th IEEE International Conference on Control System, Computing and Engineering, pp. 150–155. IEEE, Penang (2018)

8. Iryanti, E., Pandiya, R.: Evaluating the quality of E-learning using consistent fuzzy preference relations method. In: 6th International Conference on System Engineering and Technology, pp. 61–66. IEEE, Bandung (2016)
9. Wang, T.C., Chen, Y.H.: Applying consistent fuzzy preference relations to partnership selection. Omega **35**(4), 384–388 (2007)
10. Herrera-Viedma, E., et al.: Some issues on consistency of fuzzy preference relations. Eur. J. Oper. Res. **154**(1), 98–109 (2004)
11. Dominic, P.-D.-D., Jati, H.: A comparison of Asian airlines websites quality: using a non-parametric test. Int. J. Bus. Innov. Res. **5**(5), 499–522 (2011)

A Case Study of Customers' Payment Behaviour Analytics on Paying Electricity with RFM Analysis and K-Means

Fakhrul Hazman Bin Yusoff[✉]
and Nur Liyana Asyiqin Binti Rosman[✉]

Universiti Teknologi MARA (UiTM), Shah Alam, Malaysia
fakhrul@tmsk.uitm.edu.my,
2017107061@isiswa.uitm.edu.my

Abstract. The lack of management control and not aware about their potential customer is becoming a growing concern of the service provider. The problem of the late payment cannot be solved simply by providing more penalty and spending more money on management. There is an urgent need for upgrading, for better understanding of the current and potential electricity customers to meet the needs in modern urban life. Hence, this study proposed an integrated data mining and customer behaviour scoring model to manage existing tenants at Empire Damansara. This segmentation model was developed to identify groups of customers based on their electricity payment transaction background of history. Thus, the developer or provider can develop its intensive actions that can maintain its incomes and keep high customers' satisfaction.

Keywords: Data mining · Paying behaviour · Customer segmentation · RFM analysis · K-Means

1 Introduction

In spite of the fact that power is a basic product with for all intents and purposes no substitute, poor bill instalment with respect to power utility customers is compelling the exertion of power service organizations, particularly those in creating nations to continue and extend its arrangement. Populace and salary development are the key drivers behind the developing interest for vitality. Interest for power in Malaysia is continually developing couple with its Gross Domestic Product (GDP) development. The development for power in Malaysia anticipated by Economic Planning Unit (EPU) has demonstrated an expansion of 3.52% in 2012 contrasted with 3.48% in 2011. This development has been driven by solid interest development from business and local parts. The offer of power utilization to add up to vitality utilization has expanded from 17.4% in 2007 to 21.7% in 2012.

It is getting torment when working together and been hoping to be paid for the items or administrations however those instalments are late – or it most noticeably terrible when it doesn't come by any stretch of the imagination. As indicated by Caramela (2018), this circumstance comes up frequently to the organizations which

© Springer Nature Singapore Pte Ltd. 2019
M. W. Berry et al. (Eds.): SCDS 2019, CCIS 1100, pp. 40–55, 2019.
https://doi.org/10.1007/978-981-15-0399-3_4

legitimately influence the progression of their organizations as they need to battle onto the organizations' income. The charging will be higher than the instalment gotten, thus the customer who is declined of postponed to pay caused an obligation and diminished the pay of the supplier organization.

Concerning this investigation, been the outsider of the vitality supplier, it is plainly appeared on the supplier organization gets issue with customer who will not pay subsequent to utilizing their administrations. It is getting more enthusiastically to recognize dissolvability customer as they don't do the customer money check. In spite of the fact that there is punishment which will be given, however it is as yet being the issue. Distinguishing and counteracting of customer conduct who will not pay is one of target which need to fathom by industry. Subsequently, with respect to this exploration intends to profiling their customers so as to show signs of improvement understanding on the customers paying conduct – for controlling and checking the defaulter customers to profiling the customers productively. Additionally, thus, the organization can make vital move dependent on their customer bunch later.

The aim of this paper is described as to provide an accurate description of the specific actions that will be taken in order to reach these aims. The objectives are:

(a) To analyse the patterns of paying behavior of the customers.
(b) To identify the features or variables of customers on paying electricity.
(c) To determine the most suitable data mining model based on identified variables.

There are huge amounts of study concentrating on purchasing behaviours of the customers next to the market crate customers profiling. Notwithstanding, for this investigation will give more thoughts and commitment on how as academicians can investigate more on paying examples of the customers instead of useful for the organizations.

In degree and restriction segment, information utilized in this examination is explicitly taken from the central station of Mammoth Empire Estate Sdn. Bhd., which the engineer of the Empire Damansara from year 2015 April until 2018 December. The information taken is explicitly utilizing the power solicitations and the power instalment receipt which incorporates the subtleties of the customers and the date of each exchange.

In Table 1 demonstrates the subtleties on the units in Empire Damansara, which the 1,535 units are as of now completely involved. It likewise demonstrated that they have four squares with different kind of units, for example, little office/home office (SOHO) and studio – allude to the unit that has everything in a solitary space. Even though there are numbers of the unit, the transaction data taken is only in three-year duration, which is from year 2015 April until 2018 December, which the duration is 45 months. It is due to the business has just started on April 2015. Hence, there was a limitation on this data as the data was not big enough for every customer.

Table 1. Details on units in Empire Damansara

No	Type of units	Number of units	Number of occupied unit
1	Business Suites	51	51
2	Detached Office	6	6
3	Office Tower	13	13
4	Jewel – Commercial Lot	1	1
5	Retail Shop	77	77
6	SOHO 1	394	394
7	SOHO 2	334	334
8	Studio	659	659
Total		**1,535**	**1,535**

The study contains five sections and a few reference sections. Notwithstanding this early on section, Sect. 2 depicts the writing survey as to clarify the progression of this examination proceed in Sect. 3 the way toward grouping is examined other than the techniques utilized all through this examination. The part begins with clarifying the general procedure of bunching by utilizing the RFM examination. Further the K-Means calculation will be complimented the divisions created by the RFM investigation. Next, in Sect. 4 clarifies the aftereffect of this investigation after experience the strategy clarified in past section while in Sect. 5 will give the end alongside the proposal in future works.

2 Literature Review

Here, division proposed by utilizing conduct information since it is usually accessible and constantly developing with time and buy history. RFM (Recency, Frequency, and Monetary) investigation is a prestigious method utilized for assessing the customers' dependent on their paying conduct. A scoring technique is created to assess scores of Recency, Frequency, and Monetary. At long last, the scores of every one of the three factors are united.

The RFM has been generally connected model for customer esteem investigation. It has been utilized by numerous researchers to achieve customer division (Spring et al. 1999; Jonker et al. 2006; Cheng and Chen 2009; Khajvand and Tarokh 2011). Since RFM breaks down the conduct of the customers, it very well may be conceivable to experience conduct based models in the writing (Yeh et al. 2008; Wei et al. 2012).

In the course of recent years, a few analysts have considered RFM models in creating expectation and order models as in Table 2.

Table 2. Overview of literature on RFM analysis and clustering techniques

Titles	Context, research design and analysis	Purposes and key findings
Sarvari et al. (2016) Performance evaluation of different customer segmentation approaches based on RFM and demographics analysis	• Context: Turkey • A data from a global pizza restaurant chain • RFM analysis + K means clustering + Association rules	• Aim to determine the best approach to customer segmentation • Different types of scenarios were designed, performed and evaluated under test condition • They showed that having an appropriate segmentation approach is vital if there are to be strong association. Also, the weights of RFM attributes affected rule association performance positively
Joy et al. (2018) RFM Ranking – An effective approach to customer segmentation	• Context: U.S. State • Retailing store • RFM analysis + K-Means + Fuzzy C-means	• Customer segmentation can be performed using a variety of unique customer characteristics and been compared on their cluster compactness and the execution time
Liu et al. (2018) Customer classification using K-Means clustering: Evidence from distribution big data of Korean retailing company	• Context: Korea • Retailing store • RFM analysis + K-Means + Association rules	• To propose CRM strategies by segmenting customers into VIPs and Non-VIPs and identifying purchase patterns using the VIPs' transaction data and data mining techniques of online shopping mall in Korea
Dogan et al. (2018) Customer segmentation by using RFM model and clustering methods: A case study in retail industry	• Context: Turkey • Retailing store • RFM analysis + K-Means	• To provide better customer understanding, well-designed strategies, and more efficient decisions by building the customer segmentation without considering the expenses of the customers

3 Methodology

3.1 Data Collection

For this examination, the information will be driven from the home office of Mammoth Empire Estate Sdn Bhd which the organization has been checked on their power utility business since 2014. There will be three separate information as appeared Table 3 which their customers' fundamental data, the solicitations and the receipt of the instalment made on power.

Table 3. The Empire Damansara dataset will be used for this study

No	Dataset	Time/period	Key variables	Details of the key variables
1	Customer Details	All	Unit No	Customers' unit number
			Name	Name of the customers
			Gender	Gender of the customers
2	Invoice	All	Unit No	Customers' unit number
			Date	Date of invoice has been produced
			Amount	Amount of invoice
3	Payment Receipt	All	Unit No	Customers' unit number
			Date	Date of the payment received
			Amount	Amount of payment made by customer
			Method	Method of payment made by customer

3.2   Data Pre-processing

The performance of the proposed methodology is evaluated by working on the transactional data set of the customers of people who are leaving in Empire Damansara, Damansara Perdana, starting from 2015 April until 2018 December. The step-by-step process of customer segmentation is presented in this section. There are three original datasets which are the datasets of customer, invoice and the payment receipt. Also, in order to achieve the objectives of this analysis, imputing and creating new attribute is necessary. For this model, there are six new attributes were created as explained in Table 4.

Table 4. New Dataset Description – Empire Damansara, Damansara

No	Variable name	Description
1	last_pay	For determining the date of the last payment made by the customers
2	InvoiceTotal	It will indicate the total of the invoices within 45 months
3	PaymentTotal	Sum of the payment made by customers within 45 months operation
4	Debit	Debit note issued to the customers
5	Credit	Credit note issued to the customers
6	Recency	Recency values are determined by the last date of 2018 December minus the last_pay (in days) Recency (in days) = last date of 2018 December – last_pay
7	Frequency	It indicates the total count of payment made which the maximum count is 45 times
8	Monetary	These monetary values are determined by the paymentTotal minus invoiceTotal Monetary = paymentTotal – invoiceTotal – debit + credit

4 Results and Findings

For this analysis, as demonstrated in Table, 1,535 units gathered in the period of observations length from 2015 April until 2018 December which were selected upon a judgemental sampling process about 1,045 persons are male (68%) and 490 others are female (32%) (Table 5 and Fig. 1).

Table 5. Gender distribution

Gender	Total	Percentage
M, Male	1,045	68%
F, Female	488	32%

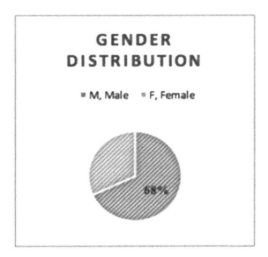

Fig. 1. Gender distribution

However, as having a limited data the descriptive statistical technique can only be applied on gender distribution. It can be explored more such as, grouping the age of the customers and their household incomes.

4.1 RFM Analysis

RFM technique is considered an exceptionally direct yet successful customer examination strategy to comprehend those sort of inquiries according to referenced beforehand. R, F, and M which develop the name of RFM strategy are identified with the accompanying 3: (RFM (Customer Value), 2016).

1. Recency – How as of late did the customer made the instalment?
2. Frequency – How regularly do they made the instalment?

3. Monetary Value – How much measure of customer's exchange that customer paid in certain period?

From the result, it has been pictured by utilizing Tableau in Figs. 2, 3 and 4 for the recovered information on Recency, Frequency and Monetary separately. It interpreted the review of the entire business. As in the recency of the customers demonstrated that the most noteworthy number of customers of 38 lays on 58 days of their recency while the least number of customers of 23 of every 8 and 23 recency days. This circumstance would give the supplier an earnestness alert as there are the greater part of their customers appears deferring their power instalment. In Fig. 4, appeared there were dissipated regularly. Additionally, it appeared a large portion of them paying less recurrence, anyway this circumstance could be taken intently as there are two potential outcomes either the customers are not paying their power bills or they wanted to pay singular amount (e.g.: paying two-months charges in a solitary instalment).

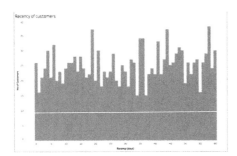 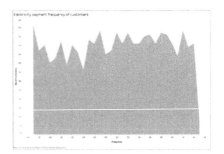

Fig. 2. Recency overview of payment behavior   **Fig. 3.** Frequency overview of payment behaviour

The overview of Monetary values of the data retrieved from Empire Damansara, Damansara Perdana as shown in Fig. 4 below. The size of the circle shown the number of customers while every circle shown the values of the monetary uphold. The colours reflect the range of the monetary values as the darker the orange shown the higher of values of unpaid bills, while the darker the blue shown the higher of surplus values made by the customers.

The biggest circle shown the monetary value of 0, which reflect that there are numbers of their customers will pay their electricity according to their bills. It also indicates that they are the best customers and by showing the gratitude for having them are crucial to the company. However, for the customers that showing the maximum unpaid bills is RM9,450.00 are should be taken seriously as their behaviour affecting the business run directly. Further in this study, it will give the provider more details on their customers which will lead to progressive action towards their various behaviours of their customers.

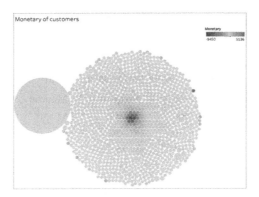

Fig. 4. Monetary overview of payment behaviour (Color figure online)

The meaning of the RFM model recently referenced has some minor change dependent on the focal point of the examination. In this way, next is part the RFM metric into three (3) fragments. As it were, giving the RFM esteems with RFM scoring. For Recency, a recency score is doled out to every customer dependent on date of latest instalment made which ran inside 2 months which 61 days. The score produced by binning the recency esteems into three classifications:

$$R = 1, 0 \leq \text{days} \leq 20$$
$$R = 2, 20 < \text{days} \leq 40$$
$$R = 3, 40 < \text{days} \leq 61$$

For Frequency, the accessible information recovered began from 2015 April until 2018 December, it demonstrates time of 45 months which identical to multiple times of instalment ought to be made by the customers. Additionally, the multiple times been isolated into three classes as pursues:

$$F = 1, 30 < \text{frequency payment made} \leq 4$$
$$F = 2, 15 < \text{frequency payment made} \leq 30$$
$$F = 3, 0 \leq \text{frequency payment made} \leq 15$$

Lastly, Monetary score is assigned on the basis of the differences payment made generated by the formula:

$$\text{Amount of difference} = \text{Amount of payment made} - \text{Amount of invoice} -$$
$$\text{Amount of Debit Note} + \text{Amount of Credit Note} \tag{1}$$

$$M = 1, \text{Amount of difference} > 0$$
$$M = 2, \text{Amount of difference} = 0$$
$$M = 3, \text{Amount of difference} < 0$$

As for simplify the RFM score, below Table 6 presents the span of RFM score that have been used in this analysis.

Table 6. Score for the span of RFM

RFM Score	Recency, R	Frequency, F	Monetary, M
1	0–20	29–45	M > 0
2	21–40	16–30	M = 0
3	41–61	0–15	M < 0

Next, creating a segmented RFM table by adding the segment numbers to the newly created segmented RFM table, which highlighted in the Table 7. Finally, adding a new column to combine RFM score: 111 is the highest score for this study, which had been highlighted in Table 8. In addition, the table only shown the first five units of the location of this case study.

Table 7. RFM table for Empire Damansara

Unit No	Recency	r_score	Frequency	f_score	Monetary	m_score
A-03-01	30	2	40	1	−73.60	3
A-03A-01	8	1	44	1	90.37	1
A-05-01	15	1	38	1	−41.01	3
A-06-01	22	2	29	2	0.00	2
A-06-02	1	1	45	1	24.95	1

Table 8. RFM score table for Empire Damansara

Unit No	r_score	f_score	m_score	rfm_score
A-03-01	2	1	3	213
A-03A-01	1	1	1	111
A-05-01	1	1	3	113
A-06-01	2	2	2	222
A-06-02	1	1	1	111

The understandings have been featured by three classifications as there was the demand made by the supplier. Be that as it may, concerning this examination, RFM scores will give us 27 Sects. (3 × 3 × 3), henceforth this RFM investigation will be reached out by utilizing K-Means grouping.

4.2   K-Means Algorithm

It has been meant to lead K-Means investigation to manufacture groups by thinking about the R, F and M pointers. As referenced previously, the quantity of group ought to be characterized in K-Means technique. Numerous estimations of k (2 to 5) has been tried as appeared in Fig. 5 and ideal arrangement have been assessed for k = 4. The groups that gotten by the K-Means bunching investigation have been qualified concurring for their RFM scores.

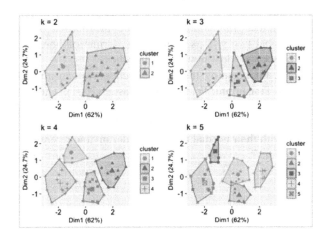

Fig. 5. Clustering evaluation model

Here we utilized K-Means grouping strategy for decide productivity of customers. K-Means grouping is the better method to fragment customers into various dimensions. The significant downside of the K-Means bunching is to decide beginning centroids. Here we utilized twisting bend to locate the ideal number of introductory centroids. As per the above we figured the subsequent bunching bending to locate a particular range that highlights a negligible lessening in normal breadth. There a few techniques for deciding the ideal bunches, which Elbow strategy, Silhouette technique and Gap insights.

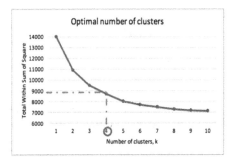

Fig. 6. Optimal number of clusters by Elbow Method

The basic idea behind cluster partitioning methods, such as K-Means clustering, is to define clusters such that the total intra-cluster variation (known as total within-cluster variation or total within-cluster sum of square) is minimized. We can implement this in R and the results suggest that 4 is the optimal number of clusters as it appears to be the bend in the knee (or elbow) in Fig. 6. Besides, the normal outline technique. As to put it plainly, the normal outline approach estimates the nature of a bunching. That is, it decides how well each article exists in its group. A high normal outline width shows a decent bunching. The normal outline technique processes the normal outline of perceptions for various qualities of k. The ideal number of clusters k is the one that augments the normal outline over a scope of potential qualities for k. We can utilize the 'silhouette' function in the group bundle to figure the normal outline width. The outcomes demonstrate that two groups amplify the normal outline esteems with four bunches coming in as second ideal number of bunches.

To wrap things up, for deciding the ideal bunches, we can utilize Gap Statistics Method, distributed by Tibishirani, Walther and Hastie (2001). The methodology can be connected to any grouping technique (for example K-Means bunching, various leveled grouping). The hole measurement looks at the all out intra-group variety for various qualities of k with their normal qualities under invalid reference conveyance of the information as in Fig. 7.

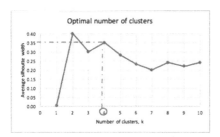

Fig. 7. Optimal number of clusters by Average Silhouette Metho

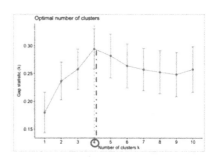

Fig. 8. Optimal number of clusters by Gap Statistics Method

To register the hole measurement technique, we can utilize the 'clusGap' work which gives the hole measurement and standard mistake for a yield. We can picture (Fig. 8) the outcomes which additionally proposes four bunches as the ideal number of groups.

Hence, it is crystal clear that we found the optimal number of initial clusters as four (4).

Before bunching the information, we taken RFM model as an info variable. RFM model is the most ideal approach to portray customers' information. As per the trial we decide RFM values for every group. To improve proficiency, we utilized weighted

RFM values for decide last estimation of every group. As indicated by the last qualities we partitioned customers into three productive dimensions. High gainful customers, Medium productive customers and Low beneficial customers.

Along these lines, getting benefit and conduct of customers, chiefs can settle on choices to improve their administration and increment income of the organization. High beneficial customers additionally called devotion customers, since they are essential to the organization. They straightforwardly influence to the organization's pay. Organizations need serve better administration handle above talked about customer class. They need to deal with them with consideration. On the off chance that one of the high productive customer leave organization it will impact to the organization. Along these lines, most significant customer class is high gainful customers.

We clearly can't simply disregard medium and low gainful customers. Since they are a piece of the organizations' benefit. One day they can be steadfast customer. In this way, organizations need to worry about them. They can propel them for increment their salary. They can orchestrate advancements, limits to fulfill them. They can send a few letters by recollecting advancements and limits.

4.3   RFM + K-Means Clustering Analysis

After a progression of treatment forms, we at that point assess the outcomes acquired for the significance of division understanding formed. As a consequence of a procedure of customer reliability division calculation K-Means and RFM examination. The elucidation of each bunch would having the spot presently directly after the groups are distributed. The rating of the group dependent on RFM separated into four phases, which have been chosen as ideal number of bunches through K-Means investigation. Each property of RFM scoring is finished beginning from the number 3, 2 and 1. Every customer can be distinguished by their unit number. Tables underneath present the aftereffects of RFM scoring each quality of the information group. From 1,535 units have been bunches by utilizing K-Means investigation after experience the RFM examination already. As allude previously, which featuring the scope of scores for each quality R, F, and M. In view of the rating and the consequences of bunching has been acquired, it tends to be deciphered on the attributes of customers Recency, Frequency and Monetary according to show in Table 9 and further will be explained in Table 10.

Table 9. Final clustering and the RFM values

Cluster, k	Total of customers	R	F	M
1	419	2	1	1
2	277	1	1	2
3	430	3	2	3
4	407	2	2	3

Table 10. Interpretation of every cluster

Cluster, k	Interpretation
1	In cluster 1, there are 419 customers that hold the characteristics: Recency, R with a score of 2, which shows that the last exchange was completed in the no so distant past, it gives potential customers to make exchanges back (Soeini 2012). Likewise, customers in this gathering have the best score in scoring their Frequency, F-score by scoring 1, it implies that the customers in this gathering oftenly make instalments at a specific time. While, the score of M, Monetary additionally generally excellent in which customers in this gathering have spending level more in paying their bills
2	In cluster 2, there are 277 customers with the following characteristics: Recency, R esteem with a score of 1, implying that the customers in this gathering has making exchange in a brief period before this examination directed. So also, the quantity of the exchanges on Frequency, F demonstrated that they are making exchange much of the time as showing the score of 1. Be that as it may, the spending level for the organization simply gone through beneficial enough with score 0 as they are paying what should they be paid for consistently
3	In cluster 3 there are 430 customers data with the following characteristics: Score Recency, R is awful, it demonstrates that the gathering of those customers' season of the last exchange was completed quite a while in the past, it high-help that those customers are the potential default customers. Likewise, the estimation of Monetary, M correspondingly to the R as it demonstrated that those customers will pay lesser than the sum they ought to have been paid. In any case, the Frequency, F has a score 2 which depicts a decent recurrence of the instalment made by the customers
4	In cluster 4, there are 407 data of customers that hold the characteristics of: Recency, R-score is 2 which shows that the last exchange was done not very far in the past, it gives potential customers to make exchanges back (Soeini 2012). Thus to the Frequency, F has score by 2, which depicts a decent recurrence of the instalment made by the customers. Nonetheless, the estimation of Monetary, M is so terrible (score by 3) as it demonstrated that those customers will pay lesser than the sum they ought to have been paid. Accordingly it will end up being a torment for the supplier which they need to pay more than they ought to pay as to cover the less sum paid by the customers

Those elucidation has been clarified as and finished up in Table 11, which gives the estimation of every bunch to their customers – from this last worth the company would now be able to concentrate on the best way to design and to get ready on the most proficient method to taking care of their customers as indicated by their groups.

Table 11. Final value for each cluster

Cluster, k	Values, V_k
1	108.016
2	37.456
3	4.277
4	8.462

5 Conclusion and Recommendations

Associations should better comprehend their segments. The orders of customer examination under the various sort of prerequisites are testing task in administrations industry today. An expanding number of customers make this issue increasingly confused. K-Means bunching with RFM model is a superior method to address the above issue. Fundamental issue of the K-Means grouping is to discover beginning bunch centroids. Here we proposed three different ways of distinguishing ideal number of bunches, by utilizing; Elbow strategy, Average Silhouette technique and furthermore Gap Statistics strategy.

Information digging for customer profile is utilized for customer division. To structure the customer division model, this examination take RFM model and K-Means Clustering. The outcome demonstrates that the technique is practical to be fabricated. The aftereffects of this investigation, the customer can be distinguished steadfast customers, productive customers, customers who conceivably to be a defaulter and furthermore customers who are as of now turned into a defaulter. After customers are gathered, the organization can give treatment to every customer gathering.

As a future work we can propose idea of enormous information to deal with colossal customers' information. By the manner in which we can improve RFM model into progressively entangled model like RFMDC model. By including extra two factors into RFM model for improving effectiveness of the info variable of the K-Means calculation. D and C represent Diversity and Continuousness. The assortments of administrations we can use for assessments are, particularly, volume of customer's defaulter. Consistency is the ceaseless after succession in a specific period.

References

Abirami, M., Pattabiraman, V.: Data mining approach for intelligent customer behavior analysis for a retail store. In: Vijayakumar, V., Neelanarayanan, V. (eds.) Proceedings of the 3rd International Symposium on Big Data and Cloud Computing Challenges (ISBCC–2016) Smart Innovation, Systems and Technologies, vol. 49, pp. 283–291. Springer, Cham (2016). https://doi.org/10.1007/978-3-319-30348-2_23

Ansari, A., Riasi, A.: Taxonomy of marketing strategies using bank customers' clustering. Int. J. Bus. Manag. **11**(7), 106–119 (2016)

Caramela S.: How to Handle Non-Paying Customers (2018). https://www.business.com/studys/overdue-and-over-you-what-actions-can-you-take-when-a-customer-hasnt-paid/

Cheng, C.H., Chen, Y.S.: Classifying the segmentation of customer value via RFM model and RS theory. Expert Syst. Appl. **36**(3), 4176–4184 (2009)

Cho, Y.S., Moon, S.C., Jeong, S., Oh, I.B., Ryu, K.H.: Clustering method using item preference based on RFM for recommendation system in U-commerce. In: Han, Y.H., Park, D.S., Jia, W., Yeo, S.S. (eds.) Ubiquitous Information Technologies and Applications. Lecture Notes in Electrical Engineering, vol. 214, pp. 353–362. Springer, Dordrecht (2013). https://doi.org/10.1007/978-94-007-5857-5_38

Coussement, K., Van den Bossche, F.A.M., De Bock, K.W.: Data accuracy's impact on segmentation performance: benchmarking RFM analysis, logistic regression, and decision trees. J. Bus. Res. **67**(1), 2751–2758 (2014)

Dogan, O., Aycin, E., Bulut, Z.: Customer segmentation by using RFM model and clustering methods: a case study in retail industry. Int. J. Contemp. Econ. Adm. Sci. **8**(1), 1–19 (2018). (ISSN 1925 – 4423)

Dursun, A., Caber, M.: Using data mining techniques for profiling profitable hotel customers: an application of RFM analysis. Tourism Manag. Perspect. **18**, 153–160 (2016)

Hu, Y.H., Yeh, T.W.: Discovering valuable frequent patterns based on RFM analysis without customer identification information. Knowl.-Based Syst. **61**(3), 76–88 (2014)

Jonker, J.J., Piersma, N., Van den Poel, D.: Joint optimization of customer segmentation and marketing policy to maximize long-term profitability. Expert Syst. Appl. **27**(2), 159–168 (2006)

Joy, C., Umamakeswari, A., Priyatharsini, L., Neyaa, A.: RFM ranking – an effective approach to customer segmentation. Knowl.-Based Syst. **61**(3), 76–88 (2018)

Khajvand, M., Tarokh, M.J.: Estimating customer future value of different customer segments based on adapted RFM model in retail banking context. Proc. Comput. Sci. **3**, 1327–1332 (2011)

Khajvand, M., Zolfaghar, K., Ashoori, S., Alizadeh, S.: Estimating customer lifetime value based on RFM analysis of customer purchase behaviour: case study. Proc. Comput. Sci. **3**, 57–63 (2011)

Kumar, M.V., Chaitanya, M.V., Madhavan, M.: Segmenting the banking market strategy by clustering. Int. J. Comput. Appl. **45**, 10–15 (2012)

Liu, R.Q., Lee, Y.C., Mu, H.L.: Customer Classification and Market Basket Analysis using K-Means Clustering and Association Rules: Evidence from Distribution Big Data of Korean Retailing Company. Dongguk University, Korea (2018)

RFM (Customer Value): 8 November 2016. https://en.wikipedia.org/wikiRFM_(customer_value)

Sarvari, P.A., Ustundag, A., Takci, H.: Performance evaluation of different customer segmentation approaches based on RFM and demographics analysis. Kybernetes **45**(7), 1129–1157 (2016)

Soeini, R.A.: Customer segmentation based on modified RFM model in the insurance industry. In: International Conference on Machine Learning and Computing, Singapore (2012)

Spring, P., Leeflang, P.S., Wansbeek, T.: The combination strategy to optimal target selection and offer segmentation in direct mail. J. Mark.-Focused Manag. **4**(3), 187–203 (1999)

Sualihu, M.A., Rahman, M.A.: Defaulting on water utility bills: evidence from the greater Accra region of Ghana. Indian Journal of Finance **8**(3), 22–34 (2014a)

Sualihu, M.A., Rahman, M.A.: Payment behaviour of electricity consumers: evidence from the greater Accra region of Ghana. Glob. Bus. Rev. **15**, 477–492 (2014b)

Wei, J.T.: A review of the application of RFM model. Afr. J. Bus. Manag. **4**(19), 4199–4206 (2010)

Wei, J.T., Lin, S.Y., Weng, C.C., Wu, H.H.: A case study of applying LRFM model in market segmentation of a children's dental clinic. Expert Syst. Appl. **39**(5), 5529–5533 (2012)

Yeh, I.C., Yang, K.J., Ting, T.M.: Knowledge discovery on RFM model using Bernoulli sequence. Expert Syst. Appl. **36**(3), 5866–5871 (2008)

You, Z., Si, Y.W., Zhang, D., Zeng, X., Leung, S.C., Li, T.: A decision-making framework for precision marketing. Expert Syst. Appl. **42**(7), 3357–3367 (2015)

Zalaghi, Z., Varzi, Y.: Measuring customer loyalty using an extended RFM and clustering technique. Manag. Sci. Lett. **4**(5), 905–912 (2014)

Visual Data Science

Amniotic Fluid Segmentation by Pixel Classification in B-Mode Ultrasound Image for Computer Assisted Diagnosis

Desiana Wulaning Ayu[1,2], Sri Hartati[1(✉)], and Aina Musdholifah[1]

[1] Universitas Gadjah Mada, Sekip Utara, Yogyakarta 55281, Indonesia
{shartati,aina_m}@ugm.ac.id
[2] Institute of Technology and Business STIKOM Bali, Bali 80226, Indonesia
wulaning_ayu@stikom-bali.ac.id

Abstract. B-mode ultrasound imaging segmentation is facing a challenge in the artifacts such as speckle noise, blurry edges, low contrast, and unexpected shadow. This study proposed a model segmentation considering the local information from each pixel based upon its neighborhood information. The features used are a statistical texture (mean intensity, deviation standard, skewness, entropy, and property) taken based upon the 3×3 and 5×5 window. Random forest was used to classify each pixel into three regions: the amniotic fluid, uterus, and fetal body. An evaluation was carried out by calculating the comparison between the ground truth area and the segmentation results of the proposed model. The experimental results showed that the proposed model has an average accuracy of 81.45% in the 3×3 window and 85.86% in the 5×5 window on 50 tested images.

Keywords: Statistical texture · Pixel classification · Random Forest

1 Introduction

Examination using ultrasonography by a sonographer in pregnancy used as the investigation and diagnosis from the organ anatomy figure. One of the quite essential examinations in pregnancy is an examination volume of amniotic fluid. This fluid divided into three categories: Oligohydramnios (low volume of amniotic fluid), Polyhydramnios (an excess of amniotic fluid) and Normally. One of the techniques to observe the Amniotic Fluid category is using the SDP (Single Deep Pocket) method. This method is carried out by measuring the vertical dimension or perpendicular pull on the largest uterus quadrant (Maxima Vertical Pocket), by placing a caliper between the deepest dimension with the highest area of amniotic fluids, without any umbilical cord and with a fetal body part with a minimum distance of 1 cm in width [1]. On SDP method, each category has the size of vertical dimension in which oligohydramnios is <2 cm, polyhydramnios is >8 cm, and normal or plenty is in the range of 2–8 cm [2]. The volume of amniotic fluids is increasing until the age of pregnancy reaches 14–20 weeks. During the 20–37 week period, the volume of fluid will be stable [3]. In the 30–37-week period of pregnancy, the cavity of the gestational sac is getting narrower as the

© Springer Nature Singapore Pte Ltd. 2019
M. W. Berry et al. (Eds.): SCDS 2019, CCIS 1100, pp. 59–70, 2019.
https://doi.org/10.1007/978-981-15-0399-3_5

fetal body has filled it. To distinguish the area of amniotic fluids, fetal body, and uterus in ultrasound have a difficulty issue, need more times, knowledge, and experience of a radiologist. Segmentation proses can solve this issue.

Some ultrasonography image segmentation is challenging because they had an artifact such as speckle noise, blurry edges, low contrast, and unexpected shadow [4]. The ultrasonography image segmentation, it can be done through several method approaches such as active contour [5, 6] this model requires initialization to obtain the accurate segmentation. Subsequently, the procedure of pixel grouping based upon similarity intensity [7, 8]. Another method as a learning-based approach proven robust to the noise and artifact [9–11], one of them is a pixel classification method. Pixel classification for segmentation is robust but often slow [12], these problems solved by selection of classifier method and modeling label strategy. In this study, labeling sampling carry of using window sampling marker in the classification amniotic fluid object is using Random Forest. This will be a novelty.

The dataset used this study is 50 B-mode ultrasound amniotic fluid images from Surya Husadha Hospital. From the dataset, randomly selected ten imagery as reference images. The reference image generates the dataset pixel represent each class based on the statistical feature texture. The three regions clustered are obtained by learning-based model use the random forest classifier, based on the similarity of the statistical feature from the dataset pixel. Amniotic fluid region obtained through by morphology operation. The result of the segmentation compared with ground truth by measuring the similarity correlation.

2 Material

In this section, explained detail about how to take and prepare the dataset and the data acquisition process.

2.1 Image Acquisition

All dataset used were taken from Surya Husada Hospital in Bali. The age of pregnancy used as the research subject was in the range of 30 to 37 weeks consisting of 50 B-mode ultrasonography image taken using the Accuvix XG machine and transducer with a frequency of 3, 5 Hz, lateral resolution corresponding to 3 mm to 0.2 mm. The size of each image in the dataset was 800×600 pixels, as shown in Fig. 1(a). Each image is cropped to uniform the image dimension in the size of 744×522 pixel. Cropping was also aimed to take the area of interest for the amniotic fluids and to separate other information such as the names of patients, age of the pregnancy, the distance of amniotic fluids, as shown in Fig. 1(b).

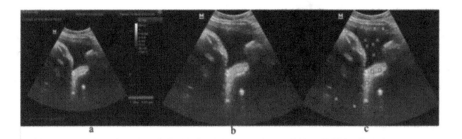

Fig. 1. Ultrasound Amniotic Fluids B-mode. (a) Image as the result of examination from the patients, (b) image as the cropping result and (c) illustration of a manual marker in each class ('blue' - uterus, 'red'- amniotic fluid, 'orange'- fetal body). (Color figure online)

3   Proposed Model

Proposed model segmentation of amniotic fluid region presented in Fig. 2; it consists of 3 phases: (1) dataset construction to obtain the pixel dataset in each class; (2) pixel classification to get the segmentation of the amniotic fluid region; (3) evaluation phase for the segmentation result. Each of the process phases explained as follows.

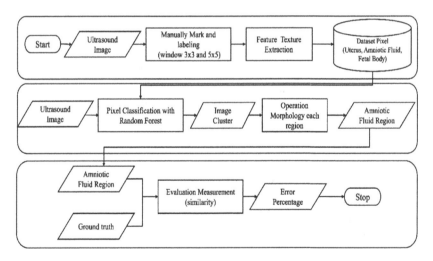

Fig. 2. The proposed model to obtain the region of amniotic fluids.

3.1   Dataset Construction

The proposed model in the process of dataset construction consisted of three steps: manually marker, labeling, and feature extraction.

(1) *Manually Marker and Labeling*

The study used ten randomly images as reference. Each image taken as five samples in each region with the window size of $l \times l$ were ($l = 3$ and 5) thus, obtained 150 datasets of pixels representing each class (amniotic fluid, fetal body, uterus). Retrieval sample and labeling of each class done by radiologist expert. Sampling marker each class in image reference illustrated in Fig. 1(c).

(2) *Feature Extraction*

The feature value of each reference image result obtained through the texture feature approach using the statistical texture (histogram). Statistical textures are the first orde that uses statistical measurements based on the pixel value of the original image without concern with the environmental pixel correlation. In this study, measured five features: mean intensity, standard deviations, skewness, entropy, and properties. Mathematical expressions as listed in Table 1.

Table 1. Statistical features and mathematical expressions

Number	Feature name	Mathematical expression
1	Mean	$m = \sum\limits_{l=1}^{L=1} z_i\, p(z_i)$
2	Variance	$\sigma^2 = \sum\limits_{l=1}^{L=1} (z_i - m)^2 p(z_i)$
3	Skewness	$\mu_3 = \sum\limits_{l=1}^{L=1} (z_i - m)^3 p(z_i)$
4	Entropy	$e = -\sum\limits_{l=0}^{L=1} p(z_i)\, log2\, p(z_i)$
5	Smoothness	$R = 1 - \frac{1}{1+\sigma^2(z)}$

The equation notation in Table 1 is explained as follows:

L = number of distinct gray level
z_i = random variable denoting gray levels
$p(z_i)$, i = 0, 1, 2 ...... $L-1$
n^{th}= moment of z
$\sigma^2(z)$ = the variance in z_i.

3.2   Segmentation by Pixel Classification

Random Forest classifier has been implemented widely for detection, classification and segmentation on the medical images such as MRI [13, 14], CT [15], US [11, 16–18], X-Ray [19]. It has high accuracy, non-parametric method, and establishes variable importance. Based on the advantaged of Random Forest, the problem in the segmentation of the amniotic fluids is solved using the model classification approach, which each pixel will learn to be the membership in one of three classes (amniotic fluid, fetal body, uterus). A different approach, such as clustering technique as K-Means based on image intensities give reasonable success for good quality data. However, in this study, the K-Means method not represented the region of amniotic fluid, because a characteristic of data image is poor, blurry edge, and many artifacts, as shown in Fig. 3(b). Thus, the pixel classifier model with Random Forest using the feature space such as the statistical textures can give a better result in the cluster separation. In this paper, pixel classification was carried out using Random Forest by classifying each pixel based upon the existing pixel dataset $(\bar{x})$, thus resulting in the image divided into three regions, as shown in Fig. 3(c).

Three classes (uterus, amniotic fluid, fetal body) were classified using Random forest feature classifier. This classifier was used to find patterns based on statistical features into the three classes technically, Random Forest classifier does the training by taking the sample randomly in the dataset to result in the subset for each decision tree. Training in each tree was given using 2/3 of total data of training and validated using the samples used to build that tree (out-of-bag-error-estimation) [20]. The stages of Random Forest are as follows:

Step 1: Pick at random K data points from the Traning set.
Step 2: Build the Decision Tree associated to these K data points.
Step 3: Choose the Number Ntree of trees to build and repeat Step 1 and 2.
Step 4: For a new data point, make each one of Ntree trees predict the category to which the data points belong, and assign the new data point to the category that wins the majority vote.

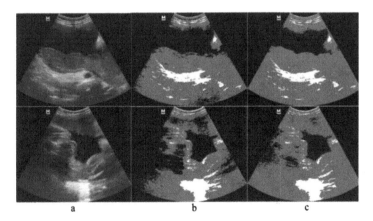

Fig. 3. (a) Image test; (b) clustering result with K-Means; (c) result of the pixel classification with Random Forest.

In this paper, Random Forest using the maximum tree-depth of 5, and the number *Ntree* set in 60 trees as the optimal number from tree to obtain the state-of-bag-error-estimation as shown in Fig. 4. The test on each pixel with Random Forest based on the neighborhood pixel value from the 3×3 and 5×5 window (n-neighborhood) as shown in Fig. 5.

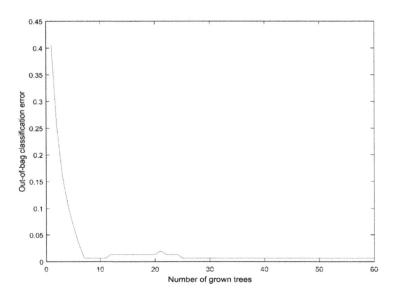

Fig. 4. The optimal number from tree to obtain the state-of-bag-error-estimation.

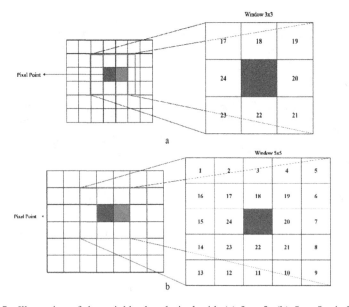

Fig. 5. Illustration of the neighborhood pixel with (a) 3×3; (b) 5×5 window.

3.3 Operation Morphology

As shown in Fig. 6, closing operation and opening operation make the contour smooth and to remove the small holes from the result of pixel classification. Opening operation is the erosion operation followed with the dilation using the similar structuring elements. This operation has a function to make object surface smooth and to eliminate all pixels in the very narrow and small area to be filled by the elements. As a result, all smaller areas from the structuring elements finally will be removed using the erosion operation. The operation morphology placed a role in the smoothing process. The opening operation requires a strel function as the particular structuring element in its operation, as shown in Eq. (1).

$$AoB = (A\theta B) \oplus B \tag{1}$$

The closing operation is useful to smoothen the contour and remove the small holes as defined in Eq. (2).

$$A \bullet B = (A \oplus B)\theta B \tag{2}$$

where A = image pixel A, and B = structuring element pixel B.

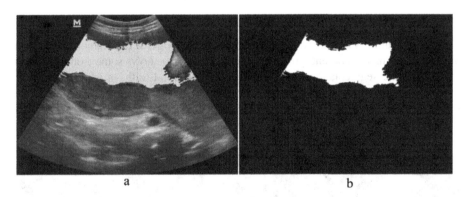

a   b

Fig. 6. (a) Yellow shows the amniotic fluid region resulted from the pixel classification process by Random Forest; (b) result of morphology with the opening and closing operation. (Color figure online)

4 Experiment Result

The proposed segmentation method had evaluated for quantitative comparison results. Quantitative results were obtained based on testing with groundtruth. This validation or testing conducted by similarity of pixels between the groundtruth and segmentation proposed model based on the matrix equation, as follows; $T(A, B) = |A \cap B|/|A \cup B|$ in which A and B were the pair of the ground truth pixel, and the result of segmentation from the model proposed. Similarity method measures the similarity of the pixel does not providing the error of segmentation. At the first phase proposed segmentation method, the proses manually marking utilized in 3 × 3 and 5 × 5 window. In this study also indicated the result of the comparison by using the K-means clustering method. The use of two sizes of the window was to find out the difference in the accuracy of the segmentation results. Table 2. Shows the results of pixel similarty between groundtruth and segmentation of the K-means clustering method and Table 3. shows the results of the pixel similarity between ground truth and the segmentaion of the proposed model.

Table 2. Result of similarity pixel (ground truth and K-means clustering)

Testing window	Similarity pixel average (%)	Dissimilarity pixel average (%)
3 × 3	50.07	49.93
5 × 5	55.53	44.47

Table 3. Result of similarity pixel (ground truth and proposed model)

Testing window	Similarity pixel average (%)	Dissimilarity pixel average (%)
3 × 3	82.39	17.60
5 × 5	85.86	14.13

Figures 7 and 8 show some result examples of the edge of the amniotic fluid area with the K-means method and proposed model. Figure 10 shows some examples from the result of the segmentation test compared to the ground truth, in which green line shows the ground truth, and the yellow line shows the result of the segmentation by the proposed model, and Fig. 9 shows with K-means clustering method.

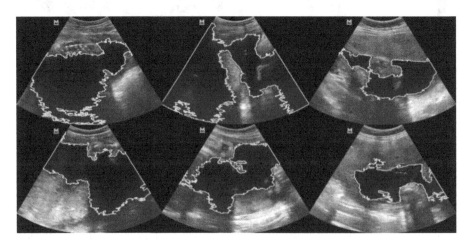

Fig. 7. Result of the edge detection of amniotic fluid with the K-means clustering.

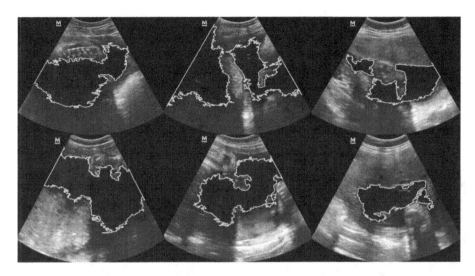

Fig. 8. Result of the edge detection of amniotic fluid is with the proposed model.

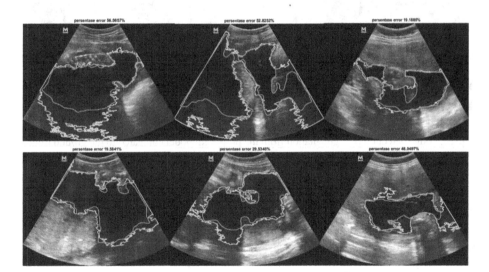

Fig. 9. Edge detection of segmentation from the K-means clustering method compared to the ground truth.

The evaluation was carried out using 50 B-mode to take reference sample use 3 × 3 and 5 × 5 window. The experimental result using the proposed model in a 3 × 3 window reached the average accuracy of 81.45%. On the other hand, using the 5 × 5 window reached total average accuracy of 85.86%. In contrast with the K-means clustering method only produces a total average accuracy of 50.07% using 3 × 3 and 55.53% using 5 × 5. Differences significant results indicate that the method of K-means clustering with less appropriate in the image of the amniotic fluid of the object that has the characteristics of uneven pixel intensity in one region. A very significant difference in accuracy between the Random Forest and K-means is probably due to the Random Forest using an assembly technique that uses multiple classifier learning to determine the class label. With the bootstrap technique, the sample in the Random Forest makes predicting the class of a pixel more accurate because based on voting on several classifiers. It shows that amniotic fluids ultrasound B-mode image are more suitable to use assembly learning techniques, like Random Forest.

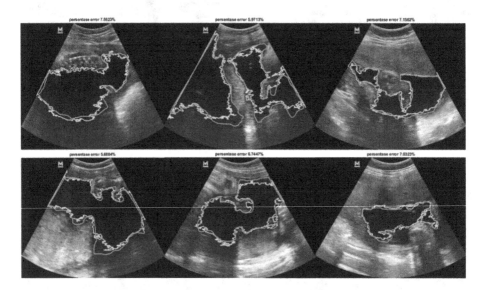

Fig. 10. Edge detection of segmentation from the proposed model compared to the ground truth.

5   Conclusion

From this research, it can be concluded that a segmentation model using the method of pixel classification based on the Random Forest results reach average accuracy result of 85.86% using a 5 × 5 window. In contrast, K-means clustering method only produces total average accuracy 55.53%. This gap occurs because of random forest method adopted an assembly technique that uses multiple classifier learning to determine the class label. With the bootstrap technique, the sample in the Random Forest makes predicting the class of a pixel more accurate because based on voting on several classifiers. Maker size in 5 × 5 window produce better result because the higher window size makes the value of the feature increasingly general in each region but not too good on the pixel located between the edge of each region because it can happen to overlap information between different classes. Overall result indicates by using the random forest as a classifier give the more successful result and size 5 × 5 widows as a recommendation for marker sample.

Acknowledgment. The author would like thank to the Research Directorate of Gadjah Mada University for funding this research in the RTA (Rekognisi Tugas Akhir) 2019 scheme. The author also would thank the Surya Husada Hospital, Bali, for supporting this research in providing data.

References

1. Edwards, A.: 3-D Ultrasound in Obstetrics and Gynecology, vol. 42, no. 2 (2004)
2. Magann, E.F., Hess, L.W., Martin, R.W., Whitworth, N.S., Morrison, J.C., Nolan, T.E.: Measurement of amniotic fluid volume: accuracy of ultrasonography techniques. Am. J. Obstet. Gynecol. **167**(6), 1533–1537 (2013)
3. Magann, E.F., Sanderson, M., Martin, J.N., Chauhan, S.: The amniotic fluid index, single deepest pocket, and two-diameter pocket in normal human pregnancy. Am. J. Obstet. Gynecol. **182**(6), 1581–1588 (2000)
4. Meiburger, K.M., Acharya, U.R., Molinari, F.: Automated localization and segmentation techniques for B-mode ultrasound images: a review. Comput. Biol. Med. **92**, 210–235 (2018)
5. Fang, L., Qiu, T., Zhao, H., Lv, F.: A hybrid active contour model based on global and local information for medical image segmentation. Multidimens. Syst. Signal Process. **30**(2), 1–15 (2018)
6. Chaudhry, A., Hassan, M., Khan, A., Kim, J.Y.: Automatic active contour-based segmentation and classification of carotid artery ultrasound images. J. Digit. Imaging **26**(6), 1071–1081 (2013)
7. Yang, M.-C., et al.: Robust texture analysis using multi-resolution gray-scale invariant features for breast sonographic tumor diagnosis. IEEE Trans. Med. Imaging **32**(12), 2262–2273 (2013)
8. Cai, L., Wang, X., Wang, Y., Guo, Y., Yu, J., Wang, Y.: Robust phase-based texture descriptor for classification of breast ultrasound images. Biomed. Eng. Online **14**(1), 1 (2015)
9. Liu, B., Cheng, H.D., Huang, J., Tian, J., Tang, X., Liu, J.: Fully automatic and segmentation-robust classification of breast tumors based on local texture analysis of ultrasound images. Pattern Recognit. **43**(1), 280–298 (2010)
10. Ye, C., Vaidya, V., Zhao, F.: Improved mass detection in 3D automated breast ultrasound using region based features and multi-view information. In: 2014 36th Annual International Conference EEE Engineering in Medicine and Biology Society, EMBC 2014, pp. 2865–2868 (2014)
11. Qian, C., Yang, X.: An integrated method for atherosclerotic carotid plaque segmentation in ultrasound image. Comput. Methods Programs Biomed. **153**, 19–32 (2018)
12. Fang, H., Kim, J.-W., Jang, J.-W.: A fast snake algorithm for tracking multiple objects. J. Inf. Process. Syst. **7**(3), 519–530 (2012)
13. Gray, K.R., Aljabar, P., Heckemann, R.A., Hammers, A., Rueckert, D.: Random forest-based similarity measures for multi-modal classification of Alzheimer's disease. Neuroimage **65**, 167–175 (2013)
14. Qian, C., et al.: In vivo MRI based prostate cancer localization with random forests and auto-context model. Comput. Med. Imaging Graph. **52**, 44–57 (2016)
15. Criminisi, A., et al.: Regression forests for efficient anatomy detection and localization in computed tomography scans. Med. Image Anal. **17**(8), 1293–1303 (2013)

16. Li, Y., Ho, C.P., Toulemonde, M., Chahal, N., Senior, R., Tang, M.X.: Fully automatic myocardial segmentation of contrast echocardiography sequence using random forests guided by shape model. IEEE Trans. Med. Imaging **37**(5), 1081–1091 (2018)
17. Abdel-Nasser, M., Melendez, J., Moreno, A., Omer, O.A., Puig, D.: Breast tumor classification in ultrasound images using texture analysis and super-resolution methods. Eng. Appl. Artif. Intell. **59**, 84–92 (2017)
18. Ni, D., et al.: Standard plane localization in ultrasound by radial component model and selective search. Ultrasound Med. Biol. **40**(11), 2728–2742 (2014)
19. Ko, B.C., Kim, S.H., Nam, J.Y.: X-ray image classification using random forests with local wavelet-based CS-local binary patterns. J. Digit. Imaging **24**(6), 1141–1151 (2011)
20. Hartati, S., Harjoko, A., Rosnelly, R., Chandradewi, I.: Soft Computing in Data Science, vol. 545. Springer, Singapore (2015)

Machine Learning Assisted Medical Diagnosis for Segmentation of Follicle in Ovary Ultrasound

Eliyani[1,2], Sri Hartati[1(✉)], and Aina Musdholifah[1]

[1] Universitas Gadjah Mada, Sekip Utara Bulaksumur,
Yogyakarta 55281, Indonesia
shartati@ugm.ac.id
[2] Universitas Muhammadiyah Gresik, Gresik 61121, Jawa Timur, Indonesia

Abstract. Machine learning can be applied to the diagnosis of polycystic ovarian syndrome (PCOS), one of criteria for PCOS patients is presence polycystic ovary (PCO). PCO is the presence of the least 12 follicles in the ovary or follicular diameter between 2 and 9 mm and/or increased ovarian volume 10 cm^3. In this research, a computational model for the detection of follicles of various sizes and extracting relevant features of the follicle and calculate the diameter and the number of follicles is proposed. The proposed model consists of pre-processing, speckle noise reduction, follicular segmentation, feature extraction, feature selection, and calculate the diameter of number follicles. The segmentation method uses active contour to divide objects base on the similarity of follicle shape feature so that it is more accurate in calculating the number and diameter of follicles. The performance of this method is tested on a dataset of ovarian ultrasound images of patients at Sardjito Hospital, Yogyakarta using Probabilistic Rand Index (PRI) and Global Consistency Error (GCE).

Keywords: Segmentation · Active contour · Feature extraction

1 Introduction

Machine learning has developed rapidly in various fields to help humans make decisions. This development has great potential for medical imaging technology, medical analysis, medical diagnostics, general health care, one of which is machine learning-assisted medical diagnosis. Some machine learning studies that have helped in the health sector, such as diagnosing leukemia [1], classification of malaria [2] and evaluation of the administration of hypertension drugs [3]. This paper highlights the direction of new research on machine learning to help detect follicles in ovarian ultrasound images. Detecting the number of follicles will be useful as one of means in determining the diagnosis of patients with polycystic ovary syndrome.

Polycystic ovary syndrome (PCOS) is experienced women of reproductive age with excessive androgen production, usually ovulation is irregular, sometimes even does not have ovulation, ovulation disorders cause difficulty in getting pregnant. Women experiencing PCOS have higher risk of type 2 diabetes mellitus, hypertension, cardiovascular, ovarian hyperstimulation syndrome (OHSS) and endometrial cancer [4, 5].

© Springer Nature Singapore Pte Ltd. 2019
M. W. Berry et al. (Eds.): SCDS 2019, CCIS 1100, pp. 71–80, 2019.
https://doi.org/10.1007/978-981-15-0399-3_6

PCOS can be diagnosed by various biochemical criteria such as blood tests to check hormone levels, glucose and thyroid levels. Another method is to see the presence of follicles in the ovary by ultrasonography (USG). This method is often used because it is cheap, fast and low risk for patients [6, 7]. Ultrasound images of ovarian morphology with polycystic ovary (PCO) are characterized by 12 or more follicles measuring 2–9 mm and/or increasing ovarian volume of more than 10 cm^3 [8] Diagnosis of PCOS patients with calculating the number of the follicle is more reliable than calculating ovarian volume [6]. Follicles such as fluid-filled bags to form eggs, follicles produce estrogen which is needed in the development of eggs, so monitoring follicles in the ovary are very important for women for planning pregnancy. The characteristic appearance of the follicle by ultrasound [9] (see Fig. 1).

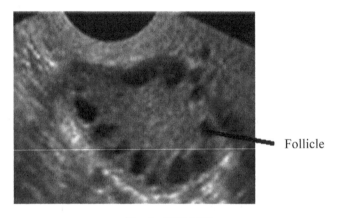

Follicle

Fig. 1. PCOS [9]

PCOS is the most common female endocrine disorder, affecting 10% of women in the reproductive age [10, 11]. Diagnosis of PCOS is a pleasant experience because it has an explanation of the ovarian state and gives hope for treatment, but sometimes makes a woman frightened by the results of the diagnosis. Therefore it is essential to make an accurate diagnosis so that there is no diagnostic error [9, 12].

2 Data Collection

Follicle images are taken from ultrasound tests in accordance with specific rules, in women with regular menstrual cycles, examinations are carried out in the initial follicular phase (days 3–5), for women with irregular menstrual cycles (oligo-amenorrheic) can choose random days, or more precisely, the first 3–5 days after bleeding caused by progesterone. Ultrasonography should be arranged to achieve the best contrast to distinguish follicular fluid from the ovarian stroma [7, 13]. This research uses digital images of ovarian ultrasound from Permata Hati patients in Sardjito Hospital by converting analog images into digital images using the scanner.

3 Method

In general, machine learning for detection and calculation of follicle's number and diameter in ovarian ultrasound images consists of several stages, including pre-processing, reduce speckle noice, segmentation, feature extraction, feature selection, calculate number of follicle, calculate diameter of follicle and model performance evaluation.

3.1 Preprocessing

The initial stage of image processing a pre-processing is an essential stage that needs to be done before image processed; it serves to improve image quality to produce better features in the next step. This research will use pre-processing method of histogram equalization and reduce speckle noise.

3.2 Segmentation

This research uses an active contour segmentation method, and active contour divides ovarian ultrasound image into several regions and separate object area and background from an ovarian ultrasound. Segmentation of active contour has the advantage of being able to adjust to object pattern according to its input parameters so that it can move broader or narrower to look for ovarian ultrasound objects.

3.3 Features Extraction

Feature extraction is a step to create a feature and reduce dimensions of dimensions from high dimensions to lower dimensions. Reliable feature extraction techniques are the main key in solving pattern recognition problems. Chain code is one shape feature extraction algorithm whose value does not change with the treatment of rotation, translation, reflection, and scaling. This method results in a value that shows the direction of the pixel, making up the object. This research will use geometric feature extraction, such as area, perimeter, major axis, minor axis, eccentricity, extent, circularity, and tortuosity produce measurements to recognize correct follicles.

3.4 Feature Selection

Once the feature extraction process is complete, then the next process is the selection of important feature that able to differentiate between very similar objects. The number of elements from the feature extraction process is too much possible to have the same values. The Feature selection is needed to improve accuracy. This research uses a chi-square feature selection technique to extract the most relevant features.

4 Proposed Method

In general, a computational model for detection and calculation of the number and diameter in ovarian ultrasound images are shown in Fig. 2 which consists of stages of image acquisition, pre-processing, follicle detection, feature extraction, feature selection, calculating follicle counts, calculating follicular diameter and model performance evaluation.

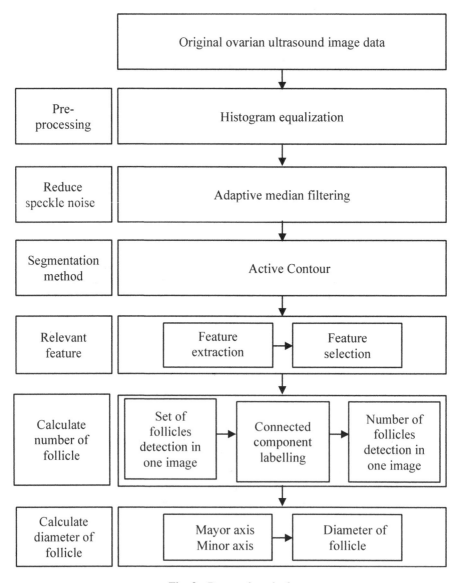

Fig. 2. Proposed method

The initial pre-processing is carried out to improve the quality of the image so that it can produce better features for the next stage. Pre-processing method used here is histogram equalization and speckle noise reduction.

One attempt to improve the distribution of ovarian ultrasound images that irregular shape of circles, is using histogram equalization method. There is a lot of speckle noise in ovarian ultrasonography, so to reduce the speckle noise used with the adaptive median filtering. This method can be used to handle filter operations in damaged images with impulse noise and smooth noise so that the image output is much better than the median filtering result.

The adaptive filter works on a rectangular region S_{xy}. The adaptive median filter changes the size of S_{xy} during the filtering operation depending on specific criteria as listed below. The output of the filter is a single value which replaces the current pixel value at x, y the point on which S_{xy} centered at the time. The Adaptive Median Filter classifies pixels as noise by comparing each pixel in the image with its neighboring pixels around it. The neighborhood size is adjustable, and the comparison threshold is adjustable. The Adaptive median filters work on two levels, namely [14].

Level A: $A1 = Z_{median} - Z_{min}$
                $A2 = Z_{median} - Z_{max}$
                If $A1 > 0$ and $A2 < 0$ then do level B
                Else increase window size
                If $window\ size \le maximum\ allowed\ size\ of\ S_{xy}$ repeat level A
                Else, $output = Z_{xy}$

Level B: $B1 = Z_{xy} - Z_{min}$
                $B2 = Z_{xy} - Z_{max}$
                If $B1 > 0$ and $B2 < 0$ then $output = Z_{xy}$
                Else, $output = Z_{median}$

With
Z_{min} = Minimum gray level in S_{xy}
Z_{max} = Maximum gray level in S_{xy}
Z_{median} = Median gray level in S_{xy}
Z_{xy} = Gray level at coordinates (x, y)

5 Proposed Method Segmentation of Follicle

This research uses an active contour segmentation method to divide the ovarian ultrasound image into several regions and separate the object area and background from an ovarian ultrasound. Active contour for dividing ovarian ultrasound images into several regions and separating the object area and background from an ovarian ultrasound. The segmentation steps using active contour are shown in Fig. 3.

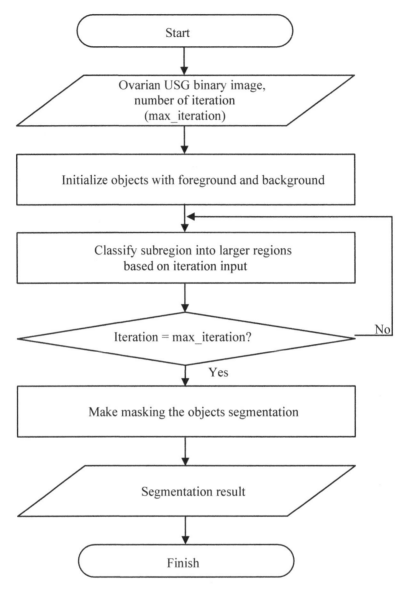

Fig. 3. Active contour

This method has the advantage of being able to adjust the object's pattern with its input parameters so that it can move broader or narrower to look for ovarian ultrasound objects. The active contour is developed to form a contour curve, the edge of the object in the image. Active contour model can be roughly classified as parametric active contour model, with snake parametrically by $z(s) = (a(s), b(s))$.

Active contour model is the snake model proposed by Kass et al. [15]. The snake model is an energy functional is in Eq. (1) which consists of two part called the inner and outer energy.

$$
\begin{aligned}
E^*_{snake} &= \int_0^1 E_{snake}(z(s))ds \\
&= \int_0^1 E_{internal}(z(s)) + E_{imageforce}(z(s)) + E_{constraint}(z(s))ds
\end{aligned}
\tag{1}
$$

Where $E_{internal}$ represent the internal energy of spline due to bending, $E_{imageforce}$ gives rise to the image force, and $E_{constraint}$ gives rise to the external constraint force [15].

The segmentation of the Chan-Vese model is usually dependent on the placements of initial contours, Chan Vese was formulated by minimizing the energy functional is in Eq. (2) [16].

$$
\begin{aligned}
F(c_1, c_2, C) = {}& \mu \, Length(C) + v.Area(inside(C)) \\
&+ \lambda_1 \int_{inside(C)} |u_0(x, y) - c_1|^2 dxdy \\
&+ \lambda_2 \int_{outside(C)} |u_0(x, y) - c_2|^2 dxdy
\end{aligned}
\tag{2}
$$

Parameters for determining evolution are used μ, C is any variable curve, v is the parameter to increase speed, and the constants c_1, c_2, depending on C, are the averages of u_0 inside C and respectively outside C, λ_1 and λ_2 parameter to adjust the intensity inside c_1 and c_2 Active contour model with $v = 0$ and $\lambda_1 = \lambda_2 = \lambda$ is a particular case of the minimal partition problem, the best approximation u of u_0 as a function taking only two value, namely [16].

$$
u = \begin{cases} average\,(u_0)\ inside\ C \\ average\,(u_0)\ outside\ C \end{cases}
\tag{3}
$$

Calculate the related Eular-Lagrange equation for unknown functions ϕ, slightly regularized from functions H and δ, denoted here by H_ε and δ_ε. F_ε is represented associated regularized functional [16].

$$
\begin{aligned}
F_\varepsilon(c_1, c_2, \phi) = {}& \mu \int_\Omega \delta_\varepsilon(\phi(x, y))|\nabla \phi(x, y)| dx\, dy \\
&+ v \int_\Omega H_\varepsilon(\phi(x, y)) dx\, dy \\
&+ \lambda_1 \int_\Omega |u_0(x, y) - c_1|^2 H_\varepsilon(\phi(x, y)) dx\, dy \\
&+ \lambda_2 \int_\Omega |u_0(x, y) - c_2|^2 (1 - H_\varepsilon(\phi(x, y))) dx\, dy
\end{aligned}
\tag{4}
$$

The segmentation method will be evaluated using Probabilistic Rand Index (PRI) and Global Consistency Error (GCE) for evaluating the performance of segmentation method. The PRI value must be higher than the GCE value.

Probabilistic Rand Index (PRI) is rand index that is a fasimilarity function that converted the problem of comparing two partitions with possibly differing number of classes into a problem of computing pairwise label relationships.

PRI calculates pixel pair fractions whose labels are consistent between calculated and actual segmentation. A measure of similarity between two groups of data. PRI gives a value between zero and one. If two segmented results do not have a similarity, the results are zero and if segmented images are identical, result is one. Formula for PRI is defined in Eq. (5) [17].

$$PR(S_{test}, \{S_{GT}\}) = \frac{1}{\binom{n}{2}} \sum_{\substack{i,j \\ i<j}} \left[c_{ij}p_{ij} + (1 - c_{ij})(1 - p_{ij}) \right] \tag{5}$$

Where S_{GT} is ground truth images which are grouped manually $\{S_1, S_2 \ldots S_{GT}\}$ according to segmented ovary images by the algorithm. Set of all perceptually correct segmentation is defined by random numbers p_{ij}, c_{ij} shows the event that a pair of pixels i and j have the same label on S_{test} image [17].

$$c_{ij} = I\left(L_i^{S_{test}} = L_j^{S_{test}} \right) \tag{6}$$

Global Consistency Error (GCE) is a measure to the extent of the results of segmentation. If one segment is the appropriate subset of the other, then the pixel is located in the area of repair and the error must be zero. If there is no subset relationship, then it indicates that the two regions overlap inconsistently. Let S and S' be two segmentations of an image $X\{x_1, x_2 \ldots x_N\}$ consisting of N pixels. The Local Refinement Error (LCE) allows fo refinement to occur in either direction at different locations in the segmentation [17].

$$GCE(S, S') = \frac{1}{N} \min \left\{ \sum_i LRE(S, S', x_i), \sum_i LRE(S', S, x_i) \right\} \tag{7}$$

6 Experiment Results

Proposed segmentation method will be tested on 100 images of ovarian ultrasound from Sardjito Hospital Yogyakarta, Indonesia. This research right now is in the stage of implementation of the segmentation method using active contour. The Fig. 4(a) depicts an original ultrasound image of the ovary and the result adaptive median filtering images are shown in Fig. 4(b).

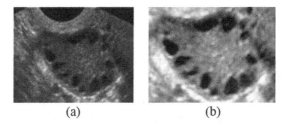

(a)   (b)

Fig. 4. Original ultrasound image of the ovary and result image of proposed method. (a) Original image, (b) Adaptive median filtering image

Images may have poor contrast and can't be used directly. Therefore, it is necessary to get rid of speckle noise presented in an image. In order to remove speckle noise, adaptive Median filters are commonly used. An additional benefit of the adaptive median filter is that it seeks to preserve detail while smoothing speckle noise, the adaptive algorithm performed quite well. The Fig. 5(a) depicts an original ultrasound image of the ovary and the result images of proposed method are shown in Fig. 5(b) and (c).

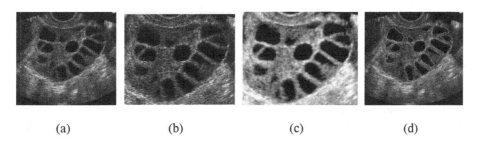

(a)   (b)   (c)   (d)

Fig. 5. Original ultrasound image of the ovary and result image of proposed method. (a) Original image, (b) Histogram equalized image (c) Adaptive median filtering image, (d) Manual segmentation of follicles by medical expert.

7 Conclusion

Previous stage has detected presence of follicles in ovarian ultrasound images, then next step in calculating the number of follicles using connected component labeling. The expected result of this process is to be able to label and color each follicle in an ovarian ultrasound image making it easier to calculate number of follicles in one ovarian ultrasound image. Calculation of follicular diameter that is shaped like a circle is produced from area of follicle. Area is the number of pixels that make up a follicle (region), result of calculating the diameter in pixel values will be converted into millimeters.

Acknowledgment. The author would like to thank the Research Directorate of Universitas Gadjah Mada for funding this research in the RTA (Rekognisi Tugas Akhir) scheme 2019 program.

References

1. Bagasjvara, R., Candradewi, I., Hartati, S., Harjoko, A.: Automated detection and classification techniques of Acute leukemia using image processing: a review. In: 2016 2nd International Conference on Science and Technology-Computer (ICST), pp. 35–43 (2016)
2. Hartati, S., Harjoko, A., Rosnelly, R., Chandradewi, I., Faizah, : Performance of SVM and ANFIS for classification of malaria parasite and its Life-Cycle-Stages in blood smear. In: Yap, B., Mohamed, A., Berry, M. (eds.) Soft Computing in Data Science, SCDS 2018, Communications in Computer and Information Science, vol. 937, pp. 110–121. Springer, Singapore (2019). https://doi.org/10.1007/978-981-13-3441-2_9
3. Soetanto, H., Hartati, S., Wardoyo, R., Wibowo, S.: Hypertension drug suitability evaluation based on patient condition with improved profile matching. Indones. J. Electr. Eng. Comput. Sci. 11(2), 453 (2018)
4. Bilal, M., Haseeb, A., Rehman, A.: Relationship of polycystic ovarian syndrome with cardiovascular risk factors. Diab. Metab. Syndr. Clin. Res. Rev. 12(3), 375–380 (2018)
5. Fernandez, E.D.T., Adams, K.V., Syed, M., Maranon, R.O., Romero, D.G.: Long-lasting androgen-induced cardiometabolic effects in polycystic ovary syndrome. J. Endocr. Soc. 2(8), 949–964 (2018)
6. Ali, H.I., Elsadawy, M.E., Khater, N.H.: Ultrasound assessment of polycystic ovaries: ovarian volume and morphology; which is more accurate in making the diagnosis?! Egypt. J. Radiol. Nucl. Med. 47(1), 347–350 (2016)
7. Coelho Neto, M.A., et al.: Counting ovarian antral follicles by ultrasound: a practical guide. Ultrasound Obstet. Gynecol. 51(1), 10–20 (2018)
8. Azziz, R.: Definition, diagnosis, and epidemiology of the polycystic ovary syndrome. In: Azziz, R. (ed.) The Polycystic Ovary syndrome, Current Concepts on pathogenesis and Clinical care, Endocrine Updates, vol. 27, pp. 1–15. Springer, Boston (2007). https://doi.org/10.1007/978-0-387-69248-7_1
9. Karakas, S.E.: New biomarkers for diagnosis and management of polycystic ovary syndrome. Clin. Chim. Acta 471, 248–253 (2017)
10. Al-Shaikh, S.F.M.H., Al-Mukhatar, E.J., Al-Zubaidy, A.A., Al-Rubaie, B.J.U., Al-Khuzaee, L.: Use of clomiphene or letrozole for treating women with polycystic ovary syndrome related subfertility in Hilla city. Middle East Fertil. Soc. J. 22(2), 105–110 (2017)
11. Caburet, S., Fruchter, R.B., Legois, B., Fellous, M., Shalev, S., Veitia, R.A.: A homozygous mutation of GNRHR in a familial case diagnosed with polycystic ovary syndrome. Eur. J. Endocrinol. 176(5), K9–K14 (2017)
12. Skiba, M.A., Islam, R.M., Bell, R.J., Davis, S.R.: Understanding variation in prevalence estimates of polycystic ovary syndrome: a systematic review and meta-analysis. Hum. Reprod. Update. 24(6), 694–709 (2018)
13. Porru, C., Fulghesu, A.M., Canu, E., Cappai, A.: Ultrasound diagnosis of polycystic ovarian syndrome : current guidelines, criticism and possible update. Austin J. Obstet Gynecol. 4(2), 1074 (2017)
14. Gonzalez, R.C., Woods, R.E.: Digital image processing. p. 976, Nueva Jersey (2008)
15. Kass, M., Witkin, A., Terzopoulos, D.: Snakes: active contour models. Int. J. Comput. Vis. 1(4), 321–331 (1988)
16. Chan, T.F., Vese, L.A.: Active contours without edges. Br. Dent. J. 142(2), 73 (2001)
17. Unnikrishnan, R., Pantofaru, C., Hebert, M.: Toward objective evaluation of image segmentation algorithms. IEEE Trans. Pattern Anal. Mach. Intell. 29(6), 929–944 (2007)

Malaysian Budget Visualization Using Circle Packing

Nur Atiqah Sia Abdullah[1,2(✉)], Nursyahira Zulkeply[1],
and Zainura Idrus[1]

[1] Faculty of Computer and Mathematical Sciences,
Universiti Teknologi MARA, 40450 Shah Alam, Selangor, Malaysia
atiqah@tmsk.uitm.edu.my
[2] Research Interest Group of Knowledge and Software Engineering,
Universiti Teknologi MARA, 40450 Shah Alam, Selangor, Malaysia

Abstract. Data visualization is used to analyze the patterns and trends of data, including budget distributions, business analytics and so forth in a form of diagrams, charts, graphs and so on. In Malaysia, the budget is still in the form of a speech text and infographics. In the previous study, a treemap visualization technique is used to visualize the Malaysia budget, but it has resulted in data congestion as there are too many ministries and its programs in Malaysia. Besides, the visualization failed to compare previous and current budgets. Therefore, this study uses the circle packing technique to visualize Malaysia's budget where it provides a more presentable and interactive way to explore the budget that can compare the current and previous budget on a single website. In order to construct this data visualization, it starts with the preparation of a JSON file to reorganize the budgets data and imported as an input file. Then, circle packing algorithm is used, which includes creating a new pack layout, followed by packing the root node by assigning coordinate x, y, and radius. Besides, it needs to pack the radius, and then set the size using two elements of an array. It follows by setting the padding and packing the array of siblings circles. Lastly, it encloses the circles to the packing. This algorithm is then integrated with JSON file and HTML to visualize it interactively on the web. This dynamic budget visualization webpage is a better option of exploration and it is beneficial to Malaysians as it helps the citizen in seeing a clearer picture of the distributions of their money. Furthermore, it is handy to understand the government future financial planning for Malaysia in an interactive way. This algorithm can be reusable to visualize any other financial data in a hierarchy structure.

Keywords: Data visualization · Budget distribution · Hierarchical data visualization · Circle packing · Malaysia budget

1 Introduction

Data visualization, in general, is a graphical representation of data in the form of a diagram, chart, and graph, which is the art of making people understand data by converting them into visuals [1]. The main advantage of data visualization is not that it

© Springer Nature Singapore Pte Ltd. 2019
M. W. Berry et al. (Eds.): SCDS 2019, CCIS 1100, pp. 81–90, 2019.
https://doi.org/10.1007/978-981-15-0399-3_7

makes data increasingly attractive, yet it gives insight into the complex data to users by using information graphics with a specific end goal to help them effortlessly comprehend and break down information, which generally may be difficult to grasp [2] and literally enable them to take smarter actions and make a better decision [3].

Data visualization can be used to represent data in various domains including emotion [4, 5], online news [6], Malay pronunciation [7], election data [8, 9], budget distribution [10] and etc. Currently, most of the countries used data visualization techniques to represent their data. This kind of visualization eases the viewers to have a quick glance of the overview of the data and it is more interactive, which provides more insight into the data. However, in Malaysia, the budget is still in the form of speech text and Touchpoints through a PDF that is downloadable from Malaysia Treasury and Prime Minister Office (PMO) official websites [11]. Some of the newspapers, websites or even government agencies' websites actually took the initiatives to present the budget through info graphics. For example, iMoney [12] and Bernama [13] had published a simplified 2019 budget on their websites. However, these websites and info graphics are static information and not interactive enough for the viewers.

This study aims to visualize the Malaysia Budget 2019 in a more dynamic approach. This study helps the citizen to view the information at first glance. It is also imperative for them to understand the budget distribution through data visualization.

2 Reviews on Budget Data Visualization

Data visualization is being used in multiple industries including the domains of politics, business and research. Since the study focuses on the budget data, the literature reviews are mostly reviewed previous studies in budget distribution-related representations, which include Singapore [14], Oakland [15], Arlington [16], United Kingdom [17], United State of America [18], and Malaysia Budget 2018 [10].

Singapore has visualized their government revenue and expenditure budget using a stacked area chart as shown in Fig. 1 [14], in order to, gives a sense of the long-term trends in the budget to the readers. Colors are used to differentiate between the government's departments in the country. Even though it is very aesthetically pleasing to view the budget using a stacked area graph, it is impossible to see all data and hard to make a comparison between the years accurately, as shown in Fig. 1.

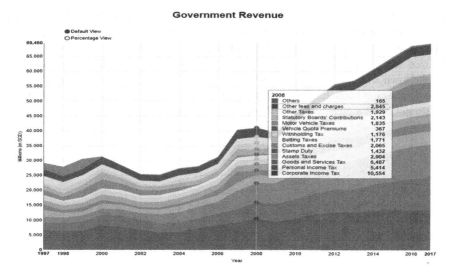

Fig. 1. Singapore government revenue budget visualization. (Color figure online)

Oakland uses a Sankey diagram to visualize the stream of money through Oakland's financial plan [15]. Figure 2 shows the revenue sources, the General Fund, and various non-discretionary funds are in the center, and the rightmost is the various city departments' expenses. It provides hover over and clickable interactivity to give more insights into the money flow and allocation. However, this technique cannot show the absolute figures or the total sum and the comparisons of the budget. Additionally, the texts are clustered and the graphic condensed if there are too many data.

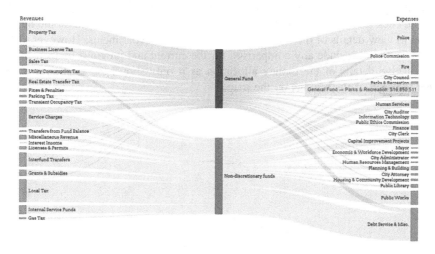

Fig. 2. Oakland budget visualization.

Arlington, MA had taken the initiative to visualize their budget by using a stacked area chart and treemap as shown in Fig. 3 [16]. The stacked area chart is used to compare the total budget for the previous and current year, while the treemap is used to visualized four types of funds. It is always convenient to use various techniques to show the town's budget. However, the budget visualization is not clear enough for the users as it does not show the flow and distribution of the money in detail. Therefore, the users do not have a clear view of where the money is being allocated.

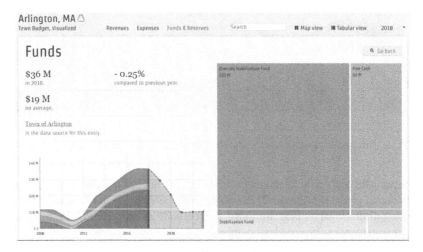

Fig. 3. Arlington, MA 2018 budget.

United Kingdom budget is visualized using bubble tree and mapping technique as shown in Fig. 4 [17]. These visualization techniques provide an interaction where the users can click the regions on the map and see the money allocation and on the bubble treemap to view details of the money distribution for the departments. However, these techniques do not include any comparison for the current and previous budget. It also does not give details of the money allocation as it only shows the total sum of the budget for each of the regions on the map.

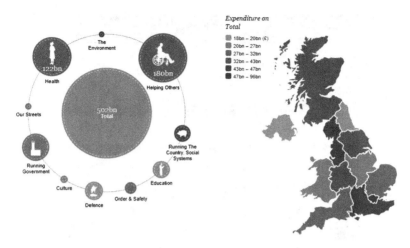

Fig. 4. United Kingdom budget visualization.

USA budget is visualized using the circle packing technique as shown in Fig. 5 [18]. Its objective is to explore the Obama budget for the year 2013 and furthermore to compare the budget with the previous year, in an interactive way. The details of the money allocations are being visualized by using different colors and sizes of the circles. However, this technique does not implement space-efficient as the user can see the white un-used space on the diagram.

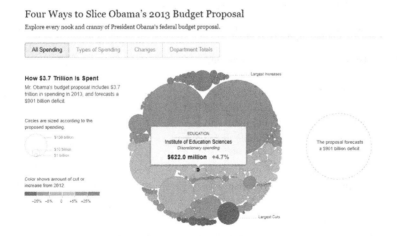

Fig. 5. USA 2013 budget visualization.

Malaysia budget is visualized using a treemap technique as shown in Fig. 6 [10]. However, by using the treemap, the information is too congested because there are too many ministries and departments in Malaysia. Additionally, there is no comparison between previous budgets with the current budget by using this type of visualization.

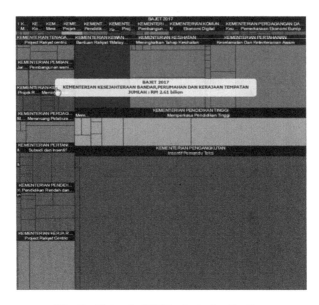

Fig. 6. Malaysia 2017 budget visualization.

From the review of different budget visualization, there are several alternatives in visualizing budget and its distribution, as shown in Table 1.

Table 1. Comparison of budget visualization techniques.

Country	Category	Technique	Budget comparison	Budget allocation
Singapore	Temporal	Bar chart and stacked area chart	Yes	Color
Oakland	Hierarchical	Sankey diagram	No	Size
Arlington	Temporal and hierarchical	Stacked area chart and treemap	No	Size and color
UK	Hierarchical and mapping	Bubble tree and map	No	Size and color
USA	Hierarchical	Circle packing	Yes	Size and color
Malaysia	Hierarchical	Treemap	No	Size and color

From Table 1, it is suggested that hierarchical data visualization is the most used technique in visualizing the budget for different countries. Sankey diagram, treemap, bubble tree, and circle packing, are used to represent budget data. Besides, to ease the comparison of two consequence years of budgets, visualization using a bar chart, stacked area chart, and circle packing are used. Budget allocation is mostly differentiated by using the sizing and colors approach.

From the comparison table, it shows that the circle packing technique is the most suitable technique of visualization as it can make comparison and shows details of money allocation clearly. Besides, it can avoid data congestion because Malaysia has too many ministries and programs to be visualized in a single webpage. It also provides insights regarding the programs of each of the ministries in Malaysia. Circle packing is also able to make a comparison between the current and previous budget allocation in order to show the differences that the government had made for Malaysia financial plan.

3   Circle Packing Representation for Malaysian Budget

After the reviews, the circle packing visualization technique is chosen for visualizing the Malaysia 2019 budget. The original budget data are analyzed and clustered based on ministries in Malaysia. Then, the processed budget data is converted into JSON file. The process is repeated until there is no error before finalizing the correct JSON file. The process continues with constructing the JavaScript based on the circle packing algorithm. The algorithm is explained in details in the next sub-sections. Then the integration of the JSON input file and circle packing JavaScript into HTML are implemented before visualizing it in a 2D dynamic page. This study uses a combination of HTML, Cascading Style Sheet (CSS), Javascript (JS) and Data-Driven Document (D3.js). The HTML and CSS are used to setup the foundation of the User Interface (UI) of the system while JS and D3.js are used to implement the algorithm flow of the system. All the outputs in the HTML document are in Simple Vector Graphic (SVG) format produced by D3.js.

3.1   Data Representation

The first step in this study is to prepare the data. This study retrieved the raw data from the official website of Budget 2019 [11]. There are two years of budget data, which are Budget 2018 and Budget 2019. In Malaysia, the budget distribution is based on pri-oritized programs. The raw data is then reorganized according to the programs and 23 ministries. The differences between budgets are calculated. Then the processed data is converted into Java script in the format of JSON file for each ministry.

3.2   Set Attributes of Circle

The logic of creating a circle is assuming its position by identifying the x and y coordinates and determining the radius, which is r. This logic has been implemented at the beginning of the development by using JavaScript and HTML. The Math.random() function is also used to randomly generate circles position in the canvas. The radius is set randomly from the range of 5 to 50. The array of data will generate random sizes and number of circles, which only stop until it fully fit in the canvas.

3.3 Set Distance

Then, Pythagoras theorem is used to ensure that the circles will not overlap among themselves. This theorem is used to calculate the distance between two points. The result of creating random coordinates and circles using Pythagoras theorem will display a random set of circles without overlapping.

3.4 Set Scales of Circle

Next, the scales of each circle are defined to control the size of the circles to become over-sized. Each of the nodes is defined by using the input data size, which are the size of the circles, following the current year's budget allocation, the colors of the circles following the total difference of the two years' budget.

3.5 Set Force Function

The forces function is added in order to force all the circles to be relocated into designated locations. Lastly, the nodes are separated and clustered into its own ministry by defining the target for its coordinates x and y.

4 Results and Discussion

This study has created circle packing visualization for Malaysia 2019 Budget. The budget overview of the overall budget is as shown in Fig. 7. The amount of money allocation is represented directly proportional by the size of the circles. The increasing money allocation form previous years are represented by the green colored bubble and red means decreasing money allocation, the tone of the color represent the amount of the allocation, the darker color means bigger allocations and lighter color means smaller money allocations compared to the previous year and each of the circle represent various government departments. Details of the budget allocation are shown in the tooltip whenever users hover over the mouse over the circles.

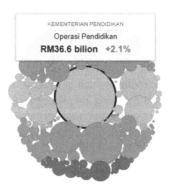

Fig. 7. Malaysia 2019 budget visualization.

The budgets are separated and clustered to its ministry in order to provide more insights regarding the distribution of the money. Figure 8 shows the circles that have been clustered according to their own ministry. Users can view how each of the ministries distributes the budget allocation according to the programs available in the ministry.

Fig. 8. Budget distribution based on ministries.

The project evaluates the accuracy of the budget by validating the data from the actual budget for two years. The evaluation process is conducted by evaluating part of the system and the main objective of this process is to validate the accuracy of the visualized data with the given dataset, which are stored in the JSON input file. The result between the input data and the output data is compared to validate the system correctly visualized on the webpage. These validation processes are done manually by cross-checking the input data with the output data.

5   Conclusion

Data visualization can be categorized into several categories. Finding the most suitable techniques usually depends on the type of data to be visualized. Hierarchical visualization technique is found to be the most suitable technique to represent budget data. The literature review can conclude that circle packing is suitable to be used as it can make a comparison between two years of budget data. The circle packing is created based on the logic of a circle that needs to have three important variables, which are the coordinates x, y, and radius. Different color schemes are useful to clearly show the difference between the budget allocation between two years. Therefore, with the help of visuals, the viewer can interpret easily the budget allocation in Malaysia.

Acknowledgments. The authors would like to thank Faculty of Computer and Mathematical Sciences, Universiti Teknologi MARA for sponsoring this paper.

References

1. Ephrati, A.: Buyers Beware: Data Visualization is Not Data Analytic. https://www.sisense.com/blog/buyers-beware-data-visualization-not-data-analytics/
2. Friendly, M.: Milestones in the history of thematic cartography, statistical graphics, and data visualization. Engineering **9** (2008). https://doi.org/10.1016/S1360-1385(01)02193-8
3. Saltin, J.: Interactive visualization of financial data: development of a visual data mining tool. http://www.diva-portal.org/smash/record.jsf?pid=diva2:555392
4. Idrus, Z., Bakri, M., Noordin, F., Lokman, A.M., Aliman, S.: Visual analytics of happiness index in parallel coordinate graph. In: Lokman, A., Yamanaka, T., Lévy, P., Chen, K., Koyama, A. (eds.) KEER 2018. AISC, vol. 739, pp. 891–898. Springer, Singapore (2018). https://doi.org/10.1007/978-981-10-8612-0_94
5. Montanez, A.: Data Visualization and Feelings – Psychology project explores human emotions through interactive visualization. https://blogs.scientificamerican.com/sa-visual/data-visualization-and-feelings/
6. Yafooz, W.M.S., Abidin, S.Z.Z., Omar, N., Hilles, S.: Interactive big data visualization model based on hot issues (online news articles). In: Berry, M.W., Hj. Mohamed, A., Yap, B.W. (eds.) SCDS 2016. CCIS, vol. 652, pp. 89–99. Springer, Singapore (2016). https://doi.org/10.1007/978-981-10-2777-2_8
7. Noh, Z., Abidin, S.Z.Z., Aliman, S., Ardi, N.: Visual design for Malay consonants pronunciation. Int. J. Eng. Technol. (UAE) **7**(4.36 Special Issue 36), 212–215 (2018)
8. Abdullah, N.A.S., Mohamed Idzham, M.N., Aliman, S., Idrus, Z.: Malaysia election data visualization using hexagon tile grid map. In: Yap, B.W., Mohamed, A.H., Berry, M.W. (eds.) SCDS 2018. CCIS, vol. 937, pp. 364–373. Springer, Singapore (2019). https://doi.org/10.1007/978-981-13-3441-2_28
9. Krum, R.: Landslide for the "Did Not Vote" Candidate in the 2016 Election!. http://coolinfographics.com/blog/2017/8/9/landslide-for-the-did-not-vote-candidate-in-the-2016-electio.html
10. Abdullah, N.A.S., Wahid, N.W.A., Idrus, Z.: Budget visual: Malaysia budget visualization. In: Mohamed, A., Berry, M.W., Yap, B.W. (eds.) SCDS 2017. CCIS, vol. 788, pp. 209–218. Springer, Singapore (2017). https://doi.org/10.1007/978-981-10-7242-0_18
11. Kementerian Kewangan Malaysia, Belanjawan Tahunan 2019. http://www.treasury.gov.my/index.php/belanjawan/belanjawan-tahunan.html
12. Gazi, F.: Budget 2019 By The Numbers. https://www.imoney.my/articles/budget-2019-infographic
13. Budget 2019. http://budget.bernama.com/iglist.php
14. A data-centric look: Singapore Budget 2017. https://viz.sg/viz/budget2017/
15. MacFarland, I., Betancourt, F., Stiles, A.: Open Budget: Oakland (2019). http://openbudgetoakland.org/2017-19-proposed-budget-flow.html
16. Arlington, MA Town Budget Visualized (2019). http://arlingtonvisualbudget.org/funds/2018/t/d42b2bb7
17. UK Budget 2019: Where Does My Money Go?. http://app.wheredoesmymoneygo.org//bubbletree-map.html#/~/total
18. Carter, S.: Four Ways to Slice Obama's 2013 Budget Proposal. https://archive.nytimes.com/www.nytimes.com/interactive/2012/02/13/us/politics/2013-budget-proposal-graphic.html?hp

An Overview of Visualization Techniques: A Survey of Food-Related Research

Nurfarah Mazarina Mazalan$^{(\boxtimes)}$, Zainura Idrus,
Nur Atiqah Sia Abdullah, and Zaidah Ibrahim

Faculty of Computer and Mathematical Sciences,
Universiti Teknologi MARA (UiTM), 40450 Shah Alam, Selangor, Malaysia
nurfarah.mazarina@gmail.com

Abstract. The variety of food including food choice and preference, food production, dietary intake and other have been done by previous researchers. Other than that, the implementation of visualization techniques has been used in visualizing the result and its findings regarding on food-related research. However, there is lack of understanding towards food-related research in Malaysia. Furthermore, the graphical view that had been illustrated is very limited which lead into the difficulty of understanding the pattern of food-related studies. Therefore, this paper purposely is to do an overview regarding on the visualization techniques that have been used in displaying the results and its finding. The methods that have been used in this study is by constructing the comparison table of visualization techniques such as tree map, heat map, canonical map, word cloud and semantic tree. The classification of visualization techniques also has been differentiated into four type of structures of visualization which are hierarchical, relational, textual and spatial. The aim of this study is to analyze the variety of visualization techniques. This study also is to identify the criteria of each visualization techniques such as size, color, latitude and longitude, vertices, edges and others. Lastly, this paper produces the conceptual framework of visualization techniques as expected outcome due to assist people in making decision in order to get the suitable of visualization technique based on the criteria that matches with its requirement.

Keywords: Visualization techniques · Information visualization · Food-related · Social media · Criteria of visualization first section

1 Introduction

The previous researchers have done many studies towards food-related such as dietary food patterns, food choices and preferences, food intake and others. This can be seen that food-related research have its importance in order to gain other perspectives for consumers to make decision making on what food to eat based on several criteria. The need for food is a basic need with a clear and simple goal and straight forward solution on how to be satisfied [1]. For example, food choices can be considered as common knowledge that people have different food preferences whereas food preference can be indicated as a consumers' choice of food product over another.

© Springer Nature Singapore Pte Ltd. 2019
M. W. Berry et al. (Eds.): SCDS 2019, CCIS 1100, pp. 91–104, 2019.
https://doi.org/10.1007/978-981-15-0399-3_8

The identification of food-related research will facilitate our understanding of consumer choices, with important implications for marketing strategies and efforts to improve human and environmental health with dietary modifications [2]. The graphical representation in visualizing the overall patterns of food-related studies in tasks such as healthcare and business purposes have been contributed to the understanding of information to describe the food choices, dietary patterns and capture hidden information towards food-related and emotions, public health and even its location.

However, most of the finding towards food-related research in Malaysia have not been visualized properly which can be ease the understanding of consumers. For instance, the study of efficiency and productivity of selected food processing industries had been developed in comparative analysis which was to evaluate the performance and change in the technical towards 34 food processing industries in Malaysia [3]. Anyhow, the graphical view that had been illustrated is very limited which lead into the difficulty of understanding the pattern of studies. For example, a few papers regarding on Malay traditional faced the similar problems as it failed to visualize overall pattern towards the variety of food heritage and culture in Malaysia [4, 5]. Furthermore, the visualization that have been developed was only used for results and its finding without stressing the criteria of visualization in representing the data.

The researchers have conducted the several of visualization techniques in food-related research in the past few years in order to some purposes such as to enable the unique reflection of food practices [6], offers real-time business overviews [7] and identify patterns the potential relationships between food and contextual factors [8]. The main purpose in visualization techniques is to identify all the criteria that have been used to visualize the findings in food-related research. To overcome the above weakness and fill the gap in current research, this study proposes an overview of the visualization techniques that had been used in food-related domain.

Therefore, the aims of this research are to analyze the variety of visualization technique in order to get the overall pattern or idea towards food-related research. Then, this study also intended to identify the criteria of visualization technique that had been matched with its findings towards food-related research. Lastly, this study also is to conceptualize the framework of visualization techniques that have been used based on the domain of food-related researches.

2   Food-Related Visualization

There are numerous studies regarding of visualization of food-related research that have been found in the past few years. This section will be explained the visualization techniques that had been used in previous studies with its criteria. There are numerous studies regarding of visualization of food-related research that have been found in the past few years. This section will be explained the visualization techniques that had been used in previous studies with its criteria.

2.1   Food-Related Research

Food can be described as the core of individual's daily survival which can effect on personal health, emotions and even its location of food. One of the studies that have been found in food-related domain was the visualization of food named FoodMood (Fig. 1) is data visualization project that capture the content and sentiment of global English-language tweets about food. Since FoodMood considered as global food sentiment measurement, this project aimed to gain better understand towards food consumption patterns and the impact on daily emotional well-being of people such as Gross Domestic Product (GDP) and obesity levels involved.

Through this research, FoodMood has been developed where all the lists of food choices were in western foods and level of happiness indicates the emotion taken from the tweets at a time [6]. Furthermore, the major task in search interface that provides an engaging means of exploration, the purpose of FoodMood is to utilize and visualize the digital social data to lend important insights into citizen behavior and urban living patterns and practices. This can be resulted as an immersive interaction with the data for user experience in order to find the overall pattern of food.

Another studies regarding on food-related research is collecting and visualizing the dietary behavior and reasons for eating using Twitter platform [8]. Instead of testing the feasible and acceptable of Twitter in capturing young adults' dietary behavior and reasons for eating, another aim of this study was to visualize data from Twitter by using novel analytic tool designed which helped in identifying relationships among dietary behaviors, reasons for eating and contextual factors. This can be concluded as a first step toward understanding the complexity of eating behavior is developing and testing a method to capture these critical data. This study proposed by tracking typical dietary choices and reasons for making food-related decisions simultaneously and in real time, relationships between these factors would emerge, and the addition of contextual factors, such as when, where, and with whom a person eats, would enrich these data providing additional information related to diet quality and potential mediators of dietary behavior.

One of the case studies regarding on Noodle Company in visualizing the results of opinion mining from social media contents have been done in food-related research domain. The study is purposely focused on formulation on comprehensive and practical approach to conduct opinion mining with visual deliverables. Despite to describe the cycle of opinion mining practical by using social media, the presentation of visualization output such as domain specific lexicons, volume and sentiment graphs, information visualization and others so the potential users can consider the tools and techniques for immediate adoption. Thus, the method of opinion mining with had been conducted on with a leading Korean instant noodle company.

The study regarding on the location of food-related research in tourism purpose has been conducted from perspectives such as regional geography [9]. Over the past decade, tourism and hospitality research has studied means to examine consumer perceptions of the food safety of restaurants and improve the safety of the food supply chain. Some studies have investigated the impact of health inspectors on food safety scores and health inspection violations. This study also investigated food safety violations in chain and non-chain restaurants and found that the location of restaurants within a state is important. This finding indicates that the standards of health inspections and food safety violations may differ depending on location. Thus, exploring the spatial patterns of food safety may be a prerequisite to effective food safety planning and policy-making in tourism.

As the purpose of study was to explore the spatial patterns of food safety, some of findings can help tourism or hospitality managers and local destination management organization (DMOs) better understand the local patterns of food safety violations in tourism destination and better cultivate location-based food safety planning, policy and strategy to develop a destination.

2.2   Visualization Techniques

The visualization can be made in the variety of techniques to create tables, diagrams, images and other intuitive display ways in order to reveal the data. The basic criteria of visualization techniques will produce the pattern to get better understanding of data collection.

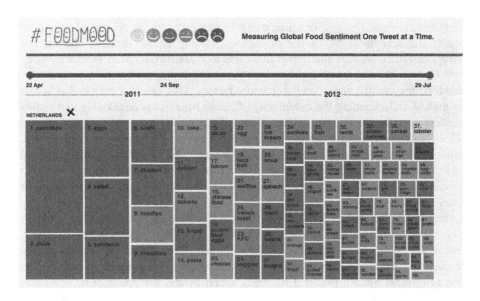

Fig. 1. Example of tree-map of FoodMood [6] (Color figure online)

The use of treemap as the visual structure of FoodMood enables enable users to explore emotion in an interactive play space, where various sorting options require users to search, analyze and compare by time, country, food, emotions and number of tweets [6]. Figure 1 shows the example of tree map visualization. Given the multiple outcome of datasets, the various areas of the treemap were assigned visual marks such as color which indicates range of emotion whereas size indicates the quantity of tweets.

The treemap design of FoodMood empowers users to first gain an overview and broad awareness of food landscape, then zoom in and filter for more granularity and finally, analyze the visual data with details on demand. This can be seen that user can see which foods are most tweeted with highest sentiment expression and view the most obese countries and least GDP countries and select an interval or a point in time on the timeline.

Another tree map that have been developed in food-related research study was on the visualizing the result from case study of noodle company at Korea by using valence tree maps techniques [7]. In generating this tree map, the criteria that have been used were the size of block which represents main category and sub-category, the sentiment elements represented by using three colors where green indicates positive, red indicate negative and black indicates natural. Figure 2 shows the example of valence tree map for Noodle Company in Korea.

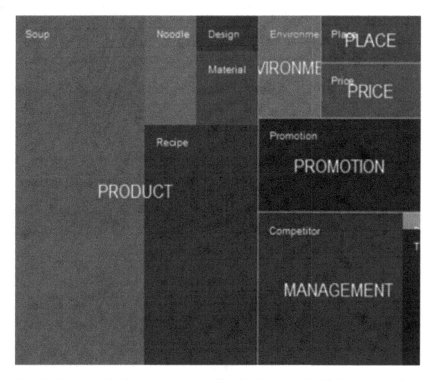

Fig. 2. Example of valence tree map for Noodle Company [7] (Color figure online)

Next, the spatial structure of visualization have commonly used in visualizing the location of food-related research. The studies regarding on dietary behavior and reasons for eating has been developed behavioral maps that consists of canonical maps and heat maps [10]. In this study, the canonical maps have been developed by using GMap software application. GMap maps have been used for a variety of behavioral applications. Other than that, using these maps is also able to see behavioral patterns and

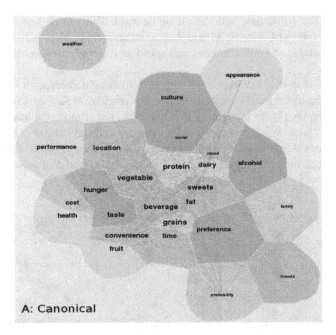

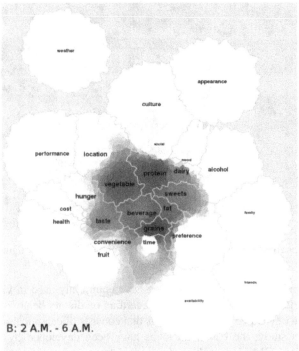

Fig. 3. (a) Example of canonical maps (Behavior maps) [10]. (b) Example of heat maps (Behavior maps) [10] (Color figure online)

infer relationships. The criteria of visualization that have been implemented in this map were vertices represents hashtag usage in Twitter, edge weight indicates the value of occurrence and co-occurrence which have been merged into matrix form. The variety colors to represent "countries" indicates the most frequency co-occurring hashtag and the size of hashtag represents the frequency of hashtag received.

Other than that, the heat maps also have been produced its criteria to visualize the behavioral patterns as well. The heat maps represent by time of day which explored over 24 h. The frequency of hashtag reported during 4-h time blocks which represented by the intensity of color where the darkest colors represent more frequent hashtag. Meanwhile, the light blues showed the progressively in lower frequencies and white color indicates none of frequency activity. Figure 3(a) and (b) shows the example of canonical maps and heat map in visualizing the dietary pattern by using Twitter.

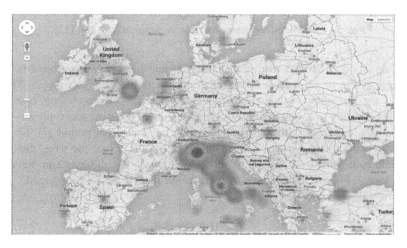

Fig. 4. Example of heat maps [15] (Color figure online)

The usage of heat maps also had been implemented in findings as the output deliverables which have been done by previous research. By using the same features in developing the heat maps, the intensity of colors had been shown the highest frequency of entities. Figure 4 shows the Italian food topics has high intensity in Italy as reflective of immigration patterns to worldwide in related to topics for foods of various geographic origins [15].

Other than heat map visualization had been produced as graphical deliverable in analyzing the language of food from social media, another map had been discovered as well within top terms by state [15] as shown in Fig. 5. The implementation of term frequency- inverse document frequency, *tf-idf* in reducing words that occur frequently across all states have been done for ranking purposes. Thus, the terms with highest ranking in each state visualized on the state in the maps.

Fig. 5. Example of top terms by state by geographical map [15]

The tag cloud is another textual visualization that have been developed in food-related research in to get the overview regarding on the frequency of topics. The tag cloud was also one of the visualization outputs in visualizing the result from case study of Noodle Company at Korea in order to see the frequent hot issue and topic keywords that have been extracted with high frequency during a targeted period [7]. Size and color of the words in the cloud reflect volume of a topic. From the word cloud, the study could gain some overview intuitively which reveals the difference in perception between topics. Figure 6 is example of word cloud that have been developed.

Fig. 6. Example of word cloud [7]

Another tag cloud that had been used in food related research in this study where it offers a space-efficient way to summarize text by highlighting important words. The tag cloud was focused on comparing two or more groups of texts which are the words are scaled by importance, related words are close to each other, and important words that occur in both groups are in the same locations in all clouds. Figure 7 shows the words that appear frequently in sets of tweets are black and those that are frequent on weekend related word were blue [15].

Fig. 7. Example of parallel word cloud [15] (Color figure online)

The semantic tree of generic food taxonomy that have been illustrated in semantic aware food recognition in order to produce classification [14]. The criteria needed in implemented this tree were weight indicates the network parameters from input layers to high-level of outmost layer of semantic tree and node represents the initial probability scores which had considered as initial values. Figure 8 shows the example of semantic tree which consists of classification of food classes.

Table 1 shows the comparison of visualization types that have been used in presenting the food-related research based on previous researchers. Additionally, the factors that influenced towards food related research were also mentioned in comparison Table 1. Some papers do not carry out visualization techniques in illustrating the result and finding. However, the content regarding on the food-related studies had been implemented where the relationship of food and other factors such as emotion, healthcare and location were studied in the last few years. Based on the tables, FR for Food Related, E for Emotion, PH for Public Health, LB for Location or Business, TM for Tree Map, HM for Heat Map, CM for Canonical Map, WC for Word Cloud and ST for Semantic Tree.

Fig. 8. Example of semantic tree [14]

Table 1. Background study of visualization technique past research

Literature review	Food-related	Emotion	Public health	Tree-map	Word cloud	Semantic tree
Data visualization and retrieval based convenience store location model of fresh products [10]	✓					
Using Twitter data for food-related consumer research: a case study on "what people say when tweeting about different eating situations" [11]	✓	✓				
You tweet what you eat: studying food consumption through Twitter [12]	✓		✓			
FoodMood: measuring global food sentiment one tweet at a time [6]	✓	✓	✓	✓		
Data visualizing the story of food and emotion [13]	✓	✓		✓		
Visualizing the results of opinion mining from social media contents: case study of a Noodle Company [7]	✓	✓		✓	✓	
Learning to make better mistakes: semantics-aware visual food recognition [14]	✓					✓
Analyzing the language of food on social media [15]	✓				✓	

3   Food-Related Visualization

This section, this paper will be explained on the research method. The chosen information visualization for this study have been categorized in the structure of visualization as Fig. 9.

In this study, there are six types of information visualization that have been developed in the previous researches. The structure of visualization has been carried out from the "design for information" [16]. Based on the figure, the tree map has the hierarchical structure in order to visualize the big picture of information based on area size and color.

Heat map and canonical map and have been undergo spatial structure of visualization. By visualizing the geographical map, the visualization towards location factors have been clearly illustrated to gain the information by region. The information needed such as frequency of entities will be visualized by using features such as intensity of color at each region of state involved instead gaining the accurate location through latitude and longitude in the maps.

Word clouds are the visualization that have been in textual visualization. The color and size of word plays a major role in visualizing the information needed such as the frequency of entities. Unlike tree map, the visualization of word does not need block. The size of word can be merged by the highest frequency of topic related and vice versa.

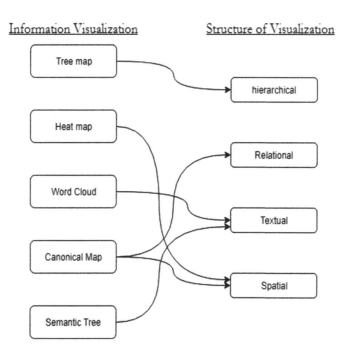

Fig. 9. Information visualization and structure of visualization

Semantic tree also can be labelled in textual structure of visualization. This is because the usage of vertices and node are connected to each other in order to produce the layers from inner to outmost. The classification of variables is made by clustering. However, the hierarchy of classification of classes have been visualized by node with its child in order to gain "big picture" of the topic related.

4   Result and Discussion

The conceptual framework is the expected outcome of this paper. Based on the conceptual framework in Fig. 10, the criteria that have been listed in visualization technique have been stated in previous researcher regarding on its information visualization. The criteria listed based on visualization techniques have been included such as area size, color scheme, latitude and longitude, vertices, edges, time-varying, color intensity, ranking, parallel, space-efficient, weight and node to visualize the result and its finding.

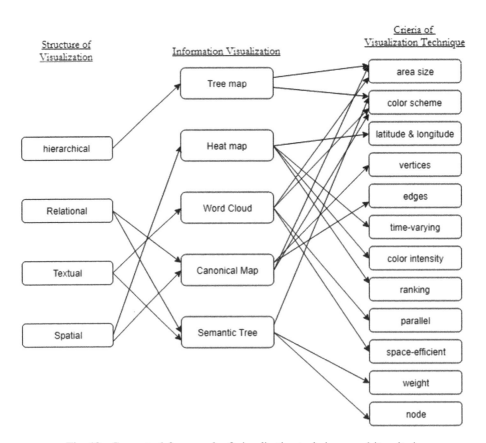

Fig. 10. Conceptual framework of visualization techniques and its criteria

Through the conceptual framework, this can be seen the most features that have been used in this developing visualization techniques are the color and size. Based on studies, the color and size indicate the frequency. From the conceptual framework, the tree map has the least of combination features which only consist of area size and colors. This can be seen that the characteristic of visualization is hierarchical ways and it directly straight forward. However, some of visualization that had been used did not particularly explained on the criteria of visualization. Thus, it is difficult to identify the accurate features that have been used that have been matches the data of food-related topics.

Thus, the conceptual framework has been illustrated in order to assist people in making decision in order to get the suitable of visualization technique based on the criteria that matches with its requirement.

5   Conclusion

Based on the aims of this paper, the first aim of this study is to analyze the variety of visualization techniques is achieved where the list of information visualization analyzed in this paper are tree map, behavioral maps such as heat map and canonical map, spatial pattern, word cloud and semantic trees based on food-related research. The second aims which is to identify the criteria of visualization techniques have been illustrated in the conceptual framework where all the criteria are based on each information visualization that have been analyzed in food-related research. In conceptual framework, the study has been carried out the overview of information visualization, structure of visualization and criteria of visualization techniques This study will assist people in order to find the most suitable of visualization techniques in visualizing the commonly food mentioned in social media.

Acknowledgement. The authors would like to thank Faculty of Computer and Mathematical Sciences, as well as Universiti Teknologi MARA for the facilities and financial support.

References

1. Vabø, M., Hansen, H.: The relationship between food preferences and food choice: a theoretical discussion. Int. J. Bus. Soc. Sci. **5**(7), 145–157 (2014)
2. Pula, K., Parks, C.D., Ross, C.F.: Regulatory focus and food choice motives. Prevention orientation associated with mood, convenience, and familiarity. Appetite **78**, 15–22 (2014)
3. Afzal, M.N.I., Lawrey, R., Anaholy, M.S., Gope, J.: A comparative analysis of the efficiency and productivity of selected food processing industries in Malaysia. Malays. J. Sustain. Agric. (MJSA) **2**(1), 19–28 (2018)
4. Raji, M.N.A., Karim, S.A., Ishak, F.A.C., Arshad, M.M.: Past and present practices of the Malay food heritage and culture in Malaysia. J. Ethn. Foods **4**(4), 221–231 (2017). LNCS Homepage, http://www.springer.com/lncs. Accessed 21 Nov 2016
5. Ng, C.Y., Karim, S.A.: Historical and contemporary perspectives of the Nyonya food culture in Malaysia. J. Ethn. Foods **3**(2), 93–106 (2016)

6. Dixon, N., Jaki, B., Lagerweij, R., Mooij, M., Yudin, E.: FoodMood: measuring global food sentiment one tweet at a time. In: 2012 AAAI Workshop - Semantic Cities, no. Card, pp. 2–7 (2012)
7. Kim, Y., Jeong, S.R.: Visualizing the results of opinion mining from social media contents: case study of a noodle company. J. Intell. Inf. Syst. **20**(4), 2288–4866 (2014)
8. Hingle, M., et al.: Collection and visualization of dietary behavior and reasons for eating using Twitter, vol. 15, pp. 1–16 (2013)
9. Lee, Y.J., Pennington-Gray, L., Kim, J.: Does location matter? Exploring the spatial patterns of food safety in a tourism destination. Tour. Manag. **71**, 18–33 (2019)
10. Li, L.: Data visualization and retrieval based convenience store location model of fresh products, vol. 15, no. 3 (2017)
11. Vidal, L., Ares, G., Machín, L., Jaeger, S.R.: Using Twitter data for food-related consumer research: a case study on 'what people say when tweeting about different eating situations'. Food Qual. Prefer. **45**, 58–69 (2015)
12. Abbar, S., Mejova, Y., Weber, I.: You tweet what you eat: studying food consumption through Twitter (2014)
13. Yudin, E.: Data visualizing the story of food and emotion, pp. 1–6 (2011)
14. Merler, M., Smith, J.R.: Learning to make better mistakes: semantics-aware visual food recognition, pp. 172–176 (2016)
15. Fried, D., Surdeanu, M., Kobourov, S., Hingle, M., Bell, D.: Analyzing the language of food on social media, no. Section II (2014)
16. Meirelle, I.: Design for Information: An Introduction to the Histories, Theoris and Best Practices Behind Effective Information Visualizations. Rockport Publishers, Osceola (2013)

Machine and Deep Learning

A Hybrid TSR and LSTM for Forecasting NO_2 and SO_2 in Surabaya

Suhartono[1(✉)], Hendri Prabowo[1], and Soo-Fen Fam[2]

[1] Department of Statistics, Institut Teknologi Sepuluh Nopember,
Jl. Raya ITS Sukolilo, Surabaya 6011, Indonesia
suhartono@statistika.its.ac.id
[2] Faculty of Technology Management and Technopreneurship,
Universiti Teknikal Malaysia Melaka, Melaka, Malaysia

Abstract. This study proposes a hybrid model that combining time series regression (TSR) and long short-term memory (LSTM) to forecast NO_2 and SO_2 in Surabaya City, Indonesia. TSR is one of the linear statistical methods to capture deterministic patterns, both are trend and seasonal, whereas LSTM is a neural network method that has a memory block in its hidden layer to handle the nonlinear pattern. Data about half-hourly NO_2 and SO_2 at three SUF stations in Surabaya City are used as cased study. These data have double seasonal pattern, i.e. daily and weekly seasonality. The performance of this hybrid model was compared to several individual models both linear and non-linear methods, i.e. TSR and ARIMA as linear model and FFNN and LSTM as nonlinear model. Based on the smallest RMSEP and sMAPEP, the results showed that hybrid TSR-LSTM yielded more accurate forecast at two datasets, whereas LSTM as an individual method produced more accurate forecast at three datasets. Hence, it is in line with the results of M3 and M4 forecasting competition, i.e. more complex methods do not necessary yield better forecast than simpler ones.

Keywords: Hybrid · TSR · LSTM · NO_2 · SO_2 · Forecasting

1 Introduction

Recently, air pollution becomes a main problem of all regions in the world [1]. Air pollution can affect humans directly or indirectly. The direct effect of air pollution is disease and death [2], while the indirect effect is the disruption of various natural resources that are important for human life and well-being. The information used to indicate the condition of air quality in the city of Surabaya is the Air Pollution Index (API). The parameters used to calculate API are PM_{10}, CO, SO_2, NO_2 and O_3. For high concentrations, SO_2 and NO_2 can cause various respiratory diseases such as bronchitis, pneumonia and emphysema [3]. Most sources of NO_2 come from the transportation sector and fossil fuels. The main source of SO_2 is combustion of sulfur-containing fuels such as coal and petroleum. Due to SO_2 and NO_2 are very dangerous for humans and the environment [4], it is very necessary to forecast the concentration of air pollutants of NO_2 and SO_2.

© Springer Nature Singapore Pte Ltd. 2019
M. W. Berry et al. (Eds.): SCDS 2019, CCIS 1100, pp. 107–120, 2019.
https://doi.org/10.1007/978-981-15-0399-3_9

The forecasting method that commonly be used for solving practical problem is statistical methods [5]. Time series regression (TSR) in general is the same as the linear regression model [6]. Other linear forecasting model is Autoregressive Integrated Moving Average or ARIMA method. ARIMA method is one of the most popular methods in time series forecasting [7, 8]. Furthermore, TSR and ARIMA can be used for forecasting seasonal and non-seasonal data. However, many real data not only follow linear pattern. Thus, nonlinear model is needed to handle this nonlinearity pattern. Recently, many nonlinear methods were proposed and applied for time series forecasting. Neural Network (NN) is one of nonlinear methods that frequently used for solving forecasting problems [9]. Moreover, there are many models on NN such as feed forward neural networks (FFNN) and long short-term memory (LSTM) that usually be applied for forecasting [10].

Due to time series data are rarely found only having linear or nonlinear pattern, currently many hybrid methods that combining linear and nonlinear models have been proposed and applied for forecasting. Data frequently have both linear and nonlinear patterns, so it is necessary to build a model that combining individual models for handling these patterns [11]. However, it must be noted that more complex methods do not always yield better forecast than simpler ones [12].

Many researches in prediction of NO_2 and SO_2 have been carried out. Mishra and Goyal [13] used multiple linear regression and ANN to predict NO_2 levels at the Taj Mahal. The inputs of ANN are significant variables in linear regression and principal component analysis (PCA). It was found that the ANN method yielded better result than multiple linear regression in predicting NO_2 at the Taj Mahal. Reisen et al. [14] used a seasonal autoregressive fractional integrated moving average (SARFIMA) to predict SO_2. Cheng et al. [8] forecasted $PM_{2.5}$ in five cities in China by comparing individual methods, i.e. ARIMA, ANN, SVM, with the hybrid method. The results showed that hybrid methods produced better results than individual methods for predicting $PM_{2.5}$ in five cities in China.

In this paper, hybrid TSR-LSTM method will be proposed and applied for forecasting half-hourly NO_2 and SO_2 at three SUF stations in Surabaya City. The accuracy of this hybrid model will be compared with several individual methods, i.e. TSR, ARIMA, FFNN and LSTM. The rest of paper is organized as follows: Sect. 2 reviews the methodology, i.e. TSR, ARIMA, FFNN, and LSTM, and hybrid method; Sect. 3 presents the dataset and methodology; Sect. 4 presents the results and analysis; and Sect. 5 presents the conclusion from this study.

2 Time Series Analysis

Time series analysis is an analysis to predict future data based on past data. The purpose of time series analysis is to find the patterns and characteristics of data in the past that are used to predict future [15]. In general, there are two main time series methods, i.e. statistical methods and machine learning. TSR and ARIMA are popular statistical methods for linear time series forecasting. Whereas, NN methods such as FFNN and LSTM are popular machine learning techniques for nonlinear time series forecasting.

2.1 Time Series Regression

Time series regression or TSR is one of the classical time series models. In general, model of TSR is same as the linear regression model, i.e. consists of predictor that affects the response [6]. In forecasting, the focus of TSR slightly different from general linear regression, i.e. on accuracy of forecast [16].

2.2 ARIMA

Autoregressive Integrated Moving Average or ARIMA is one of the most popular methods in time series forecasting. ARIMA(p, d, q) is an ARMA(p, q) model after implementing differencing order d to time series data. This ARIMA model can be used for both seasonal and non-seasonal patterns. In general, ARIMA(p, d, q) for non-seasonal data can be written as follows [17]:

$$\phi_p(B)(1 - B)^d Y_t = \theta_0 + \theta_q(B)a_t. \tag{1}$$

If there are seasonal and non-seasonal patterns in time series, then ARIMA multiplicative model, i.e. ARIMA$(p, d, q)(P, D, Q)^S$ can be applied. The general form of ARIMA(p, d, q) $(P, D, Q)^S$ model is

$$\phi_p(B)(1 - B)^d \Phi_P(B^S)(1 - B^S)^D Y_t = \theta_q(B)\Theta_Q(B^S)a_t \tag{2}$$

where p and q are the non-seasonal order of AR and MA, respectively, P and Q are the seasonal order of AR and MA, respectively.

2.3 Feed Forward Neural Network

Neural Network (NN) is an information processing system that has characteristics similar to biological neural networks, namely the human brain consisting of a number of neurons that perform simple tasks to process information. NN is one of the nonlinear models that can be used to model and predict time series data [9]. NN can estimate almost all linear or nonlinear functions in an efficient and stable way, if the underlying data is unknown.

The NN architecture most widely used in engineering applications is the feed forward neural network (FFNN) or also called multi perceptron layer [18]. In the FFNN structure there are several layers, i.e. the input layer, one or more hidden layers and the output layer as Fig. 1. Each neuron will receive information only from the neurons in the previous layer where the input neurons come from the output weights in the previous layer [19]. In FFNN the process will be initiated when neurons receive input grouped into an input layer then information is directed through the hidden layer then sorted until it reaches the output layer [20] as follows:

$$\hat{Y}_{(t)} = f^o\left[\sum_{j=1}^{q}\left[w_j^o f_j^h\left(\sum_{i=1}^{p} w_{ji}^h X_{i(t)} + b_j^h\right) + b^o\right]\right], \tag{3}$$

where b^o, $b_j^h(j = 1, \ldots, q)$, $w_{ji}^h(j = 1, \ldots, q; i = 1, \ldots, p)$, $w_j^o(j = 1, \ldots, q)$ are parameters of the FFNN model, $f^o, f_j^h(j = 1, \ldots, q)$ is an activation function used in FFNN, p is the number of input variables and q is the number of neurons in hidden layer.

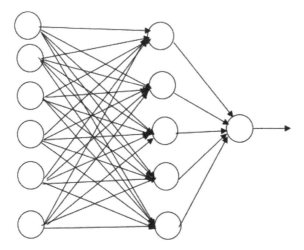

Fig. 1. The FFNN architecture with 6 neurons in the input layer, 6 neurons in the hidden layer and 1 neuron in the output layer

2.4 Long Short-Term Memory

Long short-term memory or LSTM is a development of the recurrent neural network (RNN) method. LSTM was developed due to there are many weaknesses in the RNN method, namely long-term delay that is not accessible in an architecture [21]. The structure of LSTM consists of several layers, i.e. input layer, recurrent hidden layer and output layer [10]. The difference with ordinary NN is that the hidden layer used is a memory block. This hidden layer of the LSTM network contains a memory cell and a pair of adaptive, multiplicative gating units that direct input and output to all cells in the block. An LSTM cell consists of four gates, i.e. input gate, input modulation gate, forget gate and output gate as in Fig. 2 [21].

Based on Fig. 2 the equation in LSTM is obtained as follows:

Forget gate:

$$F_t = \sigma\left(W_{fh}h_{t-1} + W_{fx}X_t + b_f\right), \tag{4}$$

Input gate:

$$\hat{C}_t = tanh(W_{ch}h_{t-1} + W_{cx}X_t + b_c), \tag{5}$$

$$U_t = \sigma(W_{uh}h_{t-1} + W_{ux}X_t + b_u), \tag{6}$$

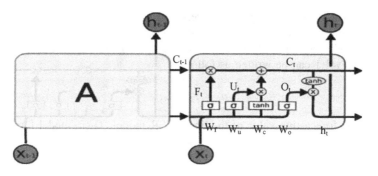

Fig. 2. Structure of memory block LSTM

Memory cell unit:

$$C_t = F_t C_{t-1} + U_t \hat{C}_t, \tag{7}$$

Output gate:

$$O_t = \sigma(W_{oh}h_{t-1} + W_{ox}X_t + b_o), \tag{8}$$

$$h_t = O_t \, tanh(C_t), \tag{9}$$

where W_{fh}, W_{fx}, b_f, W_{ch}, W_{cx}, b_c, W_{uh}, W_{ux}, b_u, W_{oh}, W_{ox}, b_o are the parameters of the LSTM model, $\sigma(.)$ is a sigmoid activation function that shown in Eq. (10) and $tanh(.)$ is a tanh activation function as shown in Eq. (11), i.e.:

$$\sigma(x) = \frac{1}{(1 + e^{-x})} \tag{10}$$

$$tanh(x) = \frac{e^x - e^{-x}}{e^x + e^{-x}}. \tag{11}$$

2.5   Hybrid Model

Hybrids are combination of two or more systems in one function. In this paper, the hybrid model is a combination of TSR as statistical model and LSTM as neural network model. The statistical model can produce good forecasting for linear patterns, but it will be bad if it encounters a nonlinear condition. Then, the TSR model is combined with a neural network that yields good performance if the data is nonlinear. Thus, a hybrid model can overcome the complex structure of data due to the existence of linear and nonlinear patterns [22]. In general, the combination of TSR and LSTM can be written as follows:

$$Y_t = L_t + N_t + \varepsilon_t, \tag{12}$$

where L_t showing linear components represented by TSR and N_t showing nonlinear components represented by LSTM. This hybrid model consists of two levels modeling, i.e. TSR in level 1 and then LSTM for the residuals of TSR at level 2. In general, the residuals for level 1 can be written as follows:

$$a_t = Y_t - \hat{L}_t, \tag{13}$$

where L_t is the forecasting value of the linear TSR model. Then, model for level 2 is done by applying LSTM. In general, the model at level 2 is as follows:

$$a_t = f(a_{t-1}, a_{t-2}, \ldots, a_{t-k}) + \varepsilon_t \tag{14}$$

with f is a non-linear function obtained from the LSTM. If $\hat{N}_t$ is the result of forecasting from the LSTM, then the forecast of hybrid model is obtained as follows [11]:

$$\hat{Y}_t = \hat{L}_t + \hat{N}_t. \tag{15}$$

Thus, the proposed hybrid TSR-LSTM method is combination between TSR at level 1 and LSTM at level 2. The scheme for building this hybrid TSR-LSTM model is illustrated as in Fig. 3.

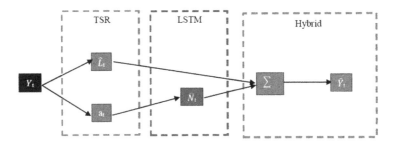

Fig. 3. Hybrid TSR-LSTM

3 Dataset and Methodology

3.1 Dataset

The data used in the study is secondary data about half hourly SO_2 and NO_2 which is the result of air quality monitoring data at three SUF stations in the city of Surabaya, Indonesia. Three SUF stations are SUF 1, SUF 6, and SUF 7 as shown in Fig. 4. Air quality monitoring data is recorded every half hour. The data about half hourly SO_2 and NO_2 were observed in 2018. The time series plots of these data are shown at Fig. 5. Based on Fig. 5, it is found some missing values in the data. Thus, imputation is needed to fill these missing values. Due to the seasonal pattern at the data, the missing values are replaced by the median at the same period, i.e. the median for the same half hour and day data.

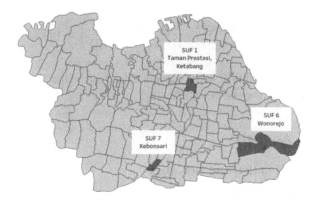

Fig. 4. SUF station map in Surabaya

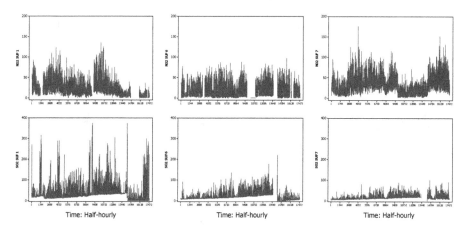

Fig. 5. Time series plot of SO$_2$ and NO$_2$ in SUF 1, SUF 6 and SUF 7

3.2 Methodology

The data period used is data from 1 January 2018 to 20 December 2018. Data is divided into data in-sample (training dataset) and out-of-sample (testing dataset). Data from January 1, 2018 at 00.30 to December 20, 2018 at 19.00 becomes in-sample data and data from December 20, 2018 at 19.30 until 00.00 to become out-of-sample data. In-sample data is used for modeling, whereas out-of-sample data is used for validating and selecting the best model. To evaluate the forecast accuracy, root mean square prediction or RMSEP [17] and symmetric mean absolute percentage error or sMAPE [23] are used and the formulae is as follows:

$$RMSEP = \sqrt{\frac{1}{L}\sum\nolimits_{l=1}^{L}\left(Y_{n+l} - \hat{Y}_n(l)\right)^2}, \qquad (16)$$

$$sMAPEP = \left(\frac{1}{L} \sum\nolimits_{l=1}^{L} \frac{2|Y_{n+l} - \hat{Y}_n(l)|}{(|Y_{n+l}| + |\hat{Y}_n(l)|)} \right) \times 100\%, \tag{17}$$

where Y_{n+l} is the actual value at out-of-sample data, $\hat{Y}_n(l)$ is the forecast value at out-of-sample data and L is the out-of-sample size.

4   Result and Analysis

4.1   Characteristics of SO$_2$ and NO$_2$ Data

The analysis begins with visualizing the data. By applying interval plot as shown at Fig. 6, the seasonal pattern of the data can be identified clearly. It shows that half hourly data of NO$_2$ and SO$_2$ at three SUF stations in the city of Surabaya tend to have double seasonal patterns, i.e. daily (period 48) and weekly (period 336). The highest NO$_2$ level occurred at SUF 7, whereas the highest SO$_2$ level occurred at SUF 1.

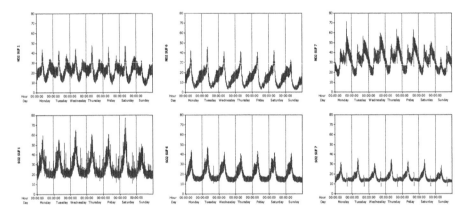

Fig. 6. Interval plot of NO$_2$ and SO$_2$ in SUF 1, SUF 6 and SUF 7

The levels of NO$_2$ and SO$_2$ tend to have high condition at around 08.00 am on both working day and weekend day. In working day, it is caused by many vehicles in the main road at rush hour in the morning. In general, the levels of NO$_2$ and SO$_2$ on Monday to Friday at three stations are almost the same. But these values on Sundays are lower than other days due to few people in Surabaya do activities outside compare with working day.

4.2   Forecasting NO$_2$ and SO$_2$

The modeling process is done after identifying the pattern of data. The identification step shows that the data tend to have small trend and double seasonal patterns, i.e. daily and weekly seasonality. Hence, TSR model that be proposed is involving two main predictors, i.e. trend and seasonal dummy variables. The dummy variables for these double seasonal are dummy for every half hour in a day and dummy for every day in one week. The proposed TSR model is a multiplicative model that consists of 336 dummy variables for capture double seasonality. After fitting the TSR model, the model evaluation showed that the residual of TSR model did not fulfill white noise condition. Hence, the modeling process can be continued by applying LSTM as nonlinear mode to these residuals.

Moreover, ARIMA model based on the Box Jenkins procedure was also applied for forecasting these data. The identification step showed that data of NO_2 and SO_2 were not stationary in the mean and variance. Thus, transformation and differencing are applied to achieve stationary condition. The results of identification and estimation steps showed that the best ARIMA model for each data was double seasonal and subset ARIMA models as shown in Table 1.

Table 1. The best ARIMA model for each dataset

Variable	ARIMA Model
NO_2 SUF 1	$([6,9,13,33,36,47,48],1,[1,2,3,4,5,7,8,10,11,18,19,20,21, 24,27,29]) (0,1,1)^{48}$ $(0,1,1)^{336}$
NO_2 SUF 6	$([8,15,16,17,33,35,43,47,48],1,[1,2,3,4,5,6,7,9,11,21,22, 45,47]) (0,1,1)^{48}$ $(0,1,1)^{336}$
NO_2 SUF 7	$([8,10,14,15,21,29,40,47],1,[1,2,3,4,5,6,7,9,13,16,22,26, 27,33,38,46])$ $(0,1,1)^{48} (0,1,1)^{336}$
SO_2 SUF 1	$([1,7,15,17,18,19,20,42,43,44,45,46,47],1,[2,3,4,5,6,8,9,21, 27,43,45,46,47])$ $(0,1,1)^{48} (0,1,1)^{336}$
SO_2 SUF 6	$([5,6,7,9,10,11,12,13,14,18,23,31,38,47,48],1,[1,2,3,4,8, 15,16,17,20,21,22,28,29,30,36,45,46,47]) (0,1,1)^{48} (0,1,1)^{336}$
SO_2 SUF 7	$([16,23,28,38,48],1,[1,2,3,4,5,6,7,8,9,10,15,18,27,43,46]) (0,1,1)^{48} (0,1,1)^{336}$

NN modeling is done based on the ARIMA model, particularly about the determination of the inputs. The AR lags of the ARIMA model are used as inputs of both FFNN and LSTM. FFNN with one hidden layer and number of neurons from 1 to 5 are applied to the data. The sigmoid activation function is used in the hidden layer of FFNN. Moreover, the data used in FFNN are two dimensions, whereas in LSTM are three-dimensional data. The additional input dimensions in LSTM is done by adding one dimension with size of one until the input becomes three dimensions. The optimum number of neurons in FFNN and LSTM models are shown in Table 2.

Furthermore, hybrid TSR-LSTM method is done by applying LSTM for forecasting the residuals of TSR model. This LSTM is used to capture nonlinear patterns that cannot be captured by TSR model. The inputs of LSTM model are based on the AR lags of ARMA model in Table 3. The forecast of LSTM model, $\hat{N}_t$, is used as forecast value of nonlinear component of hybrid model. Thus, the final forecast of hybrid TSR-LSTM is as follows:

$$\hat{Y}_t = \hat{L}_t + \hat{N}_t$$

where $\hat{L}_t$ is forecast value of TSR model at level 1 as linear component and $\hat{N}_t$ is forecast value of LSTM at level 2 as nonlinear component. The optimum number of neurons in hybrid TSR-LSTM model for each dataset is shown in Table 2.

Table 2. Optimum number of neurons in FFNN, LSTM and Hybrid TSR-LSTM models

Variable	Number of neurons		
	FFNN	LSTM	Hybrid TSR-LSTM
NO_2 SUF 1	4	2	3
NO_2 SUF 6	5	5	3
NO_2 SUF 7	5	5	2
SO_2 SUF 1	3	5	3
SO_2 SUF 6	1	2	1
SO_2 SUF 7	2	3	1

Table 3. ARMA model of TSR residuals

Variable	ARMA model
NO_2 SUF 1	([1,2,4,9,10,21,23,34,45,47],[6,7,12,14,16,18,23,25,27,32,39,45, 46,47]) $(1,0)^{48}$
NO_2 SUF 6	([1,2,3,33,39,44,47],[7,8,10,11,12,17,45,46, 47]) $(1,0)^{48}$
NO_2 SUF 7	([1,2,3,6,20,34,46],[3,6,7,34,35,36,37,43,46, 47])$(1,0)^{48}$ $(1,0)^{336}$
SO_2 SUF 1	([3,6,14,15,19,21,27,40,44,46,47],[1,2,3,4,5,7,8,9,10,11,12,13,45, 46,47]) $(1,0)^{48}$
SO_2 SUF 6	([1,2,7,17,36,39,43,45,46,47],[3,4,5,6,38,46, 47]) $(1,0)^{48}$
SO_2 SUF 7	([1,2,3,4,5,6,7,21,39,44,47],[7,14,43]) $(1,0)^{48}$ $(1,0)^{336}$

The comparison results of forecast accuracy between individual and hybrid models by using RMSEP and sMAPEP are illustrated at Fig. 7. The results show that both individual linear models, i.e. TSR and ARIMA, tend to produce less accurate forecast than both individual nonlinear models, i.e. FFNN and LSTM, and hybrid TSR-LSTM model. It shows by the values of RMSEP and sMAPEP from both individual linear models are consistently higher than both individual nonlinear models and hybrid model in all datasets.

Moreover, it is also shown by the RMSEP ratio between these methods to TSR model as at Table 4. It proves that the data not only have linear pattern but also nonlinear pattern. In general, the best models are different for each dataset. Both individual nonlinear models yield more accurate forecast at four datasets, i.e. FFNN is the best model to forecast SO$_2$ in SUF 7, and LSTM is the best to predict NO$_2$ in SUF 1, NO$_2$ in SUF 7 and SO$_2$ in SUF 1. Furthermore, hybrid TSR-LSTM is the best model for forecasting NO$_2$ in SUF 6 and SO$_2$ in SUF 6. These results show that LSTM as an individual nonlinear model tend to produce more accurate forecast than TSR and ARIMA as simpler individual linear model. This conclusion is not in line with the first result of M3 forecasting competition, i.e. more complex methods do not necessary yield better forecast that simpler ones [12].

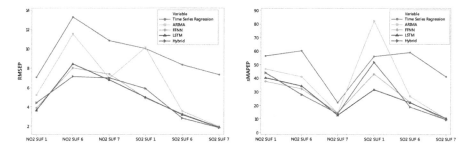

Fig. 7. Comparison of models based on RMSEP and sMAPEP

Table 4. RMSEP ratio between to TSR model

Variable	ARIMA	FFNN	LSTM	Hybrid
NO₂ SUF 1	0.74	0.55	0.51	0.63
NO₂ SUF 6	0.86	0.61	0.63	0.54
NO₂ SUF 7	0.64	0.68	0.62	0.64
SO₂ SUF 1	1.01	0.49	0.50	0.58
SO₂ SUF 6	0.43	0.39	0.38	0.34
SO₂ SUF 7	0.27	0.26	0.25	0.24

Additionally, these results also show that hybrid TSR-LSTM as a combination method tend to give better results than TSR and ARIMA as simpler individual linear model and FFNN as simpler individual nonlinear model. This result is in line with the result of M3 and M4 forecasting competition, i.e. hybrid model on average tend to produce more accurate forecast than individual methods [12, 23]. Moreover, M3 and M4 forecasting competition are big events for forecasting researchers in the world to do a competition to find the best forecasting method in many practical problems. Finally, the forecast values at testing dataset by using the best method for each series and the actual data are shown at Fig. 8.

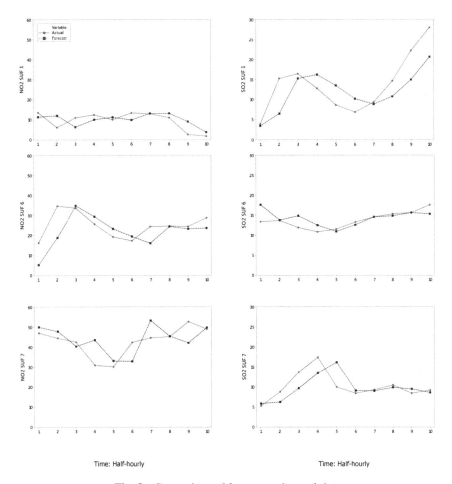

Fig. 8. Comparison of forecast and actual data

5   Conclusion

This paper proposes hybrid TSR-LSTM model for forecasting NO_2 and SO_2 at three SUF stations in Surabaya City, Indonesia, and compares the accuracy with both individual linear models, i.e. TSR and ARIMA, and both individual nonlinear models, i.e. FFNN and LSTM. In general, the results showed that both individual nonlinear models and hybrid model tend to produce more accurate forecast than individual linear models as simpler methods. This conclusion is not in line with the first result of M3 forecasting competition, i.e. more complex methods did not necessary yield better forecast that simpler ones [12]. Moreover, hybrid TSR-LSTM as a combination method tend to give better results than TSR, ARIMA, and FFNN. It means that hybrid model on average tend to produce more accurate forecast than individual methods as conclusion of M# and M4 forecasting competition [12, 23]. Additionally, these results also proved that air pollution data tend to have nonlinear pattern and only the nonlinear methods, both individual or hybrid can tackle well this pattern. Hence, further research can be proposed by developing other hybrid linear-nonlinear model, particularly by involving spatial aspect or known as hybrid spatial-temporal model.

Acknowledgements. This research was supported by DRPM-DIKTI under scheme of "Penelitian Dasar Unggulan Perguruan Tinggi ITS 2019". The authors thank to the General Director of DIKTI for funding and to anonymous referees for their useful suggestions.

References

1. Amann, M., Klimont, Z., Wagner, F.: Regional and global emissions of air pollutants: recent trends and future scenarios. Ann. Rev. Environ. Res. **38**, 31–55 (2013)
2. Tallon, L.A., Manjourides, J., Pun, V.C., Salhi, C., Suh, H.: Cognitive impacts of ambient air pollution in the national social health and aging project (NSHAP) Cohort. Environ. Int. **104**, 102–109 (2017)
3. Bernard, S.M., Samet, J.M., Grambsch, A., Ebi, K.L., Romieu, I.: The potential impacts of climate variability and change on air pollution-related. Environ. Health Perspect. **109**, 199–209 (2001)
4. Harrison, R.M., Yin, J.: Particulate matter in the atmosphere: which particle properties are important for its effects on health? Sci. Total Environ. **249**, 85–101 (2000)
5. Hanke, J.E., Wichern, D.W.: Business Forecasting, 8th edn. Pearson Practice Hall, New Jersey (2005)
6. Shummway, R.H., Stoffer, D.S.: Time Series Analysis and Its Application with R Examples. Springer, Pittsburg (2006). https://doi.org/10.1007/0-387-36276-2
7. Robles, R.A., et al.: A hybrid ARIMA and artificial neural network model to forecast particulate matter in urban areas: the case of Temuco, Chile. J. Atmos. Environ. **42**, 8331–8440 (2008)
8. Cheng, Y., Zhang, H., Liu, Z., Chen, L., Wang, P.: Hybrid algorithm for short-term forecasting of $PM_{2.5}$ in China. Atmos. Environ. **200**, 264–279 (2019)
9. Tealab, A.: Time series forecasting using artificial neural networks methodologies: a systematic review. Future Comput. Inf. J. **3**, 334–340 (2018)
10. Ma, X., Tao, Z., Wang, Y., Yu, H., Wang, Y.: Long short-term memory neural network for traffic speed prediction using remote microwave sensor data. Transp. Res. Part C Emerg. Technol. **54**, 187–197 (2015)
11. Zhang, G.P.: Time series forecasting using a hybrid ARIMA and neural network model. Neurocomputing **50**, 159–175 (2003)
12. Makridakis, S., Hibbon, M.: The M3-competition result, conclusions and implications. Int. J. Forecast. **16**, 451–676 (2000)
13. Mishra, D., Goyal, P.: Development of artificial intelligence based NO_2 forecasting models at Taj Mahal, Agra. Atmos. Pollut. Res. **6**, 99–106 (2015)
14. Reisen, V.A., et al.: Robust estimation of fractional seasonal processes: modeling and forecasting daily average SO_2 concentrations. Math. Comput. Simul. **146**, 27–43 (2018)
15. Makridakis, S., Wheelwright, S., McGee, V.: Forecasting: Methods and Applications. Wiley, New York (1983)
16. Kostenko, A.V., Hyndman, R.J.: Forecasting Without Significance Test? (2008)
17. Wei, W.W.: Time Series Analysis Univariate and Multivariate Methods. Pearson Education Inc, London (2006)

18. Azzouni, A., Pujjole, G.: A long-short term memory recurrent neural network framework for network traffic matrix prediction. arXiv preprint arXiv:1705.05690 (2017)
19. Chong, E., Zak, S.H.: An Introduction to Optimize. Wiley, Canada (2001)
20. Suhartono: New procedures for model selection in feedforward neural networks. Jurnal Ilmu Dasar **9**, 104–113 (2008)
21. Hochreiter, S., Schmiduber, J.: Long short-term memory. Neural Comput. **9**, 1735–1780 (1997)
22. Zheng, F., Zhong, S.: Time series forecasting using a hybrid RBF neural network and AR model based on binomial smoothing. World Acad. Sci. Eng. Technol. **75**, 1471–1475 (2011)
23. Makridakis, S., Spiliotis, E., Assimakopoulus, V.: The M4 competition: results, findings, conclusion and way forward. Int. J. Forecast. **34**, 802–808 (2018)

Evaluation of Pooling Layers in Convolutional Neural Network for Script Recognition

Zaidah Ibrahim[1(✉)], Dino Isa[2], Zainura Idrus[1], Zolidah Kasiran[1],
and Rosniza Roslan[3]

[1] Faculty of Computer and Mathematical Sciences,
Universiti Teknologi MARA, 40450 Shah Alam, Selangor, Malaysia
zaidah@tmsk.uitm.edu.my
[2] Crops For the Future, 43500 Semenyih, Selangor, Malaysia
[3] Faculty of Computer and Mathematical Sciences,
Universiti Teknologi MARA, Campus Jasin, 77300 Melaka, Malaysia

Abstract. This paper investigates the suitable position and number of pooling layers in Convolutional Neural Network (CNN) for script recognition from scene images. A common practice of CNN for object recognition is to position a convolve layer alternately with a pooling layer followed by a few layers of fully connected layers. We re-evaluate this basic principle by examining the position of pooling layer after every convolve layer, reducing and increasing its numbers. Experimental results on MLe2e dataset for script recognition show that a CNN with less number of pooling layers and non-overlapping pooling stride can reach excellent percentage of accuracy compared to alternating convolve layer with pooling layer.

Keywords: Convolutional Neural Network · Script recognition · MLe2e dataset

1 Introduction

In recent years, Convolutional Neural Network (CNN) has gained popularity in computer vision applications and demonstrated tremendous success including script identification [1–5]. Automatic script identification is beneficial in a multi-script environment for translation and navigational purposes. Travellers or tourists frequently encounter multi-script scene images such as signage and shop names especially in Asian countries where English is not their first language. Figure 1 illustrates some sample images of signage and shop names that consist of multi-scripts with variations of font size, colour and type. It is important to identify the script prior to the use of Optical Character Recognition (OCR) since normally OCR is script dependent. For instance, a Chinese script may produce a low recognition rate if being applied to a Roman OCR.

A lot of research has been performed to recognize scripts in documents [6–8] compared to scene images where surveys on them have been reported in [9, 10]. Machine learning approaches are popular where various hand-crafted features and classifiers are investigated. But a lot of experiments need to be conducted in order to select the significant combination of features and classifiers for good recognition accuracy. Whereas,

© Springer Nature Singapore Pte Ltd. 2019
M. W. Berry et al. (Eds.): SCDS 2019, CCIS 1100, pp. 121–129, 2019.
https://doi.org/10.1007/978-981-15-0399-3_10

for CNN, no hand-crafted feature selection is necessary since the feature extraction is being performed in the layers. Excellent performance of CNN has been reported but usually the architecture is deep and requires high-end resource capacity which is beyond the encompassment of many practical CNN applications. The accuracy of script identification from documents is higher compared to from scene images due to the less complexity and varieties of font type, size and background of the text in the documents [11, 12]. Thus, this research examines the performance of shallow CNN that requires an average hardware specification for script recognition from scene images.

Fig. 1. Some sample images of signage and shop names with multi-scripts.

Recognizing scripts from scene images provide more challenges compared to document images. Among the challenges are (i) scene images are usually made up of different font types, sizes and colours (ii) background of scene images are usually complex with various colours, not just white background as the document images (iii) word length and height of each script may varies, not just one standard aspect ratio as in document images.

In this research, MLe2e dataset [13] is being used. This dataset has been extracted from various existing scene text datasets covering four different scripts, namely Roman, Chinese, Kannada and Hangul. This dataset has been constructed for the purpose of evaluating multi-lingual scene text end-to-end reading systems including text detection, script identification and text recognition. There are 1178 cropped word text images for training and 643 for testing. Figure 2 illustrates some sample images of the different scripts from MLe2e dataset.

This paper is organized as follows: In Sect. 2, explanations about some basic definitions and the applications of CNN for script recognition are presented. Descriptions about the proposed model and result analysis are provided in Sect. 3. Finally, in Sect. 4, conclusions related to the architecture and accuracy performance of CNN is discussed and suggestion of future work is given.

Fig. 2. Some sample images of the four scripts from MLe2e dataset [13].

2 Related Work

CNN is based on the concept of Artificial Neural Network (ANN) with various hidden layers. It belongs to the deep learning domain where they are optimized to recognize two dimensional shapes or patterns. It was first introduced by Fukushima in 1988 [14] but then it was not popular due to hardware constraint. Ten years later, LeCun et al. successfully performed handwritten digit classification using CNN [15]. After that, further research on the applications and improvements of CNN for many recognition tasks has been reported.

The performance of CNN is excellent especially for image-based recognition due to the fact that CNN takes full advantage of the characteristics of the image itself [16]. An image is represented by pixels and each neighbourhood pixel is strongly related but weakly related with pixels that are far away. Furthermore, collections of pixels that are strongly related usually have similar properties. Thus, CNN applies this concept to its architecture where each neuron only links with a small local patch of neurons from the previous layer. So, not all neurons are fully connected.

Figure 3 illustrates the architecture of a basic CNN. Basically, it consists of two parts that are the feature extraction and classification. The feature extraction parts have three main layers, namely convolve, activation function and pooling. Meanwhile, the classification part consists of fully connected layer and classification layer. Normally, the number of feature extraction layers is more than the number of classification layers. The behaviour of a given layer of CNN can be considered as a linear operation where it starts with a convolution and activation function, followed by pooling or another sequence operation of convolution process, but ended with classification. The details of each layer are explained next.

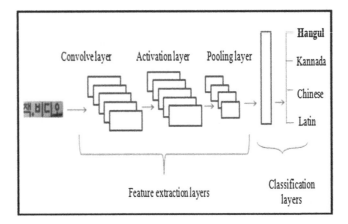

Fig. 3. Architecture of a basic CNN.

2.1 Convolve Layer

In the convolve layer, the feature maps from previous layers are convolved with learnable kernels and its purpose is to extract the features from the image. Designing of the convolve layer involves selecting the number of filters and its size. Figure 4 shows sample of two 3x3 filters to detect vertical and horizontal edges.

1	0	-1
1	0	-1
1	0	-1

1	1	1
0	0	0
-1	-1	-1

a)   b)

Fig. 4. Sample 3x3 filter for convolution (a) to detect vertical edge feature (b) to detect horizontal edge feature.

2.2 Activation Function Layer

An activation function introduces non-linearity into CNN and detects non-linear features of the input data. A popular activation function is called Rectified Linear Unit (ReLU) where the idea is that all the values above zero are maintained while all the negative values are set to zero.

2.3 Pooling Layer

Pooling layer performs the down sampled process on the input maps, depending on the size of the down sampling mask, thus reducing the computational burden. Two common types of pooling layers are max-pooling and average pooling. In max-pooling, the highest value is chosen from the NxN patches of the feature maps from the previous layer while average pooling sums the values in NxN patches of the features maps and compute the average. The conceptual diagram for average pooling and max-pooling are illustrated in Fig. 5.

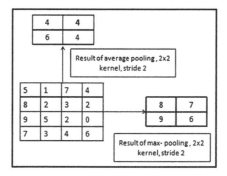

Fig. 5. Sample average and max-pooling operations.

2.4 Classification Layer

The fully-connected layer converts the 2D feature maps into a 1D feature vector. In the classification layer, a statistical classifier like softmax [17] or Support Vector Machine (SVM) [18] is being utilized for the final classification or recognition.

2.5 Data Augmentation

The main issue with CNN is that it requires a large dataset for training. Since MLe2e only has 1178 word text images for training, we have performed data augmentation to increase the number of training data without changing their natures [17]. This is where a linear transformation is conducted on the existing training images by resizing and slightly rotating them to produce new training images [19]. These transformations are chosen due to the personal observation where we tend to zoom-in or zoom-out and rotate a little bit when capturing images. After augmentation, the total number of training images becomes 8000 and testing images is 2000. Then we performed an exhaustive parameter tuning process to increase the accuracy of the proposed model.

3 Experimental Results

Designing a CNN model typically involves determining the number of layers, selecting the number of convolves filters and its size, the type of activation function and pooling method, and the classifier. In this experiment, Matlab 2018a has been used. We have conducted exhaustive experiments by varying the parameters at the feature extraction layers, that are learning rate, number of convolve layer, number of convolve filter and its size. The best parameters are selected based on the highest accuracy performance produced on the MLe2e dataset. At the classification layer, softmax is chosen since it is better than SVM [20]. The model is trained using stochastic gradient model (SGD). Besides that, we also investigate the shape of the image that is whether a square or a rectangle shape is better for classification. Pre-trained CNN models apply a square-shaped input size, for instance AlexNet uses 227x227 [21] and GoogleNet uses 224x224 [22] while most other CNN models apply a rectangle-shaped input image such as 40x480 [1]. Two datasets have been created; the images in the first dataset have been resized to 100x150 while the second dataset consists of images with the size of 150x150. Table 1 lists the results of different number of max-pooling layers, the different order between convolve and max-pooling layers and the pooling stride.

The first column in Table 1 represents the filter size and the number of convolve and max-pooling layers. Activation function layer is paired with convolve layer and considered as one layer. [3x3][3x3] in the first column means the CNN model has 2 convolve layers with filter size 3x3 for both layers and 2 activation function layers paired with the 2 convolve layers, and it is not counted as part of the number of layers in the CNN model. So, for [3x3][3x3], the total number of layers is 4 since a convolve layer is paired with max-pooling layer. An additional one fully-connected layer and one classification layer, with a total of 6 layers for that particular model with the accuracy shown in column 2. Column 3 lists the result for an additional one max-pooling layer being added before the fully connected layer. Column 4 shows the accuracy results after the elimination of the last max-pooling layer before the fully connected layer. The kernel size for the max-pooling layers for all experiments is the same which is 2x2 with stride 2. Two different learning rates and two different input sizes are also investigated in this experiment.

By referring to Table 1, we can see that in all experiments, a high accuracy is achieved in the last column where one layer of max-pooling is eliminated before the fully-connected layer. So, the total number of layers is reduced. Table 2 illustrates the accuracy performance of CNN for script recognition with different order of convolve and max-pooling layer. Referring to Table 2, C is for convolve layer while M is for max-pooling layer. By looking at the values in Table 2, we can see that max-pooling layer with stride 2 which is non-overlapping pooling process produces better accuracy compared to overlapping max-pooling process which is stride 1. Table 3 shows the best result from our experiments compared with two other CNN techniques using MLe2e which out-perform them.

Table 1. Script identification accuracy performance based on different number of convolve and max-pooling layers, and convolve filter size for MLe2e dataset.

Number of convolve layer and filter size	Accuracy (%) with LR: 0.001	Accuracy (%) with LR: 0.001, add 1 max-pooling layer	Accuracy (%) with LR: 0.001, less 1 max-pooling layer
Image size: 100x150			
[3x3]	64.6	47.04	89.91
[3x3][3x3]	80.1	92.16	**99.85**
[3x3][3x3][3x3]	96.93	95.51	99.76
[5x5]	64.83	93.17	88.28
[5x5][5x5]	96.61	97.02	83.01
[5x5][5x5][5x5]	96.06	83.54	**98.99**
[7x7]	70.93	93.95	85.89
[7x7][7x7]	98.76	97.34	**99.13**
[3x3]	84.18	91.61	77.26
[3x3][3x3]	96.61	83.68	**97.16**
[3x3][3x3][3x3]	85.69	83.22	95.64
[5x5]	67.9	89	92.34
[5x5][5x5]	96.01	89.45	**98.9**
[5x5][5x5][5x5]	92.11	73.09	96.42
[7x7]	96.74	81.8	89.78
[7x7][7x7]	88.64	80.93	**97.89**
Image size: 150x150			
[3x3]	59.65	55.48	66.8
[3x3][3x3]	61.71	79.22	91.06
[3x3][3x3][3x3]	98.54	91.66	**99.58**
[5x5]	67.49	56.49	97.43
[5x5][5x5]	92.94	96.15	84.14
[5x5][5x5][5x5]	99.5	99.36	**99.82**
[7x7]	70.93	93.95	**99.59**
[7x7][7x7]	98.76	97.34	99.13
[3x3]	76.57	68.68	61.85
[3x3][3x3]	94.54	87.16	97.8
[3x3][3x3][3x3]	92.43	84.96	**95.83**
[5x5]	96.1	86.66	71.66
[5x5][5x5]	89.45	92.21	**99.13**
[5x5][5x5][5x5]	95.46	80.65	97.11
[7x7]	96.74	81.8	89.78
[7x7][7x7]	88.64	80.93	**97.89**

Table 2. Script identification accuracy performance based on different order of convolve and max-pooling layers for MLe2e dataset.

LR: 0.0001 Image size: 100x150 Number of filter: 30 Max-pooling kernel size: 2x2; stride: 1			
Convolve filter size n order of layers	CCM	CCMM	CMCM
[3x3][5x5]	62.77	49.84	85.19
[5x5][3x3]	63.64	72.54	77.62
LR: 0.0001 Image size: 100x150 Number of filter: 30 Max-pooling kernel size: 2x2; stride: 2			
Convolve filter size n order of layers	CCM	CCMM	CMCM
[3x3][5x5]	96.72	97.62	91.2
[5x5][3x3]	96.38	98.4	96.06

Table 3. Script recognition results from different CNN methods using MLe2e dataset.

Method	Accuracy (%)
Ensemble cojoined network [2]	94.40
Attention convolutional LSTM [1]	96.70
Our method	**99.85**

4   Conclusion

We have presented the accuracy performance of script recognition for MLe2e dataset. Experiments show that shallow CNN with data augmentation can still achieve very accurate result even though with hardware constraints. Thus, deeper is not necessary better but since CNN is a data driven method, it requires a great amount of labelled data for training for the supervised approach. But this can be overcome with data augmentation. Non-overlapping max-pooling stride is better than overlapping and alternate layers of convolve and pooling layers does not really achieve a high accuracy performance. As future directions, this model will be tested for other multi-script datasets and for real-time script recognition application.

References

1. Bhunia, A.K., Konwer, A., Bhunia, Ak.K., Bhowmick, A., Roy, P.P., Pal, U.: Script identification in natural scene image and video frams using an attention based convolutional-LSTM network. Pattern Recogn. **85**, 172–184 (2019)
2. Gomez, L., Nicolaou, A., Karatzas, D.: Improving patch-based scene text script identification with ensembles of conjoined network. Pattern Recogn. **67**, 85–96 (2017)

3. Mei, J., Dai, L., Shi, B., Bai, X.: Scene text script identification with convolutional recurrent neural networks. In: IEEE International Conference on Pattern Recognition, pp. 4053–4058 (2016)
4. Sharma, N., Mandal, R., Sharma, R., Pal, U., Blumenstein, M.: ICDAR2015 competition on video script identification (CVSI 2015). In: IEEE 13th International Conference on Document Analysis and Recognition (ICDAR), pp. 1196–1200 (2015)
5. Gomez, L., Karatzas, D.: A fine-grained approach to scene text script identification. In: 12th IAPR Workshop on IEEE Document Analysis Systems (DAS), pp. 192–197 (2016)
6. Chanda, S., Pal, U., Franke, K.: Font identification – in context of an indic script. In: 21st International Conference on Pattern Recognition (ICPR2012) (2012)
7. Ul-Hasan, A., Afzal, M.Z., Shafait, F., Liwicki, M., Breuel, T.M.: A sequence learning approach for multiple script identification. In: 13th International Conference on Document Analysis and Recognition (ICDAR), pp. 1046–1050 (2015)
8. Saidani, A., Kacem, A., Belaid, A.: Co-occurrence matrix of oriented gradients for word script and nature identification. In: 13th International Conference on Document Analysis and Recognition (ICDAR), pp. 16–20 (2015)
9. Ghosh, D., Dube, T., Shivaprasad, A.P.: Script recognition – a review. IEEE Trans. Pattern Anal. Mach. Intell. 32(12), 2142–2160 (2010)
10. Ubul, K., Tursun, G., Aysa, A., Impedovo, D.: Script identification of multi-script documents: a survey. IEEE Access 5, 6546–6559 (2017)
11. Fujii, Y., Driesen, K., Baccash, J., Hurst, A., Popat, A.C.: Sequence-to-label script identification for multilingual OCR. In: 14th International Conference on Document Analysis and Recognition (ICDAR), pp. 161–168 (2017)
12. Chen, Z., Wu, Y., Yin, F., Liu, C.L.: Simultaneous script identification and handwriting recognition via muti-task learning of recurrent neural networks. In: 14th International Conference on Document Analysis and Recognition (ICDAR), pp. 525–530 (2017)
13. Gomez, L.: MLe2e multi-lingual end-to-end dataset (2016). https://www.researchgate.net/publication/297469752_MLe2e_multi-lingual_end-to-end_dataset
14. Fukushima, K.: Neocognitron: a hierarchical neural network capable of visual pattern recognition. Neural Netw. 1(2), 119–130 (1988)
15. LeCun, Y., Bottou, L., Bengio, Y., Haffner, P.: Gradient-based learning applied to document recognition. Proc. IEEE 86(11), 2278–2324 (1998)
16. Chen, L., Wong, S., Fan, W., Sun, J., Satoshi, N.: Reconstruction combined training for convolutional neural networks on character recognition. In: 13th International Conference on Document Analysis and Recognition (ICDAR) (2015)
17. Russakovsky, O., et al.: ImageNet large scale visual recognition challenge. Int. J. Confl. Violence 115(3), 211–252 (2015)
18. Cortes, C., Vapnik, V.: Support-vector networks. Mach. Learn. 20(3), 273–297 (1995)
19. Rozantsev, A., Lepetit, V., Fua, P.: On rendering synthetic images for training an object detector. Comput. Vis. Image Underst. 137, 24–37 (2015)
20. Girshick, R.: Fast R-CNN. In: IEEE International Conference on Computer Vision (ICCV), pp. 1440–1448 (2015)
21. Krizhecsky, A., Sutskever, I., Hinton, G.E.: ImageNet classification with deep convolutional neural networks. In: Proceedings of Neural Information Processing Systems (NIPS) (2012)
22. Szegedy, C., et al.: Going deeper with convolution. In: Proceedings of Computer Vision and Pattern Recognition (CVPR) (2015)

Predictive Model of Graduate-On-Time Using Machine Learning Algorithms

Nurafifah Mohammad Suhaimi[3], Shuzlina Abdul-Rahman[1,2,3(✉)],
Sofianita Mutalib[1,2,3], Nurzeatul Hamimah Abdul Hamid[1,2,3],
and Ariff Md Ab Malik[1,3,4]

[1] Research Initiative Group of Intelligent Systems, Universiti Teknologi MARA,
40450 Shah Alam, Selangor, Malaysia
{shuzlina, sofi, nurzea}@tmsk.uitm.edu.my,
ariff215@puncakalam.uitm.edu.my
[2] Faculty of Computer and Mathematical Sciences,
Universiti Teknologi MARA, 40450 Shah Alam, Selangor, Malaysia
[3] Universiti Teknologi MARA, 40450 Shah Alam, Selangor, Malaysia
afifahsuhaimi01@gmail.com
[4] Faculty of Business and Management, Universiti Teknologi MARA,
42300 Puncak Alam, Selangor, Malaysia

Abstract. In most universities, the number of students who graduated on time reflect tremendously on their operation costs. In such cases, the high number of graduate-on-time or GOT students achievement will indirectly reduce the university's annual operation cost per student. Not as trivial as it seems, to ensure most of the students able to GOT is challenging. It may vary in the perspective of university practises, academic programmes, and students' background. At the university's level, students' data can be used to identify the achievement and ability of students, interests, and weaknesses. To build an accurate predictive model, it requires an extensive study on significant factors that may contribute to students' ability to graduate on time. Consequently, this study aims to construct a predictive model that can predict students' graduation status. We applied five different machine learning algorithms (classifiers) namely Decision Tree, Random Forest, Naïve Bayes, Support Vector Machine (PolyKernel), and Support Vector Machine (RBFKernel). These classifiers were evaluated with four different k folds of 5, 10, 15, and 20. The performance of these classifiers was compared based on different measurement subject to accuracy, precision, recall, and F-Score. The results indicated that Support Vector Machine (PolyKernel) outperformed other classifiers and the best numbers of k folds for this experiment are 5 and 20. This predictive model of GOT is hopefully will beneficial to university management and academicians to devise their strategies in helping and improving the weakness of students' academic performance and to ensure they can graduate on time.

Keywords: Data mining · Graduate-On-Time · Machine learning ·
Predictive model · Supervised algorithm

© Springer Nature Singapore Pte Ltd. 2019
M. W. Berry et al. (Eds.): SCDS 2019, CCIS 1100, pp. 130–141, 2019.
https://doi.org/10.1007/978-981-15-0399-3_11

1 Introduction

The decreasing pattern of the Graduate-On-Time (GOT) students has become a major issue by The Malaysian Ministry of Education, professors, academicians and related parties despite the increasing number of student enrolment. Ensuring the students to GOT has become the biggest challenge to the university since the position of a university in the education industry relies on this indicator that acts as one of the metrics to measure institutional effectiveness [1]. The measurement of academic productivity is measured in many ways, depending on the range of academic input parameters and outcomes. The two productivity indicators used by the Malaysian universities are the Intake Graduation on Time (iGOT) and the annual cost per Full-Time Student Equivalent (Cost per FTSE) [2]. The iGOT measures the productivity based on the number of students (based on intake cohort) that graduates within the programme stipulated duration. Rather than looking at the annual graduation rate, iGOT highlights the number of students who graduated in comparison to the intake numbers. Based on this measurement, though a student may take a longer period to finish his degree, the university will bear more cost. The improvement of iGOT rate also ties with the Cost per FTSE [2]. Based on the 2013/2014 statistics, the average iGOT for public universities is 74%. The statistics also show a large gap between the universities of 27%. Though in some cases, delayed graduation is unavoidable, analysis from historical data could provide beneficial insights [1]. Hence, Education Data Mining (EDM) application in such areas could pave a way to understand the issues at hand better.

EDM devises a new exploration paradigm to analyse students' learning behaviour to gain insights. It uses machine learning techniques to explore data from educational settings such as online logs, teaching approaches, teaching resources, interim tests and examination results to predict and learn the patterns that characterize students' behaviours that affect their performance. Subsequently, the goal of the exploration is to understand and improve the educational outcomes. For example, the prediction model of the students' performance by forecasting the students' grade allows the academic management of the university to devise a proper warning mechanism for students who are at risk. Hence, able to help them to overcome difficulties in their study. Therefore, the prediction model could provide useful insights for strategic programmes to plan a suitable measure to improve the students' performance [3]. In a similar vein, research findings reveal that among the factors that contributed to students' poor performance are gender issue [4], marital status and age issue [5], family issues [6], previous academic record [7], demographics, personal, educational background, psychological, academic progress, and other environmental variables [8]. Even though most of the significant factors have been identified, the prediction of students' performance is dynamic as it varies from universities, programmes, and students' background [9]. In the EDM method, predictive modelling anticipates students' graduation time. In order to build the predictive model, there are several tasks used, which are classification, regression, and categorization. For example, classification task creates predictive models for target variable prediction based on several input variables [10]. Classification techniques are frequently applied to ease the decision-making process [11]. This paper aims to discover the ideal classifier that performs the best to predict students' graduation status or simply GOT. There are several classifiers under classification task that have been applied to predict students' graduation time. Among the classifiers used are Decision tree, Artificial Neural Networks, Naive Bayes, K-Nearest Neighbor, and Support Vector Machine [10, 12–14]. However, the performance of these classifiers is varied, as they perform differently according to the data type. Several experiments need to be carried to find which classifier works the best. We present this paper as follows: The first section summarizes the overview of this research while section two describes the related works. The third section reviews the methodology involved to carry out this research. The fourth section presents the results and discussions. Lastly, we conclude this paper in section five.

2   Related Work

The increasing number of students who are unable to graduate on time or GOT significantly affect the institution to produce the quality outputs each year and contribute the low score on the graduation rates [2]. Consequently, it gives adverse impact to the university's productivity, hence affect the university's ranking. The iGOT measures whether students finish their studies within the required time. Shariff et al. [14] defined iGOT as a state where students accomplished their studies in particular time that has been set by the university. In which most of the institutions in other countries in Europe have set the time for undergraduate students to graduate is four years, but regrettably, many students delay in finishing their degree (40% completed within four years while 60% in six years) [15].

Students who take a longer time to graduate affect the university's budget as the university has to spend more money to provide extra resources such as extra classroom to cater number of students [2]. There are about 25% of students in Malaysian universities overstay their course duration [16]. Among the suggestions proposed is to penalize students who fail to GOT. This action could act as a reminder to others to always take their study seriously. This challenging scenario worries many stakeholders especially the university's management as they have to think outside of the box and come out with a sturdy plan in solving such T as well as improving the number of graduation rates. They need to handle this issue brilliantly and proactively as the university's achievement depends highly on the graduation rate. One of the solutions to handle this issue is by analysing students' performance since it can be the indicator to predict the students' graduation time. However, analysing the performance of students is very complicated and tedious as it involves with many data that is continuously increasing year by year [17]. Alternatively, data mining is applicable to perform analysis in solving this issue.

Data mining (DM) is a process that can convert massive amount of data and turn into meaningful information or knowledge. It is about analysing and categorizing the information as well as summarizing the knowledge from different kind of data stored in database and data warehouse [9, 18]. Nowadays, the application of data mining is widely prevalent in the education system and beneficial to analyse students' performance, forecast students' graduation time and other related issue [19–21]. A predictive model to predict the graduation status of doctorate students was performed by [14]. The result shows that there is a total of 79 students who are predicted to GOT. Based on their research, they outlined that female tend to GOT compared to male as the number of female students who are predicted to GOT is 56% higher compared to male. Findings from [14] can be intensified by [22] as they underlined that from their research, the achievement level of female students in software education was higher than male students. The accuracy of DM model can be improved as well as performing a more in-depth analysis of the mined data with the use of machine learning algorithms. Machine learning algorithms are the methods usually employed by researchers to discover patterns from data sets by letting them learn on their own [23]. Besides with the advancement of extensive computing capabilities, learning through a tremendous amount of data is now seems possible.

3 Methodology

There are six phases involved in this research methodology following Cross-Industry Process for Data Mining (CRISP-DM) [24] where each phase is associated with the necessitate activities. The detail of each phase is discussed briefly in the next sections.

Business Understanding. Business Understanding is the first phase of this research. In this phase, the main area that will be examined is issues that are related to students' graduation status and the factors that contribute to students' timely graduation.

Data Understanding. The second stage of the CRISP-DM process is Data Understanding which requires the researcher to obtain the required data as well as transformed it into a format that can be mined using Data Mining Tool. In this research, the technique applied in gathering the data was through document and records, by examining UiTM students' database that contains the historical data of UiTM's students of cohort 2013. This data was then been examined carefully to identify the distribution and its range values. Based on the examination, the raw data from UiTM's database consist of 31 attributes with 74,670 instances.

Data Preparation. This is the crucial phase in CRISP-DM process as the competency of the research's model is highly depends on the quality of the dataset. Several activities involved in this phase which include data selection & cleaning, data construction, and data integrating & formatting. Several attributes from the raw dataset were selected based on the relevance to the goals of this research. The remaining of the unselected attributes were discarded as it did not give any meaning and contribution to achieve the goals of this research. Table 1 shows the list of the selected attributes, 13 attributes with GOT status as the target or class label.

Table 1. Attribute selection

Attribute	Value	Description
Prog_desc	Science programme, Engineering programme	Programme taken by the students
Study_mode	Full-Time, Extended Full-Time	The mode of study
Sponsor	Yes, No	Loan or scholarship taken by the student
Disability	Yes, No	Student's disability
Gender	Male, Female	Student's gender
Marital_status	Single, Married	Student's marital status
Race	Dusun, Iban/Sea Dayak, Jawa, Bidayuh, Melayu	Student's race
Age	18, 20, 25, 30	Student's age
Permanent_state_desc	Johor, Kedah, Kelantan, Perak, Perlis, Pulau Pinang, Sabah, Sarawak	Student's permanent address
Entry_rquirement	Diploma, Matrikulasi KPM	Student's intake mode
CGPA	0–1.99, 2.00–2.49, 2.50–2.99, 3.00–3.49, 3.5–4.0	Student's CGPA
Family_income	Nil Income RM1 - RM499.99 RM500 - RM999.99 RM1000 - RM1999.99 RM2000 - RM2999.99 RM3000 - RM3999.99 RM4000 - RM4999.99 RM5000 - RM7999.99 RM8000 - RM9999.99 RM10000 dan keatas	Student's family income
GOTstatus	GOT, Non-GOT	Student's graduation status

This research catered students' data from Science and Engineering Programme. By using "filter" task in Microsoft Excel, for *prog_desc* attribute, all programmes other than Science Programme and Engineering Programme were discarded. The total of instances of Engineering Programme is 1160 instances while Science Programme is 2575 instances.

Modelling. Modelling phase is the part to search for useful patterns in data. Within machine learning process, dataset needs to undergo the modeling process to identify the patterns from the datasets. In data mining, there are numbers of modelling algorithm or commonly known as classifier, but not all of them suit with this research's project. The list of the classifiers was narrowed, based on few rules particularly based on the business questions and the type of variables involved. The prepared data derived from data preparation process was trained on five different classifiers, namely Decision Tree, Random Forest, Naïve Bayes, SVM with PolyKernel, and SVM with RBFKernel.

Evaluation. In model evaluation phase, these classifiers were evaluated based on four performance measurements which are the value of accuracy score, precision, recall, and F-measure. The classifier that scored the best is appointed as the best classifier for this research.

Deployment. Deployment phase is the phase where all the research's progress, outcomes, results and findings as well as any problem or limitations are deployed in a report form.

The predictive models are developed using a data mining tool, called Waikato Environment for Knowledge Analysis (WEKA) [25].

4   Results and Discussions

This section presents the performance of the five classifiers as mentioned in the previous section. These classifiers were evaluated based on cross-validation with 5, 10, 15, and 20 folds. The performance of each classifier is interpreted using different performance measures such as accuracy score, precision, recall, and F-Score. The next subsections present and analyze the results gained from these classifiers.

4.1   Accuracy Score Analysis

Accuracy score is the correct number predictions made divided by the total number of predictions made, multiplied by 100 to turn it into a percentage. In other word, accuracy is the possibility that the classifier can correctly predict the positive and negative instances. Figure 1 shows the overall accuracy score for all the five classifiers.

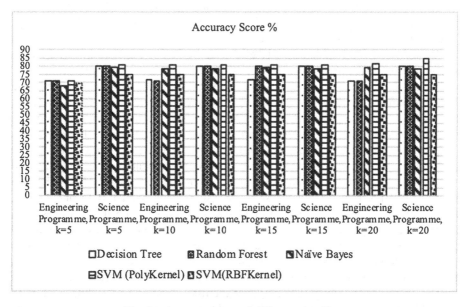

Fig. 1. Accuracy score of different classifiers

According to the chart in Fig. 1, the results of each classifier are illustrated, according to the type of data and the cross-validation fold. From the graph, the highest bar chart can be seen in Science Programme is when k = 20, which is from SVM (PolyKernel). This classifier has outperformed the others with an accuracy score of 84.78%. On the other hand, the lowest accuracy score was achieved by Naïve Bayes with k = 5, in Engineering Programme, with 68.13% only. When k = 5, we can clearly see that classifiers performed better on Science Programme, compared to Engineering Programme. However, on average, the accuracy score gained from all classifiers on Engineering Programme increased when k = 10, but there are no big differences occurred in term of accuracy score on Science Programme. The accuracy score of Random Forest rose dramatically when k = 15 for Engineering Programme, but for the other classifiers, the score remained constant for both type of data. Nevertheless, the accuracy score for Random Forest dropped back when k value is changed to 20. SVM (PolyKernel) on the other side went up steadily in the term of accuracy score when k value is changed from 15 to 20. Moreover, we can see a clear upward trend on the accuracy score of SVM (PolyKernel) for both type of data, when being tested on k = 5, 10, 15, and 20 and it reached the maximum score when k value is 20. Overall, from the graph above, we can conclude that on average, all classifiers gave better accuracy result for Science programme in the range of 75 to 85% and the ideal k value for determining the accuracy score is when k = 20.

4.2 Precision Score Analysis

In machine learning, precision is the number of positive values that are predicted correctly to the total predicted positive. Here, the correct positive prediction is highlighted out of all the positive predictions. For example, in this research, we want to highlight the real number of students who are actually GOT out of all the students who are predicted to GOT. The high precision score indicates that there is low false positive. False positive is when the classifier labels data that is actually negative with positive. In this research, non-GOT students are labeled with GOT. The bar chart in Fig. 2 shows the comparison of the precision score for all the classifiers.

Overall, from the graph above, we can see that the precision score of these five classifiers remained steady throughout the experiments when k = 5, 10, 15, and 20. Here, when k = 5, the performance of SVM (PolyKernel) when being tested on Science Programme achieved the highest precision score compared to other classifiers. The precision score of SVM (PolyKernel) decreased about 1% when k = 10, 15, 20 and the score remained unchanged throughout the experiments which are 0.82, with Science Programme. From Science Programme, it can be seen that SVM (PolyKernel) and SVM (RBFKernel) shared the same precision score when k = 15 and k = 20, which show that these two classifiers performed identically with that k values. In a similar vein, for Science Programme, Decision Tree and Random Forest also achieved the same precision score throughout the experiments, which is 0.8. On the other hand, for Science Programme, the trend of the precision score for all classifiers are the same, except for Random Forest. When k = 5 and k = 10, Random Forest scored 0.71, but the score increased slightly to 0.72 for the next cross-validation folds, which are k = 15 and k = 20. Besides, SVM (PolyKernel) scored the lowest precision score, compared to

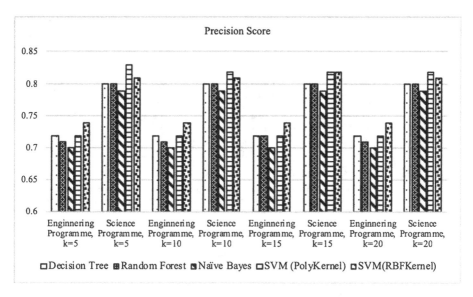

Fig. 2. Precision score of different classifiers

the other classifiers, which is 0.70. However, this classifier performed better on Science Programme, which shows that this classifier was good in classifying data in Science Programme, compared to Engineering Programme. Overall, from the graph, it can be concluded that all classifiers able to produce more precise result in predicting GOT with Science programmes and on average, the best number of cross-validation folds to measure the precision score of all the classifiers is 5. In the next section, the performance analysis on each fold is discussed in detail.

4.3 Recall Score Analysis

Recall or Sensitivity is the proportion of real positive value that is correctly predicted positive. It is also can be defined as the ratio of correctly predicted positive values to the actual positive values. Recall highlights the sensitivity of the algorithm i.e. out of all the actual positives of how many were caught by the classifier. Recall score is calculated by the real positives number divided by the real positives number plus with the false negatives number. Real positives are where data is classified as positive by the classifier that is actually positive, or in a simpler word, they are correct. False negatives on the other hand are data that the classifier marks as negatives are actually positive, or incorrect. In this research, real positives are correctly predicted GOT and false negatives are the students that the classifiers label as non-GOT that actually were GOT. Recall can be thought as of a classifier's ability to find all the data points of interest in a dataset. The bar chart in Fig. 3 shows the comparison of the recall score for all the classifiers.

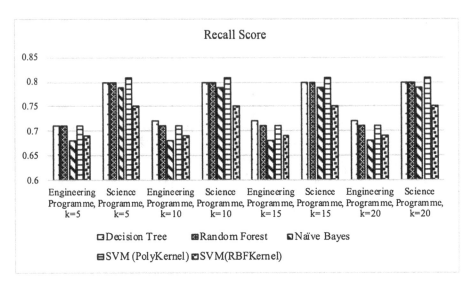

Fig. 3. Recall score of different classifiers

From the graph, for Engineering Programme, it can be seen that the performance of all classifiers was slightly low when the cross-validation folds are equal to 5. Here, Decision Tree, Random Forest and SVM (PolyKernel) shared the same level of recall score which is 0.71%, and the lowest recall score is achieved by Naïve Bayes, with only 0.68%. However, the performance of Decision Tree increased to 0.72% and remain constant for k = 10, 15 and 20. Moreover, the performance of all of the classifiers also remained constant for k = 10, 15 and 20. It can be concluded that when k = 5, the classifiers' performance was slightly lower, compared with the rest of the cross-validation folds. For Science Programme, the pattern of the recall score for all classifiers is similar when k = 5, 10, 15, and 20. Here, Decision Tree and Random Forest shared similar score, which is 0.80%, throughout the experiments. SVM (PolyKernel) has outperformed the other classifiers with a score of 0.81%, while SVM (RBFKernel) achieved the lowest recall score, which is 0.75%. Overall, from the graph, it can be concluded that all classifiers predicted GOT status better with Science Programme and the different number of folds did not affect the recall score of the classifiers.

4.4 Performance Analysis

The F score, also called as F measure, is a measure of a test's accuracy. The F score is defined as the weighted harmonic mean of the test's precision and recall. The score of F score takes the precision and recall of a test into account. As previously explained, precision is the ratio of positive results that are truly positive. This is also known as the positive predictive value. Recall on the other hand, or more known as sensitivity, is the ability of a test to correctly identify positive results to get the true positive rate. When a classifier achieved F score closest to 1, that's mean the classifier has a perfect balance of precision and recall. The F score that is closest to 0 is the worst, which means it has low score of precision as well as recall. The following Fig. 4 illustrates the trend of F score between Engineering Programme and Science Programme for all the classifiers.

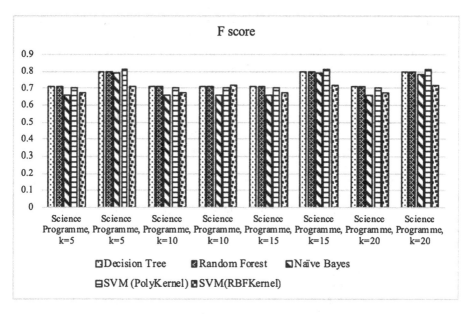

Fig. 4. F score of different classifiers

From the graph, the trend of the F score for all classifiers with Engineering Programme and Science Programme, when k = 5, 15, 20, is similar. Here, it can be interpreted that Decision Tree, Random Forest and SVM (PolyKernel) have dominated the F score for both types of data, if being compared according to the type of data. Specifically, these classifiers performed better with Science Programme, compared to Engineering Programme. When k = 10, the F score of Decision Tree, Random Forest, Naïve Bayes and SVM (PolyKernel) dropped significantly when being tested with Science Programme. However, in contra with the performance of those classifiers, the F score for SVM (RBFKernel)increased slightly when k = 10 and remained constant when k = 15 and 20. It shows that this classifier had difficulties to classify data when k = 5, but it gets better when the k values increased. For Engineering Programme, the trend of all classifiers is the same throughout the experiments. Decision Tree and Random Forest achieved the same level of F score when k = 5, 10, 15 and 20, which is the highest F score for this data type (0.71%). Naïve Bayes on the other hand achieved the lowest score, which is 0.66%, where the difference in term of F score between this classifier with Decision Tree and Random Forest is 0.05%, which is small. Overall, on the average, it can be said that classifiers performed better with Science Programme, when the cross-validation folds are equal to 15 and SVM (PolyKernel) is the best classifier that can be used to predict students' graduation status.

5   Conclusion

This paper demonstrated the use of five classifiers of machine learning algorithms to produce predictive model of Graduate-On-Time (GOT). These classifiers were modeled on Engineering Programme and Science Programme using cross-validation with 5, 10, 15, and 20 folds. The performance of each classifier was interpreted using different performance measurement such as accuracy score, precision, recall, and F-Score. The results showed that SVM (PolyKernel) outperformed other classifiers. However, Naïve Bayes had difficulties to predict students' graduation status as this classifier had produced the lowest average accuracy and score subject to precision, recall, and F Score. The model generated from Science Programme has contributed better classifiers performance compared to Engineering Programme due to its balance data distribution. The ideal k values for cross-validation folds for this research experiment were 5 and 20. This research is beneficial to many parties such as the university's academic management, academicians and students, as it can give alert about students' performance that are most likely fail to GOT and the actions can be taken to solve this problem. In addition, this approach also can improve the university's academic quality as the number of students who unable to GOT can be reduced significantly. In the future, this work can be enhanced by utilizing more datasets from different programmes particularly from non-science and technology fields.

Acknowledgment. The dataset for this research was provided by Center for Strategic Planning and Information (CSPI) Universiti Teknologi MARA. The authors would like to thank the Faculty of Computer and Mathematical Sciences and Institute of Quality and Knowledge Advancement (InQKA), UiTM for all the support given.

References

1. Ojha, T., Heileman, G.L., Martinez-Ramon, M., Slim, A.: Prediction of graduation delay based on student performance (2017)
2. Enhancing Academic Productivity and Cost Efficiency (University Transformation Programme Silver Book), Ministry of Education Malaysia (2016). http://mohe.gov.my/muat-turun/awam/penerbitan/university-transformation-programme/188-the-unitp-silver-book
3. Ibrahim, Z., Rusli, D.: Predicting students' academic performance: comparing artificial neural network, decision tree and linear regression. In: Proceedings 21st Annual SAS Malaysia Forum, pp. 1–6 (2007)
4. Dayioglu, M., Türüt-Asik, S.: Gender differences in academic performance in a large public university in Turkey. High. Educ. **53**(2), 255–277 (2007). https://link.springer.com/article/10.1007/s10734-005-2464-6
5. Amuda, B.G., Bulus, A.K., Joseph, H.P.: Marital status and age as predictors of academic performance of students of colleges of education in the Nort-Eastern Nigeria. Am. J. Educ. Res. **4**, 896–902 (2016). https://doi.org/10.12691/EDUCATION-4-12-7
6. Asif, R., Merceron, A., Ali, S.A., Haider, N.G.: Analyzing undergraduate students' performance using educational data mining. Comput. Educ. **113**, 177–194 (2017). https://doi.org/10.1016/j.compedu.2017.05.007
7. Herath, D.: Educational data mining to investigate learning behaviors : a literature review (2018). https://doi.org/10.13140/RG.2.2.20919.01446

8. Agrawal, R.S., Pandya, M.H.: Data mining with neural networks to predict students academic achievements. Int. J. Comp. Sci. Technol. **7**(2) (2016). http://www.ijcst.com/vol72/1/19-richa-shambhulal-agrawal.pdf
9. Mohammad Suhaimi, N., Abdul-Rahman, S., Mutalib, S., et al.: Review on predicting students' graduation time using machine learning algorithms. Int. J. Mod. Educ. Comput. Sci. **11**, 1–13 (2019). https://doi.org/10.5815/ijmecs.2019.07.01
10. Wah, Y.B., Ibrahim, N., Hamid, H.A., et al.: Feature selection methods: case of filter and wrapper approaches for maximising classification accuracy. Pertanika J. Sci. Technol. **26**, 329–340 (2018)
11. Akmal, E., Zaman, K., Farhan, A., et al.: Soft computing in data science. Soft Comput. Data Sci. **545**, 387–401 (2015). https://doi.org/10.1007/978-981-287-936-3
12. Cahaya, L., Hiryanto, L., Handhayani, T.: Student graduation time prediction using intelligent K-Medoids algorithm, pp. 263–266 (2017)
13. Pang, Y., Judd, N., O'Brien, J., Ben-Avie, M.: Predicting students' graduation outcomes through support vector machines. In: Proceedings - Frontiers in Education Conference FIE, October 2017, pp. 1–8 (2017). https://doi.org/10.1109/FIE.2017.8190666
14. Shariff, S.S.R., Rodzi, N.A.M., Rahman, K.A., et al.: Predicting the "graduate on time (GOT)" of Ph.D. students using binary logistics regression model. In: AIP Conference Proceedings (2016)
15. Mujani, W.K., Muttaqin, A., Khalid, K.A.: Historical development of public institutions of higher learning in Malaysia. Middle-East J. Sci. Res. **20**, 2154–2157 (2014). https://doi.org/10.5829/idosi.mejsr.2014.20.12.21113
16. Graduating on time is Malaysia's target. In: Afterschool.my (2015). https://afterschool.my/articles/graduating-on-time-is-malaysias-target. Accessed 14 Nov 2018
17. Ogwoka, T.M., Cheruiyo, W., Okeyo, G.: A model for predicting students' academic performance using a hybrid of K-means and decision tree algorithms. Int. J. Comput. Appl. Technol. Res. **4**, 693–697 (2015)
18. Jing, L.: Data mining and its applications in higher education. New Dir. Inst. Res. **2002**, 17 (2002)
19. Ma, X., Zhou, Z.: Student pass rates prediction using optimized support vector machine and decision tree. In: 2018 IEEE 8th Annual Computing Communication Workshop Conference CCWC 2018, Janua 2018, pp. 209–215 (2018). https://doi.org/10.1109/CCWC.2018.8301756
20. Athani, S.S., Kodli, S.A., Banavasi, M.N., Hiremath, P.G.S.: Student academic performance and social behavior predictor using data mining techniques. In: Proceeding - IEEE International Conference Computing Communication Automation ICCCA 2017, Janua 2017, pp. 170–174 (2017). https://doi.org/10.1109/CCAA.2017.8229794
21. Al-Shehri, H., Al-Qarni, A., Al-Saati, L., et al.: Student performance prediction using support vector machine and K-Nearest neighbor. In: Canadian Conference on Electrical and Computer Engineering, pp. 17–20 (2017). https://doi.org/10.1109/CCECE.2017.7946847
22. Lee, S.J., Kim, J.M., Lee, W.G.: Analysis of factors affecting achievement in maker programming education in the age of wireless communication. Wirel. Pers. Commun. **93**, 187–209 (2017). https://doi.org/10.1007/s11277-016-3450-2
23. Asyraf, A.S., Abdul-Rahman, S., Mutalib, S.: Mining textual terms for stock market prediction analysis using financial news. Commun. Comput. Inf. Sci. **788**, 293–305 (2017). https://doi.org/10.1007/978-981-10-7242-0_25
24. Marbán, Ó., Mariscal, G., Segovia, J.: A data mining & knowledge discovery process model. In: Ponce, J., Karahoca, A. (eds.) Data Mining and Knowledge Discovery in Real Life Applications, February 2009, pp. 438–453. I-Tech, Vienna, Austria (2009). ISBN 978-3-902613-53-0
25. Frank, E., et al.: The WEKA Workbench. Online Appendix for Data Mining: Practical Machine Learning Tools and Techniques, 4th edn. Morgan Kaufmann Press, Burlington (2016)

New Hybrid Statistical Method and Machine Learning for PM$_{10}$ Prediction

Suhartono[1(✉)], Hendri Prabowo[1], Dedy Dwi Prastyo[1],
and Muhammad Hisyam Lee[2]

[1] Department of Statistics, Institut Teknologi Sepuluh Nopember,
Jl. Raya ITS Sukolilo, Surabaya 6011, Indonesia
suhartono@statistika.its.ac.id
[2] Department of Mathematical Science, Universiti Teknologi Malaysia,
81310 Skudai, Johor, Malaysia

Abstract. The objective of this research is to propose new hybrid model by combining Time Series Regression (TSR) as statistical method and Feedforward Neural Network (FFNN) or Long Short-Term Memory (LSTM) as machine learning for PM$_{10}$ prediction at three SUF stations in Surabaya City, Indonesia. TSR as an individual linear model is used to capture trend and seasonal pattern. Whereas, FFNN or LSTM is employed to handle nonlinear pattern. Thus, this research proposes two hybrid models, i.e. hybrid TSR-FFNN and hybrid TSR-LSTM. Data about PM$_{10}$ level that be observed half hourly at three SUF stations in Surabaya are used as case study. The performance of these two hybrid models will be compared with several individual models such as ARIMA, FFNN, and LSTM by using sMAPEP. The results at identification step showed that the data has double seasonal patterns, i.e. daily and weekly seasonality. Moreover, the forecast accuracy comparison showed that hybrid TSR-FFNN produced more accurate PM$_{10}$ forecast than other methods at SUF 7, whereas FFNN yielded more accurate forecast at SUF 1 and SUF 7. These results show that FFNN as an individual nonlinear model produce better forecast than TSR and ARIMA as an individual linear model. It indicates that the PM$_{10}$ in Surabaya tend to have nonlinear pattern. Moreover, these results are also in line with the results of M3 competition, i.e. more complex method do not necessary produce better forecast than a simpler one.

Keywords: TSR · FFNN · LSTM · Hybrid · PM$_{10}$ · Surabaya

1 Introduction

Air pollution can cause various diseases and deaths for humans [1]. Moreover, it also causes disruption of various natural resources that are important for life. Due to air pollution is very dangerous for humans [2], it is very necessary to forecast air pollutants. The information used to indicate the air quality condition in Surabaya city is the Air Pollution Index (API). In Surabaya, PM10 is a pollutant in API which is often used as a main indicator for determining air quality categories. It is caused the API value of PM10 frequently becomes the highest compared to other pollutants. Since PM10 in air can be solid and liquid, it is difficult to reduce the amount [3].

© Springer Nature Singapore Pte Ltd. 2019
M. W. Berry et al. (Eds.): SCDS 2019, CCIS 1100, pp. 142–155, 2019.
https://doi.org/10.1007/978-981-15-0399-3_12

PM_{10} pollution comes from nature and human activities [4]. Natural sources that produce PM_{10} are forest fires, volcanic eruptions, and erosion. Whereas, human activities that produce PM_{10} are transportation, agriculture, and industry. Besides being caused by direct sources, PM_{10} pollution is also caused by indirect sources such as chemical processes, condensation, and coagulation involving other gas emissions [5]. Moreover, PM_{10} has impact to health problems, especially respiratory problems. The impact of PM_{10} on health will occur if exposed to long-term exposure [6].

One of the commonly forecasting methods that frequently be used in practical problems is statistical methods such as time series regression or TSR and Autoregressive Integrated Moving Average or ARIMA [8]. TSR and ARIMA can be applied for both seasonal and non-seasonal pattern data. Moreover, TSR and ARIMA are popular and easy to use for forecasting linear time series data.

In practice, data often not only have a linear pattern but also a nonlinear pattern. So, the result of using only linear statistical methods will be not good. Recently, many forecasting problems are solved by implementing machine learning which are part of artificial intelligence. These machine learning methods were used to study the structure of data in higher dimensions for predicting output in lower dimensions using certain algorithms [9]. Neural network or NN is one method in machine learning that is used in forecasting non-linear time series [10]. There are many models in NN such as Feed Forward Neural Network or FFNN [5, 11] and Long Short-Term Memory or LSTM [12, 13].

This paper proposes new hybrid forecasting methods by combining statistical method (i.e. TSR as a linear model) and machine learning (i.e. FFNN and LSTM as nonlinear model). TSR method is used to capture linear patterns from PM_{10} data, particularly the deterministic trend and seasonal patterns. Whereas, FFNN or LSTM is used to capture nonlinear patterns from PM_{10} data. Moreover, this hybrid method is proposed due to air pollution data is frequently consisted of linear and nonlinear patterns. Thus, it is necessary to build a model based on a combination of several systems [7].

There are many previous researches on PM_{10} forecasting. Hooybergs et al. [5] used FFNN to predict PM_{10} levels in Belgium. Wongsathan and Seedadan [14] used hybrid ARIMA-NN method to predict PM_{10} levels in Chiang Mai. This hybrid ARIMA-NN method compares with individual methods i.e. ARIMA and NN. The results showed that hybrid ARIMA-NN method yielded better forecast than ARIMA and NN. Robles et al. [15] also forecasted PM_{10} in Temuco Chile by comparing individual i.e. ARIMA and ANN with hybrid ARIMA-ANN. The results showed that hybrid methods produced better forecast than individual methods for predicting PM_{10}. In general, the results from several previous studies showed that the hybrid method tend to give better result than an individual method.

The rest of paper is organized as follows: Sect. 2 reviews the methodology, i.e. TSR, ARIMA, FFNN, LSTM and hybrid method; Sect. 3 presents the dataset and methodology; Sect. 4 presents the results and discussion; and Sect. 5 presents the conclusion from this study.

2 Time Series Analysis

In general, time series analysis is an analysis to predict future data based on patterns and characteristics of data in the past [16]. There are two main time series methods, i.e. statistical methods and machine learning. Statistical methods such as TSR and ARIMA are linear models. Whereas, machine learning such as FFNN and LSTM are well used for forecasting data that has a nonlinear pattern.

2.1 Time Series Regression

The general model of the TSR is same as the linear regression model, i.e. there are predictors and response [17]. Predictors used in time series regression usually are dummy variables for trend, seasonal, and other patterns. In time series forecasting, the significance of the parameter not being main considered due to more attention is the accuracy of forecasting [18, 19].

2.2 ARIMA

The ARIMA model is a combination of autoregressive (AR) and moving average (MA) models with differencing process. Determination order of AR and MA need data fulfill stationary condition both in the mean and variance. If data not yet satisfy stationary condition, transformation and differencing is done first. ARIMA modeling is usually done by implementing Box Jenkins procedure. This procedure consists of identification, parameters estimation, diagnostic check, and forecasting steps [20]. In general, ARIMA(p, d, q) can be written as follows [21]:

$$\phi_p(B)(1 - B)^d Y_t = \theta_0 + \theta_q(B)a_t. \tag{1}$$

If data consist both seasonal and non-seasonal patterns, then ARIMA multiplicative model or ARIMA(p, d, q)(P, D, Q)^S can be applied. Here is the general model of ARIMA(p, d, q)(P, D, Q)^S:

$$\phi_p(B)(1 - B)^d \Phi_P(B^S)(1 - B^S)^D Y_t = \theta_q(B)\Theta_Q(B^S)a_t, \tag{2}$$

where

$$\phi_p(B) = \left(1 - \phi_1 B - \ldots - \phi_p B^p\right)$$
$$\theta_q(B) = \left(1 - \theta_1 B - \ldots - \theta_q B^q\right)$$
$$\Phi_p(B^S) = \left(1 - \Phi_1 B^S - \ldots - \Phi_P B^{PS}\right)$$
$$\Theta_Q(B^S) = \left(1 - \Theta_1 B^S - \ldots - \Theta_Q B^{QS}\right).$$

2.3 Feed Forward Neural Network

Neural network or NN is one of the nonlinear models that frequently be used for forecasting time series data [10]. NN can estimate almost all linear or nonlinear functions in an efficient and stable way. It is caused by the NN consists of interconnected neurons that cause the exchange of data or information between neurons.

Feed forward neural network or FFNN is a form of NN architecture that is generally the most widely used in engineering [22]. In the FFNN structure there are several layers, i.e. the input layer, one or more hidden layers and the output layer as in Fig. 1. Each neuron will receive information only from the neurons in the previous layer where the input neurons come from the output weights in the previous layer [23]. In FFNN, the process starting from the input is then received by neurons which are grouped into layers called input layers, then the information is directed sequentially to the output layer [24].

$$\hat{Y}_{(t)} = f^o \left[\sum_{j=1}^q \left[w_j^o f_j^h (\sum_{i=1}^p w_{ji}^h X_{i(t)} + b_j^h) + b^o \right] \right], \tag{3}$$

where b^o, b_j^h $(j = 1, \ldots, q)$, w_{ji}^h $(j = 1, \ldots, q; i = 1, \ldots, p)$, w_j^o $(j = 1, \ldots, q)$ is the parameter of the FFNN model, f^o is the activation function in output layer, f_j^h $(j = 1, \ldots, q)$ is the activation function in hidden layer, p is the number of input variables and q is the number of neurons in the hidden layer.

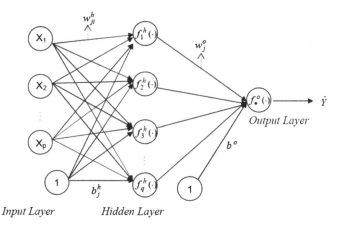

Fig. 1. FFNN architecture with p neurons in the input layer, q neurons in the hidden layer and 1 neuron in the output layer

2.4 Long Short-Term Memory

Long short-term memory or LSTM is one of the methods in the neural network that was first discovered by Sepp Hochreiter and Jurgen Schmiduber in 1997 [25]. LSTM is a development of the recurrent neural network (RNN) method. The weaknesses in the RNN method are long-term delays that cannot be accessed in an architecture [25].

The structure of LSTM consists of several layers, i.e. input layer, recurrent hidden layer and output layer [12]. Hidden layer used in LSTM is a memory block. This hidden layer of the LSTM network contains a memory cell and a pair of adaptive, multiplicative gating units that direct input and output to all cells in the block. An LSTM cell consists of four gates, i.e. the input gate, input modulation gate, forget gate and output gate as illustrated in Fig. 2 [25].

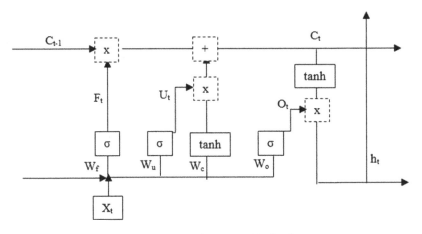

Fig. 2. Structure of memory block LSTM

Based on Fig. 2 the equation in LSTM is obtained as follows:
Forget gate:

$$F_t = \sigma\left(W_{fh}h_{t-1} + W_{fx}X_t + b_f\right), \tag{4}$$

Input gate:

$$\hat{C}_t = tanh(W_{ch}h_{t-1} + W_{cx}X_t + b_c), \tag{5}$$

$$U_t = \sigma(W_{uh}h_{t-1} + W_{ux}X_t + b_u), \tag{6}$$

Memory cell unit:

$$C_t = F_t C_{t-1} + U_t \hat{C}_t, \tag{7}$$

Output gate:

$$O_t = \sigma(W_{oh}h_{t-1} + W_{ox}X_t + b_o), \tag{8}$$

$$h_t = O_t tanh(C_t), \tag{9}$$

where W_{fh}, W_{fx}, b_f, W_{ch}, W_{cx}, b_c, W_{uh}, W_{ux}, b_u, W_{oh}, W_{ox}, b_o are the parameters of LSTM model, $\sigma(.)$ is a sigmoid activation function that shown in Eq. (10) and $tanh(.)$ is a tanh activation function that shown in Eq. (11) as follows:

$$\sigma(x) = \frac{1}{(1+e^{-x})} \tag{10}$$

$$tanh(x) = \frac{e^x - e^{-x}}{e^x + e^{-x}}. \tag{11}$$

2.5   Hybrid Statistical Method and Machine Learning

In general, hybrid is a combination of two or more systems in one function or in this paper is a combination of statistical model and machine learning. The statistical model used is TSR while the machine learning model used is FFNN and LSTM. The statistical model is used for forecasting data that has a linear pattern. While the machine learning is used to capture non-linear patterns of data. Thus, a combination of statistical method and machine learning is carried out. Hybrid models can help overcome complex structures of data [26]. In general, forecasting of hybrid models can be written as follows:

$$Y_t = L_t + N_t + \varepsilon_t, \tag{12}$$

where L_t shows the linear component represented by TSR and N_t shows the nonlinear component represented by the machine learning. Hybrid modeling starts with modeling data using TSR at level 1. Then, calculate the residual using the Eq. (13) as follows:

$$a_t = Y_t - \hat{L}_t. \tag{13}$$

Residuals from the TSR model are modeled at level 2 by using the machine learning. The residual equation of the TSR model is as follows:

$$a_t = f(a_{t-1}, a_{t-2}, \ldots, a_{t-k}) + \varepsilon_t \tag{14}$$

where f is a nonlinear function obtained from the neural network and N_t is the result of forecasting from the neural network at time-t. Hence, the forecast of hybrid model is obtained as an Eq. (15), i.e. [7]

$$\hat{Y}_t = \hat{L}_t + \hat{N}_t. \tag{15}$$

The scheme for building a hybrid model of statistical methods and machine learning is illustrated as in Fig. 3.

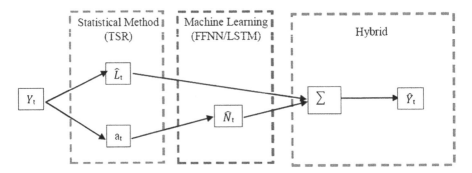

Fig. 3. Hybrid statistical method and machine learning

3 Dataset and Methodology

3.1 Dataset

The dataset used in the study is secondary data which is the result of air quality monitoring data at three SUF stations in Surabaya city, Indonesia. Air quality monitoring data is recorded every half hour. Data that be used in this paper are about PM_{10} level at SUF 1, SUF 6 and SUF 7 in 2018 as shown at Fig. 4. Based on Fig. 4 it is known that there are missing values in dataset. To overcome the missing values in the data, imputation is applied by using the median value at the same period, i.e. median at the same half hour and day.

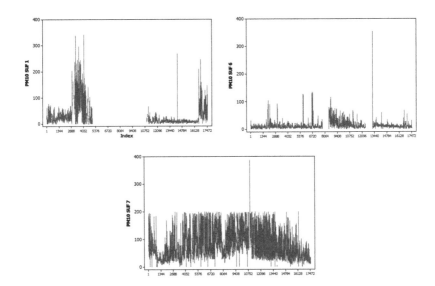

Fig. 4. Time series plot of PM_{10} in SUF 1, SUF 6 and SUF 7

3.2 Methodology

The data period is from 1 January 2018 to 20 December 2018. Data is divided into two parts, i.e. in-sample (training) and out-of-sample (testing) dataset. Data from January 1, 2018 at 00.30 to December 20, 2018 at 19.00 becomes in-sample dataset and data from December 20, 2018 at 19.30 until 00.00 to become out-of-sample dataset. To evaluate the forecast accuracy, symmetric mean absolute percentage error prediction or sMA-PEP is applied [27]. This sMAPEP is used to compare the forecast accuracy of each method and to find the best method. The formulae of sMAPEP is as follows:

$$sMAPEP = \left(\frac{1}{L} \sum_{l=1}^{L} \frac{2|Y_{n+l} - \hat{Y}_n(l)|}{(|Y_{n+l}| + |\hat{Y}_n(l)|)} \right) \times 100\%, \tag{16}$$

where L is the out-of-sample size, Y_{n+l} is the actual value in out of sample data and $\hat{Y}_n(l)$ is the forecast value in out-of-sample dataset.

4 Result and Discussion

4.1 Characteristics of PM$_{10}$ Data

Before analyzing the data, visualization was carried out to identify the pattern and characteristics at the data. Interval plot at Fig. 5 is created to find out the pattern of the data.

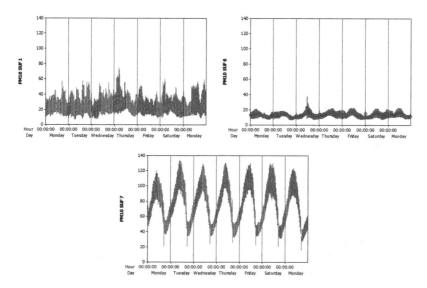

Fig. 5. Interval plot of PM$_{10}$ in SUF 1, SUF 6 and SUF 7

These graphs show that PM_{10} levels in SUF 7 is the highest compared to other SUF. Whereas, the lowest PM_{10} level is in SUF 6. SUF 6 has a low PM_{10} level due to it is located far from the highway. Moreover, the identification step shows that PM_{10} data at three SUF stations has double seasonal pattern, i.e. daily and weekly seasonality. Furthermore, PM_{10} levels at three SUF stations on Monday tend to be lower than other days.

4.2   Forecasting PM_{10}

This research compares hybrid models of statistical methods and machine learning with several individual models such as ARIMA, FFNN and LSTM. ARIMA modeling is applied based on the Box Jenkins procedure. Before doing parameter estimation, identification step is done to identify data stationarity. PM_{10} data are not stationary in the mean and variance so that transformation and differencing are applied on the data. Nonseasonal and seasonal (with period 48 and 336) differencing are employed to achieve stationary data. Then, the best ARIMA model for each SUF is obtained, i.e. double seasonal and subset ARIMA models. These best ARIMA models satisfy assumption about white noise residuals. The best ARIMA model for each SUF is shown in Table 1.

Table 1. The best ARIMA model for each SUF

Variable	ARIMA model
PM_{10} SUF 1	$([11, 13, 25, 33, 44, 47], 1, [1, 2, 3, 4, 5, 6, 7, 8, 9, 10, 15, 16, 18, 19, 22, 23, 24, 27, 28, 32, 35, 40, 46, 47]) (0, 1, 1)^{48} (0, 1, 1)^{336}$
PM_{10} SUF 6	$([1, 6, 14, 17, 33, 45, 46, 48], 1, [1, 2, 4, 8, 13, 15, 16, 20, 47]) (0, 1, 1)^{48} (0, 1, 1)^{336}$
PM_{10} SUF 7	$([7, 8, 9, 11, 15, 18, 27, 37, 38, 44, 48], 1, [1, 2, 3, 4, 5, 6, 10, 12, 14, 16, 21, 22, 23, 24, 41, 47]) (0, 1, 1)^{48} (0, 1, 1)^{336}$

The individual methods used in this paper are TSR and ARIMA as statistical methods and FFNN and LSTM as machine learning or NN methods. Modeling of NN is based on the ARIMA model, particularly determination of input. The input used on FFNN and LSTM is the significant AR lags of the ARIMA model. Furthermore, NN modeling is done by using 1 hidden layer and the number of neurons from 1 to 5. The main difference between FFNN and LSTM is the process in hidden layer, i.e. LSTM uses a hidden layer, namely memory block, whereas FFNN uses the sigmoid activation function in the hidden layer. The optimum number of neurons in the FFNN and LSTM models is shown in Table 2.

There are two hybrid models that be proposed in this research, i.e. hybrid TSR-FFNN and hybrid TSR-LSTM. The first step in hybrid modeling is applying TSR model at level 1 to PM_{10} data. This TSR model is used mainly to capture linear patterns from data. The TSR model uses dummy variable predictors for trends and seasonal. The dummy variable for seasonal is dummy every half hour in 1 day and dummy day in

1 week. Thus, the form of TSR model at level 1 is a multiplicative model that consists of 336 dummy variables for seasonal pattern. Then, residuals from the TSR model are modeled at level 2 by applying machine learning, i.e. FFNN and LSTM. This machine learning algorithm is used mainly to capture non-linear patterns that cannot be captured by TSR model. Machine learning modeling on TSR residuals is based on the ARMA model of the TSR residual in Table 3, particularly how to determine the input.

Table 2. Optimum number of neurons in FFNN and LSTM

Variable	Number of neurons	
	FFNN	LSTM
PM_{10} SUF 1	2	2
PM_{10} SUF 6	2	2
PM_{10} SUF 7	3	3

Table 3. ARMA model of time series regression residual

Variable	ARMA model
PM_{10} SUF 1	([1, 5, 6, 12, 13, 14, 16, 17, 18, 19, 21, 22, 23, 24, 26, 31, 33, 35, 40], [8, 11, 37, 38, 39, 40, 41, 42, 43, 44, 45) (1, 0)^{48}
PM_{10} SUF 6	([1, 2, 3, 12, 15, 16, 17, 45, 46], [3, 4, 12, 16, 18, 31]) (1, 0)^{336}
PM_{10} SUF 7	([1, 6, 7, 10, 11, 12, 13, 41], [1, 24, 45]) (1, 0)^{48} (1, 0)^{336}

The input for NN model is the significant AR lags of the ARMA model. After obtaining the FFNN and LSTM models from the residuals, the forecast value of hybrid TSR-FFNN and hybrid TSR-LSTM is obtained. The optimum number of neurons in hybrid TSR-FFNN and hybrid TSR-LSTM is shown in Table 4.

Table 4. Optimum number of neurons in hybrid TSR-FFNN and Hybrid TSR-LSTM

Variable	Number of neurons	
	Hybrid TSR-FFNN	Hybrid TSR-LSTM
PM_{10} SUF 1	3	4
PM_{10} SUF 6	3	5
PM_{10} SUF 7	2	5

Furthermore, Fig. 6 shows the forecast accuracy comparison between the actual data and the forecast using various methods. In general, the forecast values of ARIMA, FFNN, LSTM, hybrid TSR-FFNN, and hybrid TSR-LSTM have followed the pattern of actual data, even though some methods still have high error at certain period. For example, the forecast of ARIMA at SUF 6 has high error at period 4, 9, and 10 compared to other methods.

The sMAPEP at out-of-sample dataset of each method is presented in Table 5. To find the best method, particularly compared to the ARIMA method, it can be seen from the sMAPEP ratio in Table 6. The results show that at SUF 6 the best model for forecasting PM_{10} is hybrid TSR-FFNN method. Whereas, the results at SUF 1 and SUF 7 show that both hybrid TSR-FFNN and hybrid TSR-LSTM produce less accurate forecast than ARIMA model. These results are in line with the results of M3 competition, i.e. more complex methods do not always produce better predictions than simpler ones [28]. Moreover, the best method for forecasting PM_{10} at SUF 1 and SUF 7 is FFNN. These results are not in line with the results of M3 competition, particularly the third result, i.e. hybrid method tend to produce more accurate forecast that individual method [27, 28]. In general, the results show that PM_{10} tend to have nonlinear pattern. Thus, nonlinear method both individual or hybrid method are preferable to be used for forecasting PM_{10} at certain area.

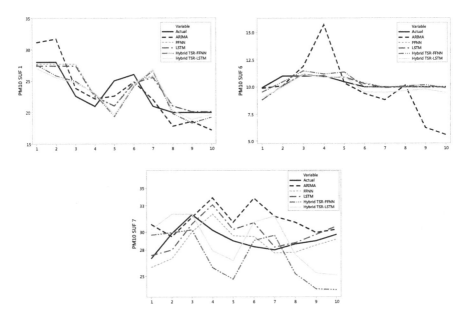

Fig. 6. Comparison between actual and forecast data

Table 5. The Results of sMAPEP for each method

Variable	ARIMA	FFNN	LSTM	Hybrid TSR-FFNN	Hybrid TSR-LSTM
PM_{10} SUF 1	8.72	7.81	7.88	9.87	9.58
PM_{10} SUF 6	17.69	3.16	3.15	2.73	4.84
PM_{10} SUF 7	7.75	3.99	4.07	11.14	9.09

Table 6. The results of sMAPEP Ratio to ARIMA method

Variable	FFNN	LSTM	Hybrid TSR-FFNN	Hybrid TSR-LSTM
PM_{10} SUF 1	0.89	0.90	1.13	1.10
PM_{10} SUF 6	0.18	0.18	0.15	0.27
PM_{10} SUF 7	0.51	0.52	1.44	1.17

5   Conclusion

This paper proposes two hybrid forecasting methods, i.e. hybrid TSR-FFNN and hybrid TSR-LSTM, for forecasting PM_{10} levels in three SUF stations in Surabaya city, Indonesia. The results then are compared to some individual methods, i.e. TSR and ARIMA as linear statistical methods and FFNN and LSTM as machine learning or nonlinear time series methods. The results showed that hybrid TSR-FFNN was the best method for forecasting PM_{10} at SUF 6, whereas FFNN yielded more accurate forecast at two other stations, i.e. SUF 1 and SUF 7. Two conclusions can be proposed based on these results. First, when hybrid TSR-FFNN produced more accurate forecast than individual methods, both linear and nonlinear methods, it is in line with the third result of M3 competition and the main result of M4 competition, i.e. hybrid statistical method and machine learning tend to give more accurate forecast than individual methods [27, 28]. Second, when FFNN produce more accurate forecast than TSR and ARIMA as linear and simpler models, it is not in line with the first result of M3 competition, i.e. more complex methods do not necessary give better forecast than simpler ones. Moreover, when FFNN produce more accurate forecast than hybrid methods, it is also not in line with the third result of M3 competition and the main result of M4 competition [27, 28]. In general, the results showed that half hourly PM_{10} data tend to have nonlinear pattern and further research could be done by proposing new individual nonlinear approach or other hybrid model by combining other statistical method (e.g. Singular Spectrum Analysis) and machine learning (e.g. Support Vector Regression) that can handle precisely both double seasonal pattern and nonlinear pattern. Additionally, other approach based on spatial-temporal model can also be proposed for forecasting PM_{10} simultaneously at three locations.

Acknowledgements. This research was supported by DRPM-DIKTI under scheme of "Penelitian Dasar Unggulan Perguruan Tinggi 2019". The authors thank to the General Director of DIKTI for funding and to anonymous referees for their useful suggestions.

References

1. Tallon, L.A., Manjourides, J., Pun, V.C., Salhi, C., Suh, H.: Cognitive impacts of ambient air pollution in the National Social Health and Aging Project (NSHAP) cohort. Environ. Int. **104**, 102–109 (2017)
2. Harrison, R.M., Yin, J.: Particulate matter in the atmosphere: which particle properties are important for its effects on health? Sci. Total Environ. **249**, 85–101 (2000)

3. Shahrayni, H.T., Sodoudi, S.: Statistical modeling approaches for PM_{10} prediction in urban areas; a review of 21st-Century Studies. Atmosphere **7**, 15 (2016)
4. Poggi, J.M., Portier, B.: PM_{10} forecasting using clusterwise regression. Atmos. Environ. **45**, 7005–7014 (2011)
5. Hooybergs, J., Mensink, C., Dummont, G., Fierens, F., Brasseur, O.: A neural network forecast for daily average PM_{10} concentrations in Belgium. Atmos. Environ. **39**, 3279–3289 (2005)
6. He, G., Fan, M., Zhou, M.: The effect of air pollution on mortality in China: evidence from the 2008 Beijing olympic games. J. Environ. Econ. Manag. **79**, 18–39 (2016)
7. Zhang, G.P.: Time series forecasting using a hybrid ARIMA and neural network model. Neurocomputing **50**, 159–175 (2003)
8. Hanke, J.E., Wichern, D.W.: Bussines Forecasting, Eight edn. Pearson Practice Hall, New Jersey (2005)
9. Lewis, N.D.: Deep Learning Made Easy with R. NigelFLewis, Australia (2016)
10. Tealab, A.: Time series forecasting using artificial neural networks methodologies: a systematic review. Future Comput. Inform. J. **3**, 334–340 (2018)
11. Suhartono, Saputri, P.D., Amalia, F.F., Prastyo, D.D., Ulama, B.S.S.: Model selection in feedforward neural networks for forecasting inflow and outflow in Indonesia. In: Mohamed, A., Berry, M., Yap, B. (eds.) Soft Computing in Data Science, SCDS 2017. Communications in Computer and Information Science, vol. 788. Springer, Singapore (2017). https://doi.org/10.1007/978-981-10-7242-0_8
12. Ma, X., Tao, Z., Wang, Y., Yu, H., Wang, Y.: Long short-term memory neural network for traffic speed prediction using remote microwave sensor data. Transp. Res. Part C: Emerging Technol. **54**, 187–197 (2015)
13. Srivastava, S., Lessmann, S.: A comparative study of LSTM neural networks in forecasting day-ahead global horizontal irradiance with satellite data. Sol. Energy **162**, 232–247 (2018)
14. Wongsathan, R., Seedadan, I.: A hybrid ARIMA and neural networks model for PM_{10} pollution estimation: the case of Chiang Mai City moat area. Procedia Comput. Sci. **86**, 273–276 (2016)
15. Robles, R.A., et al.: A hybrid ARIMA and artificial neural network model to forecast particullate matter in urban areas: the case of Temuco, Chile. J. Atmos. Environ. **42**, 8331–8440 (2008)
16. Makridakis, S., Wheelwright, S., McGee, V.: Forecasting: Methods and Applications. Wiley, New York (1983)
17. Shummway, R.H., Stoffer, D.S.: Time Series Analysis and Its Application with R Examples. Springer, Pittsburg (2006). https://doi.org/10.1007/s10182-008-0064-3
18. Kostenko, A.V., Hyndman, R.J.: Forecasting Without Significance Test? (2008)
19. Amstrong, J.S.: Significance tests harm progress in forecasting. Int. J. Forecast. **23**, 321–327 (2007)
20. Box, G.E., Jenkins, G.M., Reinsel, G.C., Ljung, G.M.: Time Series Analysis: Forecasting and Control. Wiley, Hoboken (2015)
21. Wei, W.W.: Time Series Analysis Univariate and Multivariate Methods. Pearson Education Inc., London (2006)
22. Azzouni, A., Pujjole, G.: A Long-short term memory recurrent neural network framework for network traffic matrix prediction. arXiv preprint arXiv:1705.05690 (2017)
23. Chong, E., Zak, S.H.: An Introduction to Optimize. Wiley, Canada (2001)
24. Suhartono.: New Procedures for Model Selection in Feedforward Neural Networks. Jurnal Ilmu Dasar **9**, 104–113 (2008)
25. Hochreiter, S., Schmiduber, J.: Long short-term memory. Neural Comput. **9**, 1735–1780 (1997)

26. Zheng, F., Zhong, S.: Time series forecasting using a hybrid RBF neural network and ar model based on binomial smoothing. World Acad. Sci. Eng. Technol. **75**, 1471–1475 (2011)
27. Makridakis, S., Spiliotis, E., Assimakopoulus, V.: The M4 competition: results, findings, conclusion and way forward. Int. J. Forecast. **34**, 802–808 (2018)
28. Makridakis, S., Hibbon, M.: The M3-competition result, conclusions and implications. Int. J. Forecast. **16**, 451–676 (2000)

Big Data Analytics

B-Spline in the Cox Regression with Application to Cervical Cancer

Jerry Dwi Trijoyo Purnomo$^{(\boxtimes)}$, Santi Wulan Purnami, and Sri Mulyani

Institut Teknologi Sepuluh Nopember, Surabaya, Indonesia
jerrypurnomo@gmail.com

Abstract. Recently, Cox proportional hazard (PH) models have played an important role and become increasingly famous in survival analysis. A crucial assumption of the Cox model is the proportional hazards assumption, that is the covariates do not vary over time. One way to check this assumption is to utilize martingale residuals. Martingale residual is an estimate of the overage of events seen in the data but not covered by the model. These residuals are used to examine the best functional form for a given covariate using an assumed Cox model for the remaining covariates. However, one problem that could be occurred when applying martingale residuals is that they tend to be asymmetric and the line does not fall around zero. Hence, in this paper, the main discussion will focus on the use of smoothing martingale residuals, another type of martingale residuals that give a higher rate of flexibility, by using B-spline and the relation to another smoothing technique, locally weighted scatterplot smoothing (LOWESS). An analysis of variables that probably affect the survival rate of patients with cervical cancer is used for illustration.

Keywords: Martingale residual · B-spline · Locally weighted scatterplot smoothing

1 Introduction

The use of some types of residual are the functional form for the influence of a covariate, in a model already accounting for other covariates, assessing model adequacy, with respect to proportional hazard assumption, the accuracy of the model in predicting the outcome for a particular subject, and the leverage exerted by each subject in parameter estimation [1]. The residual can be explained as the difference over time of the observed number of events minus the expected number of events under the assume Cox model (proportional hazard). Consequently, the martingale residuals are an estimate of the excess number of events seen in the data but not predicted by the model. These residuals are used to examine the best functional form for a given covariate using an assumed Cox model for the remaining covariates. However, in practice, there is often a violation of this assumption since it turns out that the covariate is a smooth function that is not linear. Therefore, a more flexible model is required. Locally Weighted Scatterplot Smoothing (LOWESS) [2] are classic smoothing methods that are still used today because they have a great degree of accuracy and flexibility ([11], and [12]). Another smoothing method that probably gives a better result is B-spline (free-knot spline approximation) ([13, 14],

© Springer Nature Singapore Pte Ltd. 2019
M. W. Berry et al. (Eds.): SCDS 2019, CCIS 1100, pp. 159–168, 2019.
https://doi.org/10.1007/978-981-15-0399-3_13

and [15]). This paper emphasizes to use smoothing martingale residual by applying B-spline and associate the result to LOWESS technique on patients with cervical cancer at Oncology hospital that experience metastasis.

2 Method

In this section, we briefly illustrate the literature review and the proposed method used in this paper.

2.1 Martingale Residual

Refer to [3], take the basis for these residuals to be the difference between the counting process and the integrated intensity function

$$M_i(t) = N_i(t) - \int_0^t Y_i(s) \, e^{\beta' Z_i(s)} d\Lambda_0(s) \qquad (i = 1, \cdots, n) \tag{1}$$

where t is survival time for each observation, $N_i(t)$ have a value 1 if this observation has experienced event, and 0 otherwise. Moreover, $Y_i(s)$ be the indicator that individual i is under study at a time just prior to time t, and $Z_i(s)$ is a vector of covariates. Let β be estimated by the maximum partial likelihood estimator $\hat{\beta}$ and the cumulative hazard Λ_0 by the [4] estimate

$$\hat{\Lambda}_0(t) = \int_0^t \frac{\sum dN_i(s)}{\sum Y_j(s) \, e^{\hat{\beta}' Z_j(s)}} \tag{2}$$

Thus the martingale residual is defined as

$$\hat{M}_i(t) = N_i(t) - \int_0^t Y_i(s) \, e^{\hat{\beta}' Z_i(s)} d\hat{\Lambda}_0(s) \tag{3}$$

with $\hat{M}_i$ as a shorthand for $\hat{M}_i(\infty)$. The residuals have some properties, $\sum_{i=1}^n \hat{M}_i(t) = 0$ for any t, and $E(\hat{M}_i) = \text{cov}(\hat{M}_i, \hat{M}_j) = 0$ asymptotically. When the data is right-censored and all covariates are fixed at the beginning of the study, then the martingale residual reduces to Eq. (4) [5]:

$$\hat{M}_i = \delta_i - \hat{\Lambda}_0(T_i) \, e^{\sum_{k=1}^p Z_{ik} b_k} = \delta_i - r_i, \quad i = 1, \cdots, n \tag{4}$$

The motivation behind martingale residual is that, if the true value of β and Λ_0 rather than the sample values from Eq. (2), the function Mi will be martingales. Suppose that the covariate vector Z is partitioned into vector Z*, for which the proper functional form of the Cox model is known, and a single covariate Z1 for which functional form of Z1 is unsure. We assume that Z1 is independent of Z*. Let f(Z1) be the best function of Z1 to explain its effect on survival. Thus the optimal Cox model is

$$\Lambda(t|Z^*, Z_1) = \Lambda_0(t) \, e^{\beta^* Z^*} \, e^{(f(Z_1))} \tag{5}$$

To determine f, fit a Cox model to the data based on Z* and compute the martingale residual. These residuals are plotted with the value of Z1 for ith observation. The smoothed fit of the scatter diagram is the indication of the function f.

2.2   Locally Weighted Scatterplot Smoothing (LOWESS)

Locally weighted scatterplot smoothing (LOWESS) was introduced by [2] and then developed by [6]. This method is more descriptively known as locally weighted polynomial regression. At each point in the data set a low-degree polynomial is fit to a subset of the data, with explanatory variable values near the point whose response is being estimated. The polynomial is fit using weighted least squares, giving more weight to points near the point whose response is being estimated and less weight to points further away. The weight (W) has the following properties:

a. $W(x) > 0$ for $|x| < 1$
b. $W(-x) = W(x)$
c. $W(x)$ is non increasing function for $x \geq 0$
d. $W(x) = 0$ for $|x| \geq 1$

The value of the regression function for the point is then obtained by evaluating the local polynomial using the explanatory variable values for that data point.

2.3   The General Proportional Hazard Model

The most general proportional hazards (PH) model is

$$\lambda(t|\mathbf{z}) = \lambda_0(t) \, \exp(g(\mathbf{z})) \tag{6}$$

where g is an unspecified function. When the Cox proportional assumption is violated then, g is a smooth function of z, g is approximated by a spline function s, where s is expressed as a linear combination of fixed knot B-spline. Spline methods are useful in proportional hazards modeling only when they provide insight into the analysis of complex data sets with a censored failure time variable. Although this model does not restrict the log hazard to be linear in z, it is usually difficult to estimate g(z) and to interpret the influence of any single covariate on survival. An additive regression model [7] provides more structure but allows a different, arbitrary function for each covariate. In the case of the Cox model, the log hazard ratio has p components, each represented by an arbitrary function:

$$\lambda(t|\mathbf{z}) = \lambda_0(t) \, \exp(g_1(z_1) + g_2(z_2) + \cdots + g_p(z_p)) \tag{7}$$

In this paper each of the p unknown functions is approximated by a spline function:

$$\lambda(t|\mathbf{z}) \approx \lambda_0(t) \exp\big(s_1(z_1) + s_2(z_2) + \cdots + s_p(z_p)\big) \tag{8}$$

Splines are capable of approximating known, smooth functions well, interpolating such functions at any number of points, and thus we expect that they will also approximate the component functions in (5) well [8].

2.4   B-Spline Function

The well-conditioned nature of B-splines makes their use preferable to that of plus functions when representing a spline function. B-splines are defined recursively: the ith B-spline of order r is a weighted sum of the ith and (i + 1)st B-splines of order r − 1, with weights depending on the breakpoints and continuity condition [9]. A B-spline basis is better conditioned than the truncated power basis, producing a more numerically stable representation of a spline function [8]. The [10] theorem implies that any spline function can be written as a unique linear combination of the element in the basis B:

$$s(x) = \sum_{j=1}^{d} \big(\alpha_j B_j(x)\big) \tag{9}$$

Three important mathematical properties of B-splines [9] are

a. $B_j(x) = 0$ if $x \notin [\tau_j, \tau_{j+r}]$.
b. $B_j(x) > 0$ if $\tau_j \leq x \leq \tau_{j+r}$.
c. $\sum_{j=1}^{d} B_j(x) = 1$.

where the knot sequence τ is defined

$$\tau = \{a = \tau_1 = \cdots = \tau_r, \, \tau_{r+1} = \xi_1, \cdots, \tau_{r+M} = \xi_m, \, \tau_{r+M+1} = \cdots = \tau_{M+2} = b\}$$

Property (a) shows that for any given x, at most r B-splines are nonzero. Property (b) defines which B-splines need to be evaluated. Property (c) shows that the basis is a partition of unity, and implies that any constant function lies in the span of the basis.

2.5   Estimation of B-Spline Coefficients

By transforming the scalar covariate Z into the d-dimensional vector $[B_1(z), \cdots B_d(z)]$ we obtain the B-spline model

$$\lambda(t|\mathbf{z}) = \lambda_0(t) \exp\left[\sum_{j=1}^{d} \alpha_j B_j(z)\right] \tag{10}$$

The log hazard ratio from (10) is estimated by $\sum\limits_{j=1}^{d} \hat{\alpha}_j B_j(z)$. Consider that the baseline covariate value is data dependent, that is, the hazard reduces to $\lambda_0(t)$ when z takes on some arbitrary value, zo, that satisfies s(zo) = 0. We could set the baseline covariate value to be a specified value z*, for example $\bar{z}$, by finding c(z*) that satisfy

$$\hat{s}(z*) = \hat{\alpha}_1 B_1(z*) + \cdots + \hat{\alpha}_d B_d(z*) + c(z*) = 0 \tag{11}$$

Thus

$$c(z*) = -\sum_{j=1}^{d-1} \hat{\alpha}_j B_j(z*) \tag{12}$$

when $\alpha_d = 0$. Consequently, as a consequence of B-spline Property (c), the estimated difference between the log hazard at z and at z* at any time t is

$$\hat{s} * (z) = \left[\hat{\alpha}_1 + c(z*)\right] B_1(z) + \cdots + \left[\hat{\alpha}_{d-1} + c(z*)\right] B_{d-1}(z) + [c(z*)] B_d(z) \tag{13}$$

2.6   Hypothesis Testing and Martingale Residual Using B-Spline

Consider in the case of a single continuous covariate, with the assumption of a non-linear covariate function, the hypothesis of no association between Z and survival is H0: $\alpha = 0$, and can be tested by using χ^2_{d-1} likelihood ratio (LR) statistics [8]. Further, recall the difference between the counting process and the integrated intensity function (2), for most general model, we obtain

$$M_i(t) = N_i(t) - \int_0^t Y_i(s) \exp\left[g(z)\right] d\Lambda_0(s) \tag{14}$$

For B-spline model, regarding Eq. (11), we transform the scalar covariate Z into the d-dimensional vector $[B_1(z), \cdots B_d(z)]$ to get

$$M_i(t) = N_i(t) - \int_0^t Y_i(s) \exp\left[\sum_{j=1}^{d} \alpha_j B_j(z)\right] d\Lambda_0(s) \tag{15}$$

Then the martingale residual for B-spline define as

$$\hat{M}_i(t) = N_i(t) - \int_0^t Y_i(s) \exp\left[\sum_{j=1}^{d} \hat{\alpha}_j B_j(z)\right] d\hat{\Lambda}_0(s) \tag{16}$$

3 Data Example

The data example to illustrate the smoothing martingale residual are the data from Oncology Hospital of Surabaya, which purpose at analyzing variables that probably affect the survival rate of patients with cervical cancer that experience metastasis. The data including survival time of 120 patients as the response variable. For the treatment, we will focus on surgery (1 = no; 2 = yes), chemotherapy (1 = no; 2 = yes), and radiotherapy (1 = no; 2 = yes), with age and cancer stage (1 = stage 1; 2 = stage 2; 3 = stage 3; 4 = stage 4) as the covariates. Table 1 provides information about patients' characteristic by group (cancer stage). Based on this table, most patients stay at stage 2 and 3, and only 18 patients out of 120 patients at stage 4. According to the Indonesian Ministry of Health, the survival rate of patients with cancer could differ among age intervals. Hence, in this paper, we divide age into three intervals, i.e. 26–45 year; 46–65 year; and >65 years as recommended by the Indonesian Ministry of Health. Among these intervals, patients aged between 46–65 tend to experience advanced cancer stage (stage 3 and 4) more than other intervals.

Table 1. Patients' characteristic by group

Covariate	Stage 1	Stage 2	Stage 3	Stage 4	Total
Cancer stage	21	41	40	18	120
Age					
26–45 yr	11	16	14	5	46
46–65 yr	10	24	23	10	67
>65 yr	0	1	3	3	7
Surgery					
No	16	41	39	18	114
Yes	5	0	1	0	6
Chemotherapy					
No	19	38	38	15	110
Yes	2	3	2	3	10
Radiotherapy					
No	20	34	37	17	108
Yes	1	7	3	1	12

Table 2 presents the estimation of parameters of the Cox proportional hazard regression model for patients with cervical cancer. Based on this table, there is no significant covariates affect the survival rate of patients as well as treatments, at 0.05 level. As mention in advance, the survival rate could differ across age interval. Hence, we can apply a proper method to face this problem. Spline, especially with truncated basis function (show by introducing knot point) is a common method to explain information that could change from one interval to other intervals. The most flexible spline usually applied is cubic spline. One important step to obtain an optimal result from the spline technique is to find the appropriate number and location of knot points. B-spline is a free-knot based spline model, that has higher flexibility compared to other spline methods. Thus, we apply cubic B-spline on age, to catch the information from different age intervals. The Cox PH model by applying B-spline method is presented in Table 3.

Table 2. Parameter estimation of Cox PH regression model for cervical cancer data

Covariate	Estimate	OR	p-value
Age	−1.915	0.981	0.340
Cancer stage (stage 1 is the reference category)			
Stage 2	1.503	4.497	0.160
Stage 3	1.312	3.712	0.225
Stage 4	2.023	7.557	0.052
Surgery	−17.410	0.000	0.997
Chemotherapy	0.171	1.186	0.749
Radiotherapy	0.153	1.165	0.789

Based on Table 3 we can find that after applying B-spline on age, there is a significant covariate that affects the survival rate of patients with cervical cancer (experience metastasis), namely cancer stage (stage 4). It means, patients with advanced cancer stage (stage 4), have more opportunity to experience metastasis. The direct relationship between stage and its survival time (time to experience metastasis) could be shown by using odds ratio (OR). The odds ratios are obtained by taking the exponential transformation on the estimates of regression coefficients, which is revealed in Table 3. Summarizing the result, we derive an important finding where patients with advanced cancer stage will experience metastasis roughly 9 times larger than patients with early cancer stage.

Finally, according to Table 3, we have one important covariate that affects survival time, i.e. advanced cancer stage (stage 4). However, from Fig. 1 we find that there are curvilinear on the martingale residual of cancer stage, surgery, chemotherapy, and radiotherapy. It means that the assumption of linear covariate on hazard is violated (the line does not fall around zero). We cannot use B-spline on these variables since all of them are categorical variable. Therefore we adopt LOWESS and applying this smoothing technique to these categorical variables. Figure 2 shows that the martingale residual of stage, chemotherapy, and radiotherapy after applying LOWESS on martingale and B-spline on age. The line of all martingale residuals of Fig. 2 fall around zero. It explains that this martingale residual is roughly better than the classic one (Fig. 1). From Fig. 2 we also be able to find that the assumption of hazard (linearity of covariate) is met. Combining both LOWESS on categorical variables that has an indication of curvilinear and B-spline on the variable that has polynomial interpolation, give the best result.

Table 3. Parameter estimation of the Cox PH regression model for cervical cancer data by applying B-spline as smoothing function

Covariate	Estimate	OR	p-value
Age			
bs (Age, df = 4) 1	−2.030	0.131	0.282
bs (Age, df = 4) 2	−2.836	0.059	0.090
bs (Age, df = 4) 3	−1.125	0.325	0.604
bs (Age, df = 4) 4	−4.290	0.014	0.181
Cancer stage (stage 1 is the reference category)			
Stage 2	1.555	4.735	0.153
Stage 3	1.278	3.589	0.241
Stage 4	2.190	8.936	0.037*
Surgery	−17.690	0.000	0.997
Chemotherapy	0.002	1.002	0.998
Radiotherapy	0.181	1.198	0.761

Abbreviations: bs, B-spline; df, degree of freedom (df = degree-intercept)

*Significant at 0.05 level.

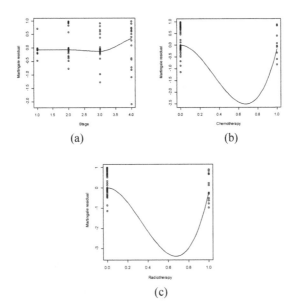

(a)

(b)

(c)

Fig. 1. The martingale residual (classic): (a) stage of cancer; (b) chemotherapy; and (c) radiotherapy.

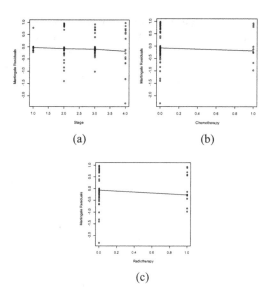

(a)   (b)

(c)

Fig. 2. The martingale residual by using LOWESS method, and by applying B-spline on Age: (a) stage of cancer; (b) chemotherapy; and (c) radiotherapy.

4 Discussion

In this paper, we analyze factors or covariates that probably affect the survival rate of patients with cervical cancer. These covariates including either categorical (cancer stage, surgery, chemotherapy, and radiotherapy) or continuous covariate (age). We emphasize to adopt the combination of locally weighted scatterplot smoothing (LOWESS) for categorical covariate and cubic B-spline for the covariate that has polynomial interpolation among interval data (age). This combination is the best way to face nonlinearity that probably happened in the Cox proportional hazard regression model.

Acknowledgment. This paper was based on the research entitle Hybrid Smoothing Method to Modeling Survival Rate and Clustering Patient with Cervical Cancer, which is supported by Institut Teknologi Sepuluh Nopember (ITS local funding, researcher contract No 1132/ PKS/ITS/2019).

References

1. Therneau, T.M., Grambsch, P.M., Fleming, T.R.: Martingale-based residuals for survival models. Biometrika **77**(1), 147–160 (1990)
2. Cleveland, W.S.: Robust locally weighted regression and smoothing scatter plot. J. Am. Stat. Assoc. **74**(368), 829–836 (1979)
3. Barlow, W.E., Prentice, R.L.: Residuals for relative risk regression. Biometrika **75**(1), 65–74 (1988)

4. Breslow, N.E.: Covariance analysis of censored survival data. Biometrics **30**, 89–99 (1974)
5. Klein, J.P., Moeschberger, M.L.: Survival Analysis-Techniques for Censored and Truncated Data, 2nd edn. Springer, New York (2003). https://doi.org/10.1007/b97377
6. Cleveland, W.S., Devlin, S.J.: Locally weighted regression: an approach to regression analysis by local fitting. J. Am. Stat. Assoc. **83**(403), 596–610 (1988)
7. Stone, C.J.: Additive regression and other nonparametric models. Ann. Stat. **13**, 689–705 (1985)
8. Sleeper, L.A., Harrington, D.P.: Regression spline in the cox model with application to covariate effects in liver disease. J. Am. Stat. Assoc. **85**(412), 941–949 (1990)
9. de Boor, C.: A Practical Guide to Splines. Springer, New York (1978)
10. Curry, H.B., Schoenberg, I.J.: On Polya frequency functions. IV: the fundamental spline functions and their limits. Journale d'Analyse Mathematique **17**, 71–107 (1966)
11. Jacoby, W.: LOESS: a nonparametric, graphical tool for depicting relationship between variables. Elect. Stud. **19**, 577–613 (2000)
12. Wilcox, R.R.: The regression smoother LOWESS: a confidence band that allows heteroscedasticity and has some specified simultaneous probability coverage. J. Mod. Appl. Stat. Methods **16**(2), 29–38 (2017)
13. Tai, C.L., Hu, S.M., Huang, Q.X.: Approximate merging of B-spline curve via knot adjustment and constrained optimization. Comput. Aided Des. **35**, 893–899 (2003)
14. Wang, Z., Wang, K., An, S.: Cubic B-spline interpolation and realization. In: Liu, C., Chang, J., Yang, A. (eds.) ICICA 2011. CCIS, vol. 243, pp. 82–89. Springer, Heidelberg (2011). https://doi.org/10.1007/978-3-642-27503-6_12
15. Hang, H., Yao, X., Li, Q., Artiles, M.: Cubic B-spline curve with shape parameter and their applications. Math. Probl. Eng. **2017**, 1–8 (2017)

Multilevel Logistic Regression and Neural Network-Genetic Algorithm for Modeling Internet Access

Wahyu Wibowo[1(✉)], Shuzlina Abdul-Rahman[2], and Nita Cahyani[1]

[1] Institut Teknologi Sepuluh Nopember, 60111 Surabaya, Indonesia
`wahyu_w@statistika.its.ac.id`
[2] Faculty of Computer and Mathematical Sciences,
Universiti Teknologi MARA, 40450 Shah Alam, Selangor, Malaysia

Abstract. Logistic regression is one of the classical methods for classification. Meanwhile, neural network is the recent method for classification. Both methods are widely used in the supervised learning and competing to be the best methods in many classifications research. This paper aims to study the performance of both methods using data of youth internet access of East Java Province of Indonesia. The first method used is Multilevel Logistic Regression, a hierarchical model which is part of Generalized Linear Mixed Model (GLMM) where the response variable is influenced by fixed and random factors. The second one is Neural Network-Genetic Algorithm in which the weight optimization is performed by selecting the relevant input variables, the optimal number of hidden nodes, and the optimal connection weights. The result shows that Multilevel Logistic Regression produced a slightly better accuracy rate of 0.873 compared to Genetic Neural Network Algorithm with an accuracy rate of 0.871.

Keywords: Multilevel Logistic Regression ·
Neural Network Genetic Algorithm · Accuracy

1 Introduction

In many fields, such as social and educational, it is often encountered the population data have hierarchical structure. The hierarchical structure is usually derived from stratified or tiered (hierarchy) population and in groups (clusters). If the data have a hierarchical structure, then there is actually an effect of the hierarchy level to the observation's units [1]. Furthermore, the development of ordinary regression to overcome the issues from a hierarchical data structure is Multilevel Modeling [2]. Multilevel model is part of Generalized Linear Mixed Model (GLMM), that is the response variable influenced by fixed and random factors. Levels of a hierarchical structure in multilevel model is defined as a level. The simplest multilevel model is a model with two levels, i.e. there are only two levels on a hierarchical data structure. The lowest level is Individual called level 1 and the higher level is District called level 2.

Researches on GLMM has been widely carried out previously. The first one is research on the status of working housewives by using multilevel regression models with binary responses and incorporating level 1 and level 2 (District) [3]. Then,

© Springer Nature Singapore Pte Ltd. 2019
M. W. Berry et al. (Eds.): SCDS 2019, CCIS 1100, pp. 169–180, 2019.
https://doi.org/10.1007/978-981-15-0399-3_14

development of multilevel model of the level of satisfaction with life in general individuals from Central and Eastern Europe grouped in the countries [4]. Another one is multilevel modeling for binary data, which contains observations on educational data in the United States using two level model [5]. Another research investigated Bayesian inference for generalized linear mixed models applied to longitudinal data with an integrated nesting Laplace approach [6]. As additional, multilevel logistic regression analysis was applied to binary contraceptive prevalence data, which contained observations on health surveys in Bangladesh using models 3 levels use the Penalized Quasi-Likelihood approach [7].

Genetic Algorithm (GA) is an optimization technique based on genetic principles and natural selection. In this technique, the population of genetics are created from many individuals who develop according to specific selection rules by maximizing fitness [8]. This algorithm is also used to obtain optimum global values by doing iterations on the Darwinian concept of evolution. In Darwin's theory of evolution, an individual is created randomly and then reproduces through the reproductive process so that a set of individuals is formed as a population. Every individual in the population has a different level of fitness/fitness. This fitness level determines how strong it is to survive in the population. Some individuals survive and others die.

Genetic Algorithms have the advantage of doing a little mathematical calculation related to the problem you want to clarify [9]. Because of the nature of its natural evolutionary changes, this algorithm will look for solutions without considering to the processes related to the problem that is solved directly. Genetic algorithms seek solutions from population points and not just from a point [10]. The search processes with a set of population points causes genetic algorithms to be less likely to be trapped at the local optimum value.

GA is a stochastic general search method, which is able to effectively explore large search spaces, which have been used with Artificial Neural Network (ANN) to determine hidden nodes and hidden layers, select relevant feature subsets, learning levels, momentum, and initialization and optimizes the network connection weights [11]. GA is used to design ANN parameters optimally, including ANN architecture, weight, input selection, activation function, ANN type, training algorithm, and a number of iterations [12, 13].

Based on the description previously, this study employs the Multilevel Logistic Regression and Neural Network-Genetic Algorithm in binary classification. As case study is real data of youth internet access of East Java Province of Indonesia. Then, the result of the two methods will be compared based on some classification criteria to get a better level of classification performance.

2 Brief Review of the Methods

This section briefly reviews the methods that are used in this study. Some basic concepts of Multilevel Logistic Regression Model, Artificial Neural Network (ANN), Genetic Algorithms (GA), Neural Network Genetic Algorithms (NNGA), and performance evaluation are explained as following:

Multilevel Logistics Regression Model. Multilevel logistic regression models are part of mixed linear model (Mixed effect model or Generalized Mixed Model (GLMM)), which is a model that combines fixed effects and random effects into one model equation. Models included in the mixed linear models are repeated measures model, longitudinal data models and multilevel models. The parameter estimates for multilevel models include using Maximum Likelihood (ML) or Restricted Maximum Likelihood (REML) [1]. The two-level logistic regression model with one predictor is written as in Eq. (1).

$$\log\left[\frac{\pi_{ij}}{(1-\pi_{ij})}\right] = \beta_{0j} + \beta_1 X_{1i} + \varepsilon_i \text{ (level 1)} \tag{1}$$

$$\beta_{0j} = \beta_0 + u_j \text{ (level 2)}$$

So, the two-level logistic regression model can be written as follow,

$$\log\left[\frac{\pi_{ij}}{(1-\pi_{ij})}\right] = \beta_0 + \beta_1 X_{1i} + u_j + \varepsilon_i \tag{2}$$

Where u_j is error at level 2, normal distributed with mean 0 and variance σ_u^2 as well as independent with ε_i.

The multilevel model is a model used in hierarchical data. Tiered data is often found in survey studies where the analysis units come from groups (clusters), or data taken through gradual sampling (cluster sampling). For example, the method of Single Stage Cluster Sampling, where the sampling units derived from the group are considered in the analysis, so in this case, the appropriate model is a multilevel model. The sampling units in the clusters are called low levels and clusters are called high levels. The number of units of analysis in clusters can be the same or different for each cluster.

Artificial Neural Network. Artificial neural networks (ANN) are information processing systems that have characteristics similar to biological neural networks. ANN were developed as mathematical models of human or biological neurology [14]. An ANN consists of a number of processing elements namely layers and neurons (Fig. 1), it appears that the network consists of 6 units (neurons) in the input layer, namely $X_1, X_2, \ldots X_6$; 1 hidden layer with two neurons, i.e. Z_1 dan Z_2 and one unit in the output layer, namely Y. Connecting weight of X_1, X_2, and X_3 with the first neuron in the hidden layer, is $V_{11}, V_{21},$ dan V_{31} (V_{ij}: the weight connecting the-i neuron to the-j neuron in the hidden layer). The notation of b_{11} and b_{12} is the weight of the bias leading to the first and second neurons in the hidden layer. The weight that connects Z_1 and Z_2 with neurons in the output layer, is W_1 and W_2. Bias weight b_2 connects hidden layer with output layer [15].

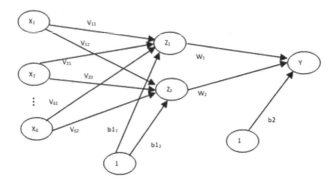

Fig. 1. Architecture of artificial neural network

A number of hidden neurons can cause problems of overfitting or underfitting. There is algorithm proposed a solution to improve hidden neurons based on statistical errors. The results show the model proposed by formula (3) can improve accuracy and minimal errors as shown below [16],

$$N_h = \frac{(4n^2 + 3)}{(n^2 - 8)} \tag{3}$$

where N_h is the number of nodes at the hidden layer and n is the number of nodes at the input layer.

Genetic Algorithms (GA). Genetic Algorithm is an optimization technique that is based on genetic principles and natural selection. In GA, genetics are formed from many individuals who develop according to specific selection rules by maximizing fitness [8]. This algorithm is also used to obtain optimum global values by doing iterations on the Darwinian concept of evolution [17, 18].

Neural Network Genetic Algorithm (NNGA). In this study, the software R with neural net packages and GA is used. Before the neural net training process is carried out, the GA process is carried out first to obtain initial weight optimization before the weights are used in the Neural Network training process. The GA operator used is as follows:

a. Encoding Schemes and Population Initialization.
b. The scheme of coding real numbers by mapping the bias and weight of ANN into the GA chromosome. Figure 1 is ANN architecture which is mapped into GA chromosome as shown in Fig. 2.

| V_{11} | V_{21} | ... | V_{61} | V_{12} | V_{22} | ... | V_{62} | W_1 | W_2 | bl_1 | bl_2 | b_2 |

Fig. 2. Chromosome coding

c. Initialize the population by generating random numbers as much as weights and biases of ANN that are mapped into the chromosomes.
d. Evaluate the value of each chromosome with the fitness value. Where the fitness value used is the value of Accuracy.
e. Conduct a selection process of N chromosomes from a number of parent P originating from the population with a roulette wheel selection. Chromosomes with high fitness values have a greater chance of being selected in randomly selecting parent pairs to reproduce.
f. Carry out the crossing process.
g. Calculate the mutation process.
h. Changing the old population with the new generation population by selecting the best chromosome from the parent and new child who has the highest fitness value after selection, crossovers, and mutations.
i. See if the solution obtained has met the criteria or not. If the solution obtained has not reached the criteria, then return step (b). The criterion is when the best fitness values are converging from the results of previous generations.

Classifier Evaluation. Actual data and predictive data from the classification model are presented using a confusion matrix, which contains information about the actual data class represented on the matrix row and the prediction data class in column [19]. The Confusion Matrix can be seen in Table 1.

Table 1. Confusion matrix

Actual group		Predicted group	
		1	0
Y	1	True Positive (n_{11})	False Positive (n_{10})
	0	False Negative (n_{01})	True Negative (n_{00})

Classification accuracy shows the performance of the overall classification model, where the higher the classification accuracy means the better the performance of the classification model.

$$Accuracy = \frac{n_{00} + n_{11}}{n_{00} + n_{01} + n_{10} + n_{11}} \quad (4)$$

The classification models can be further tested with sensitivity and specificity measures. Sensitivity is the percentage of positive data that is predicted to be positive while the specificity is the percentage of negative data predicted as negative. The formulas are shown in (5) and (6) as follows:

$$Sensitivity = \frac{n_{11}}{n_{11} + n_{01}}$$ (5)

$$Specificity = \frac{n_{00}}{n_{00} + n_{10}}$$ (6)

Performance evaluation of classification models can also be done using AUC (Area Under Curve) measures as shown in formula (7).

$$AUC = \frac{(1 + Sensitivity) - (1 - Specificity)}{2}$$ (7)

3   Data

The data used in this study are secondary data from the 2017 East Java province data of National Socio-Economic Survey (SUSENAS). This survey conducted by the Statistics Indonesia annually across Indonesia provinces.

Research Variable. The research variables used in this study are presented in Table 2. The response variable (Y) is status of youth in internet access in the last three months and the predictor variables are using computer(X_1), using mobile phone (X_2), gender (X_3), age (X_4), highest education attainment (X_5), and residence (X_6).

Table 2. Research variables

Indicator	Description	Scale
Using internet access (Y)	0: No, 1: Yes	Nominal
Using computer (X_1)	0: No, 1: Yes	Nominal
Using mobile phone (X_2)	0: No, 1: Yes	Nominal
Gender (X_3)	0: Female 1: Male	Nominal
Age (X_4)	0: <12 years old 1: 12 years old - 15 years old 2: >15 years old	Ordinal
Highest education attainment (X_5)	0: ≤primary school 1: Junior High School 2: ≥ Senior High School	Ordinal
Residence (X_6)	0: Urban, 1: Rural	Nominal

As additional, there is also information about the district origin of each sample. There is 38 districts/cities across East Java Province. The list of districts is presented as map in Fig. 3.

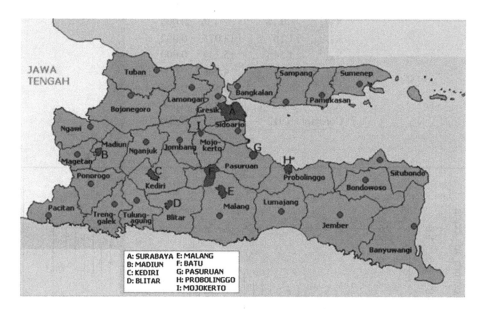

Fig. 3. Maps of East java Province (source: https://id.wikipedia.org/wiki/Daftar_kabupaten_dan_kota_di_Jawa_Timur)

For the computational process in multilevel modeling, R language is used by using library `lme4` [21]. Meanwhile, neural network and genetic algorithm used library `neuralnet` [22] and `GA` [23, 24].

4   Results

In this section, the result of Multilevel model is presented first and followed by Neural Network Genetic Algorithm model.

Multilevel Model. Multilevel regression analysis in this study is a 2-level regression model by district as random intercept at level 2. The results of the analysis can be seen in Table 3.

Based on Table 3, it can be seen that the results of testing the parameter estimation for each independent variable with the influence of level 2 (District/City), variables using computers (X_1), using mobile phones (X_2), gender (X_3), age (X_4), highest education attainment (X_5), and residence (X_6), affect the status of accessing the internet in the last three months. So, the estimated parameter of the 2-level regression model for accessing the internet in the last three months is presented in Table 3. Then, the random effect of the model is presented in Fig. 4.

Table 3. Test for estimation of multilevel regression analysis parameters

Variable	Estimation	Z	P-value
X_{11}	−2.939	−45.871	0.000
X_{21}	−2.725	−30.381	0.000
X_{31}	−0.16	−3.136	0.000
X_{41}	1.167	13.075	0.002
X_{42}	1.546	15.244	0.000
X_{51}	1.16	14.308	0.000
X_{52}	1.78	18.591	0.000
X_{61}	−0.243	-4.068	0.000
Constant	0.912	8.801	0.000

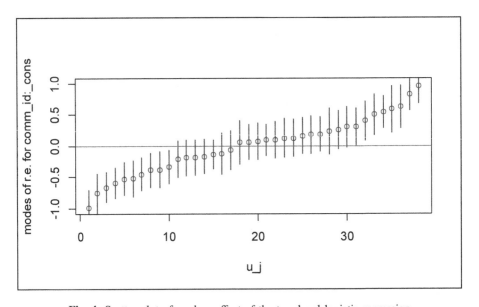

Fig. 4 Scatter plot of random effect of the two-level logistic regression

Neural Network Genetic Algorithm. In this research, the first thing to do is to determine the variables needed by Genetic Algorithm, namely the number of the population. It will be used as many as 50 chromosomes with iteration limit of 100, equal to 0.8 opportunities in crossovers and 0.1 opportunities in mutations. The selected chromosomes are as much as the weight used in accordance with the number of inputs, amounting to 6, the number of neurons in the hidden layer used is 5 neurons based on the formula (1), and the number of outputs is 1 neuron.

Furthermore, the second one is chromosome initialization. In this study, the initialization of optimization of weight parameters of Neural Network is coded using real-valued or real numbers. The process is done by representing the bias and weight on the Neural Network into the shape of the GA chromosome. Figure 1 presents an example of the structure of a Neural Network with 6 inputs and 2 neurons in the hidden layer. The formation of represented chromosomes is shown in Fig. 2. Next, 50 chromosomes are formed by generating a number of populations with variable parameter values inside. The chromosome value is generated between the range values of the parameters (bias and weight).

The third stage of the fitness function is determining the accuracy value. The 50 degenerated chromosomes will be calculated based on their accuracy value. High accuracy values produce good chromosomes and can survive. After getting a good chromosome, then the fourth stage is to select a prospective parent chromosome by using the roulette wheel. The selection of prospective parents is based on the fitness value that has been obtained at the previous stage. The frequency value is a cumulative fitness divided by total fitness. This value will be the limit of chromosome selection. Chromosomes are chosen if the value of a random number lies in the range of the previous chromosome frequency value and the chromosome.

The fifth stage is the process of crossing, given a random number uniform value (0.1) on the chromosome that has been selected as a prospective parent. If the random value of the uniform number (0.1) is less than the chance of crossing (Pc = 0.8) so that the chromosome is chosen to be a parent and there is a process of crossing that one chromosome exchanges information (parameter content) with chromosome 2 which produces a new chromosome.

The sixth stage is a mutation where mutations are carried out on the selected chromosome by changing one of the parameter values with a random number. Chromosomes are chosen by giving a uniform number (0.1) on a chromosome if uniform random numbers (0.1) are less than the chance value of mutations (Pm = 0.1) so that the chromosome becomes the chosen chromosome.

The seventh stage is by sorting the fitness values of all the chromosomes formed, from the initial 50 chromosomes, the chromosomes that are crossed, and the chromosomes at the time of mutation from the lowest fitness value to the highest fitness value. Chromosomes with the highest fitness value are stored and then used as parents in the next generation. The next stage generates 50 chromosomes which are new populations with the highest fitness value and iterations up to 100 iterations.

After obtaining the parameters for the most optimal Neural Network training, a model is then formed to determine the classification performance on internet usage status data in the last three months. Table 4 is the result of the classification performance of internet usage status data in the last three months. The following results before and after optimizing the weight with 5 neurons in 1 hidden layer.

Based on Table 4, it can be seen that initial weights and biases optimized using GA are able to improve the classification performance results in term of accuracy, AUC and Specificity. In addition, the results of NN structure before applying weight and bias parameter optimization produced the testing accuracy value of 0.8691, the AUC value of 0.9508 and the Specificity value of 0.8304. After optimization using the GA, NN produced the accuracy value of 0.8713, the AUC value of 0.9513 and the Specificity value of 0.8543. Even though it does not increase significantly but it can be said that the optimization of weight and initial bias using GA can improve the classification performance.

Table 4. Performance of classification results without optimization weight (NN) and with weight optimization (NNGA)

Classification evaluation	Method	
	NN	NNGA
Accuracy of testing	0.8691	0.8713
Accuracy of training	0.8701	0.8706
AUC of testing	0.9508	0.9513
AUC of training	0.9481	0.9478
Sensitivity of testing	0.9115	0.8899
Sensitivity of training	0.9018	0.8998
Specificity of testing	0.8304	0.8543
Specificity of training	0.8412	0.8439

Accuracy of Multilevel and NNGA Model Classifications. This subsection describes the comparison of the two methods that have been done between Multilevel Logistic Regression Model and NNGA as presented in Table 5. Based on Table 5, it can be seen that the average value of the Multilevel Logistic Regression Model classification performance values for testing data Accuracy is 0.8734, AUC amounting to 0.9518, Sensitivity of 0.8886 and Specificity of 0.8596 while for NNGA the value of testing data Accuracy is 0.8713, AUC is 0.9513, Sensitivity is 0.8899, and Specificity is 0.8543. So that in the case of this study the average performance results of the Multilevel Logistic Regression Model classification for internet usage status during the last 3 months are more appropriate to use than NNGA.

Table 5. Performance of multilevel and NNGA model classification results

Classification evaluation	Method	
	Multilevel	NNGA
Accuracy of testing	0.8734	0.8713
Accuracy of training	0.8752	0.8706
AUC of testing	0.9518	0.9513
AUC of training	0.9499	0.9478
Sensitivity of testing	0.8886	0.8899
Sensitivity of training	0.8893	0.8998
Specificity of testing	0.8596	0.8543
Specificity of training	0.8622	0.8439

5   Conclusions

The results of the study concluded that the two-level logistic regression model is slightly better than Neural Network Genetic Algorithm. This result also confirmed that the complex method doesn't always produce higher accuracy than the simple method.

It should be noted that the data characteristic also determines whether a classification method will produce good performance or not. Sometime the classical method with some modification will also produce good performance. As additional, from the two levels logistic model, it can be inferred that variables have significant impact to youth internet access, i.e., using computer, using cell phones, gender, age, educational level, and residence location.

Acknowledgment. The authors are grateful to the Ministry of Research, Technology and Higher Education of the Republic of Indonesia through Priority Fundamental Research Grant of Institut Teknologi Sepuluh Nopember with the contract number 936/PKS/ITS/2019.

References

1. Goldstein, H.: Multilevel Statistical Models, 2nd edn. E-Book of Arnold, London (1995)
2. Hox, J.J.: Multilevel Analysis: Techniques and Applications. Lawrence Erlbaum Associates Publishers, London (2002)
3. Octaviana, A.F.: Modeling of Working Status for Household Mother Using Multilevel Model With Binary Response (2017). repository.its.ac.id/2093/7/1315201029-Master-Theses.pdf
4. Hritcu, S.O.R.: A multilevel analysis of life satisfaction in Central and Eastern Europe. In: 7th International Conference on Globalization and Higher Education in Economics and Business Administration, GEBA (2013)
5. Guo, G., Zhao, H.: Multilevel modeling for binary data. Ann. Rev. Sociol. **26**, 441–462 (2000)
6. Fong, Y., Rue, H., Wakefield, J.: Bayesian inference for generalized linear mixed models. Biostatistics (2009)
7. Khan, M.H.R., Shaw, J.E.H.: Multilevel logistic regression analysis applied to binary contraceptive prevalence data. J. Data Sci. **9**, 93–110 (2011)
8. Hauput, S.E., Hauput, R.L.: Practical Genetic Algorithms. Wiley, Hoboken (2004)
9. Gen, M., Cheng, R.: Genetic Algorithms and Engineering Design. Wiley, Hoboken (1997)
10. Kolberg, D.E.: Genetic Algorithms in Search, Optimization and Machine Learning. Addison Wesley Longman Inc., Boston (1989)
11. Karegowda, A.G., Manjunath, A.S., Jayaram, M.A.: Application of genetic algorithm optimized neural network connection weights for medical diagnosis of PIMA Indian diabetes. IJSC Int. J. Soft Comput. **2**, 15–23 (2011)
12. Ahmed, O., Nordin, M., Sulaiman, S., Fatimah, W.: Study of genetic algorithm to fully automate the design and training of artificial neural networks. IJCSNS Int. J. Comput. Sci. Netw. Secur. **9**, 217–226 (2009)
13. Cahyani, N.: Deep Learning Neural Network and Genetic Algorithm in Analysis of Classification of Bidikmisi Scholarship Reception (2018). www.statistics.its.ac.id/wpp-content/uploads/advanced-cf7-upload/06211650010012_Nita-Cahyani.pdf
14. Siang, J.J.: Artificial Neural Networks and Programming Using Matlab. Andi, Yogyakarta (2005)
15. Kusumadewi, S.: Building Artificial Neural Networks Using MATLAB & EXCEL LINK. Graha Ilmu, Yogyakarta (2004)
16. Sheela, G.K., Deepa, N.S.: Review on methods to fix the number of hidden neurons in neural networks. Math. Probl. Eng. **2013**, 11 (2013)

17. Zukhri, Z.: Genetic Algorithms Evolutionary Computing Methods for Resolving Optimization Problems. Andi, Yogyakarta (2015)
18. Trevino, V., Falciani, F.: GALGO: an R package for multivariate variable selection using genetic algorithms. Bioinformatics **22**, 1154–1156 (2006)
19. Han, J., Kamber, M.: Data mining: Concepts and Techniques, 2nd edn. Morgan Kaufman, Burlington (2006)
20. Venables, W.N., Ripley, B.D.: Modern Applied Statistics with S. Springer, Heidelberg (2002). https://doi.org/10.1007/978-0-387-21706-2
21. Bates, D., Maechler, M., Bolker, B., Walker, S.: Fitting Linear Mixed-Effects Models Using lme4. J. Stat. Softw. **67**(1), 1–48 (2015). https://doi.org/10.18637/jss.v067.i01
22. Fritsch, S., Guenther, F.: neuralnet: Training of Neural Networks. R package version 1.33 (2016). https://CRAN.R-project.org/package=neuralnet
23. Scrucca, L.: GA: a package for genetic algorithms in R. J. Stat. Softw. **53**(4), 1–37 (2013). http://www.jstatsoft.org/v53/i04/
24. Scrucca, L.: On some extensions to GA package: hybrid optimisation, parallelisation and Islands evolution. Submitted to R Journal. http://arxiv.org/abs/1605.01931 (2016)

A Case Study on Student Attrition Prediction in Higher Education Using Data Mining Techniques

Syaidatus Syahira Ahmad Tarmizi[2], Sofianita Mutalib[1,2(✉)],
Nurzeatul Hamimah Abdul Hamid[1,2], Shuzlina Abdul-Rahman[1,2],
and Ariff Md Ab Malik[1,3]

[1] Research Initiative Group of Intelligent Systems, Universiti Teknologi MARA,
40450 Shah Alam, Selangor, Malaysia
{sofi,nurzeatul,shuzlina}@tmsk.uitm.edu.my,
ariff215@uitm.edu.my
[2] Faculty of Computer and Mathematical Sciences,
Universiti Teknologi MARA, 40450 Shah Alam, Selangor, Malaysia
syahiratarmizi04@gmail.com
[3] Faculty of Business and Management, Universiti Teknologi MARA,
42300 Puncak Alam, Selangor, Malaysia

Abstract. Student attrition in higher educational institutions (HEI) concern with the failure of undergraduate students who unable to complete their studies within the stipulated period. Student attrition problem relates to the resource's usage in which dropout students still use the same resources as graduated students though they do not yield any outcomes. Hence, HEI efforts to curb the percentage of student attrition numbers would have positive impact on the productivity. In a similar vein, findings from previous studies highlight numerous factors that contributed to student attrition. These factors vary from one case to another depending on the case profile. In such cases, the historical or past data can provide useful insights in understanding the factors of student attrition in an institution. In this paper, we discuss data mining techniques primarily on the supervised classification algorithms for predicting student attrition. We use the Cross-Industry Standard Process for Data Mining (CRISP-DM) that comprises of five phases for the case study. Both evaluation methods, the cross-validation and percentage split have been used to evaluate the classification methods. The study has identified the significant attributes for the student attrition prediction which are Cumulative Grade Point Average (CGPA), sponsor, family income, disability and the number of dependent. Support Vector Machine with Polynomial Kernel appeared to be the best method from the five tested algorithms.

Keywords: Classification · Data Mining · Higher education · Student attrition

1 Introduction

The need for highly-skilled graduate students with a professional career that focuses on science, technology, engineering and mathematics (STEM) fields increases yearly in line with the Industrial Revolution 4.0 (IR4.0) human capital needs [1]. Since the

© Springer Nature Singapore Pte Ltd. 2019
M. W. Berry et al. (Eds.): SCDS 2019, CCIS 1100, pp. 181–192, 2019.
https://doi.org/10.1007/978-981-15-0399-3_15

1970s, the first National Science and Technology Policy aims to achieve the 60:40 ratio of students in STEM and non-STEM enrolment which is far to be achieved until today. In contrast, the recent statistics of STEM students in the secondary education drops. It causes a chain effect on the number of STEM students in HEI [2]. In addition to the decreasing number of STEM students, HEI are facing challenges to curb student's attrition problems in STEM programs. Recent statistics from the Ministry of Education (MOE) reported a total of 22% attrition case in Malaysian Research Universities (MRU), Malaysia Technical Universities (MTUN) and Malaysian Comprehensive Universities (MCU) [3]. Out of the 22%, 9% of the total case contributed by MTUN clusters. Moreover, the need to reduce the percentage of the student attrition problem relates to the resource's usage. In such attrition cases, the dropout students use the same resources as graduated students though they do not yield any outcomes [3].

Consequently, we attempt to address two questions in this research; are there relevant data that can help us to learn the attrition pattern and based on relevant attributes of data, can we discover the pattern that contributed to student's attrition in HEI. In a similar vein, research on educational data mining (EDM) have been proposed by many researchers shows that an early identification profile of students which has tendency to attrition can be done and number of student attrition in HEI can be reduced. This paper explores data mining approaches to classify batches of students into potential to attrition or not. The rest of the paper is organized as follows: Sect. 2 provides an overview of student attrition in HEI and the contributing factors for the student to attrition in HE. Section 3 describes educational data mining in the educational sector with the supervised data classification algorithms and applications of EDM in analyzing of student attrition. In Sect. 4, the CRISP-DM approach in the methodology part will be explained. Section 5, the experimental results are presented and discussed. Finally, Sect. 6 presents future work and concludes.

2   Background Studies

2.1   Student Attrition in Higher Education Institutions

The research terms for attrition is known as withdrawal, dropout and retention [4]. A definition of attrition is referring to those persons who is withdrawing from course by canceling their candidacy, failed to progress to the next semester and failed in the examination for multiple times [5]. In general, student attrition can be identified when there is reducing the number of students who enrolled in courses per semester in comparison to the number of student intakes. Attrition can be categorized as voluntary attrition, incurred attrition and potential attrition [6]. Voluntary attrition refers to a student who is no longer wanted to continue their study within a study plan. Incurred attrition is when students breached the rules and regulation that has been set up by the institution.Potential attrition is the students with higher possibility to quit when they are no longer be able to perform any academic process for current and next semester of their studies. Students may have the possibility to return to the same institution but with a different mode of study, for example from full time to part time students; or enroll the same courses or to switch to other courses. [7] defined student attrition can be grouped as low risk students, medium risk students and highly risk students. Low risk students are students who have a high possibility of success in their study. Medium risk students are students who could succeed due to the initiative that had been taken by HEIs. Meanwhile highly risk students, who have a high probability to attrition or failing during their studies. The next section is to describe the factors of student attrition.

2.2 Factors of Student Attrition

Broad studies on attrition issues in HEI have shed lights on the contributing factors of student attritions. Reasons for attrition among undergraduate students vary from one person to another due to personal reasons for them to make the right decision. There many factors of student attrition have been identified from 25 different research articles from 2014 to 2018. From these research articles, only 15 main attributes out of 28 attributes that have been identified that can be classified as attrition related factors. The attributes are; age, race, grade in high school, loan, program code, admission type, family income, parents' education, gender, city (address), marital status, job status, study mode, Grade Point Average (GPA), and parents' occupation [7–9]. Most of these articles include first; gender (male and female) attribute, the second selected attribute is program code which is the major of student enrollment and the third attributes included are age and GPA. Later, attributes of family income and loan which is additional financial support for students. Then, the other attributes concerned are race that presents the major or minor populations of ethnicity in Malaysia, city which is the location of students from home and university, job status to identify either that particular of students are working or not, mode of study whether in full time or part time, and admission type which is qualification for student as prerequisite to further the study. Also, small portion of articles, include attributes of parents' education, parents' occupations, students grade in high school and their marital status.

Besides that, there are other factors had been identified by previous researchers that may contribute to attrition, such as due to the ability of student to study, failed to maintain academic performance and adaptation to the course enrolled, lack of appropriate qualification for the course, professional opportunities and the course offered may not suitable with their capabilities [10]. Financial reason, for example an inability to pay fees in time also might encourage a potential student to move from the institution with non-subsidized to a subsidized public or private institution. Employment issues may also be another main reason for the part-time students either involving job promotion and transfer matters or for the full-time students who are primarily working as part-time during studies to fund their studies and monthly expenses. As a result, the high rate of student attrition could be damaging the reputation level and financial of the HEIs that reflect on how these institutions arrange the funding for their enrolled students. [11] highlighted that this attrition issues could waste of taxpayers' money that was used to sponsor the students and at the same time it can lower the employment opportunity for qualifying positions. This issue could affect the students to improve their economy status if they are unable to earn a diploma or degree in the undergraduate program [12].

Based on the discussion, it shows that the attrition becomes a difficult problem and the factors might be vast. With the advance of data mining technology, it is expected that attrition can be analyzed in better way and it can help these institutions to solve their problem with better decision making.

3 Educational Data Mining

Educational Data Mining (EDM) in educational fields is used to extract the student's data to gain useful information in the educational environment. In this case, EDM can assist stakeholders in HEIs by helping educators to identify potential good and helpless students [13]. EDM can assist the researchers for better understanding of educational structures and for administrators to organize the institutions and making decisions [14]. EDM also can be used to identify what are the contributing factors for the student's success when enrolling in HEIs.

3.1 Supervised Algorithms

In data mining, classification is a supervised learning method that used the historical data. Classification algorithms are mostly used for classifying and predicting values from training dataset based on previous learned classes to build a model that can be used to classify new values [11, 15]. A successful rate for these algorithms will depend on how this variety of data is provided. There are various of classification algorithms such as Naïve Bayes (NB), K-Nearest Neighbors (KNN), Decision Tree (DT), Neural Network (NN) and Support Vector Machines (SVM).

Naïve Bayes is a conditional probabilistic classifier technique where all the attributes in a dataset are independent of one another [16, 17]. Meanwhile KNN techniques is used to find the nearest group of k instances in training set with the instance in the new dataset or training set. Problem with KNN is when dealing with a large size of data where there is lot of calculation process need to be performed [17]. At the meantime, NN is classified as a computational or mathematical model that uses to model the relationship between input and output to finds patterns in a dataset [10], though weight adjustment of nodes in layers [18, 19]. In the other hand, SVM technique is based on the idea of structural risk minimization. SVM model estimates the optimal separating hyper-plane by maximizing the margin between hyper-plane and closest points of the classes [18, 20–22]. The final technique, DT is a top-down tree model from a given dataset attributes [15, 17, 19]. This technique can provide a straightforward explanation without calculation that needs to be performed [17, 19].

3.2 Algorithms in Educational Data Mining

A review on past studies have shown that most of research papers have presented the successful of implementation of the supervised algorithms with high accuracy results, as shown in Table 1. These datasets include attributes that represent sociodemographic, enrollment, social and psychological. In sociodemographic context, the variables that have been included consists of age, gender, ethnicity, distance from home and university, marital status, parent occupation, family size, date of birth, high school results, family income, job status, sponsorships. For variables of enrollments category are consist of program code, Cumulative Grade Point Average (CGPA), Grade Point Average (GPA), semester enrolled, credits hours, study mode and student status (active, complete, attrition). Meanwhile physical activities, hobbies and socializing with friends are fall into social category. Meanwhile student personal traits and satisfaction with the enrolled course can be classified into psychological category.

Table 1. Summary of supervised algorithms.

Algorithm	Variable	Author
SVM, DT, rule induction	Sociodemographic, enrollment and social	[12]
DT	Sociodemographic, enrollment, social and physiological	[23]
NB	Sociodemographic, enrollment, social and physiological	[8]
KNN, DT, NB, RF, C5.0	Sociodemographic, enrollment and social	[24]
KNN, DT, NB, SVM and RF	Sociodemographic and enrollment	[10]

The selection of data mining tools is important in order to process the datasets for predicting of student attrition in HEI. Based on observation, R programming and WEKA are considered as the most popular open source tool had been used by researchers in their experiments because of their applicability and capability to validate the results faster compare to other tools. At the same time, DT has been identified as a popular used technique by researchers in classifying student attrition. It is because the result of this technique can be easily understood by a human with a straightforward explanation. In DT algorithms perspective, the highest accuracy measurement had been shown by the improved DT that based on combination Renyi entropy, information gain and association function with value 97.50% [24]. A SVM model also shows high accuracy result which is 89.84% [10].

4   Materials and Methods

This section describes the approach of the study: The Cross-industry Standard Process for Data Mining (CRISP-DM). The five phases of CRISP-DM are business understanding, data understanding, data pre-processing, modeling and evaluation.

4.1 Business Understanding

Business understanding phases is focusing on understanding the business problem by identifying the issues in the HEIs which is an attrition problem among undergraduate students. It involves the tasks for analyzing resources and constraints while determining the data analysis goal and project plan. Nevertheless, it is important to figure out any existing constraints in order to take prevention and avoiding any future risks that may occur.

4.2 Data Understanding

Data understanding refers to the exploration of data to gain insight into the data by looking at the dependency and its effect on the data mining tasks. Data for this study had been provided by Center for Strategic Planning and Information (CSPI) of UiTM starting from the year 2013 to 2018. A total of 74, 669 instances with 31 attributes related to graduate information were used as a set of samples for developing the models. In specific, Sciences, Engineering and Business academic program have been chosen in this study. For Science academic program, the number of instances for non-attrition was 1192 and for attrition is 111. Then, for the Engineering academic program, the number of instances for non-attrition is 1198 and the attrition is 117. For the Business academic program, the number of instances for non-attrition is 1022 and the attrition is 65.

4.3 Data Preprocessing

Data pre-processing refers to the activity to construct the pre-processed datasets from the raw data for the data mining tasks. The datasets were comprehended, transformed, cleaned, and reduced. From the total of 30 attributes, only 13 attributes are relevant and have influential for the predicting tasks. Both WEKA filters including ReplaceMissingValues and InterquartileRange were performed on the datasets [25, 26]. In the process for reducing the number of attributes, the InfoGrainAttributeEval filter in WEKA is applied to rank the attributes. The attribute selection is based on ranking; from the highest to the lowest according to the value of information gain. Only the significant attributes that has the highest information gain is selected, and the list of attributes are study mode, gender, sponsor, disability, marital status, age, race, state, CGPA, entry qualification, family income, BI (SPM), and number of dependent. The last attribute is the student status that acts as the class label. These set of attributes are shown in Table 2.

Table 2. The selected attributes for the study

Attribute	Description	Values of attributes
Study mode	Mode of study whether in full time or extended full-time	Numeric
Gender	Students gender	Numeric
Sponsor	Additional financial support such as loan or scholarship	Numeric
Disability	A physical or mental condition of student	Numeric
Marital status	Status of student either single, married or divorce	Numeric
Age	Current age of the student	Numeric
Race	Racial or ethnicity of student social group	Numeric
State	State where student currently lives	Numeric
CGPA	Current Cumulated Grade Point Average of the student per-semester	Numeric
Entry qualification	Entry qualification of the student as prerequisite to further study	Numeric
Family income	Family income of the student	Numeric
BI (SPM)	The English result during high school, Sijil Pelajaran Malaysia	Numeric
Number of dependent	The number of dependent of the student's parent has to support per-family	Numeric
Student status (class)	Status of student whether it can be classified as attrition or non-attrition	Nominal

4.4   Modeling

Modeling phase is done by testing the predictive model with cross-validation and percentage split methods. Five supervised algorithms were applied in developing the models, namely Decision Tree (J48), Random Forest (RF), Naïve Bayes, Support Vector Machine (Polynomial Kernel) and Support Vector Machine (RBF Kernel).

4.5   Evaluation

In the evaluation phase, the performance of the predictive model is evaluated to determine whether the models are successfully meet the criteria of the business or not. If it fails, it was necessarily to identify any possible reasons why the model did not satisfy the requirement. The analysis based on the selected attributes level performance is made between each classifier from three different datasets (i.e. different programmes) in order to see the effectiveness of the selected attributes. The first experiment measures 14 attributes, the second experiment measures 10 attributes and the third experiment measures 6 attributes.

5   Results and Findings

This section presents the results and findings of the study. Figure 1 presents the average of accuracy result for Sciences datasets for multiple attributes selection based on the combination between the cross-validation and percentage split method. For the 14 attributes, the best validation performance had shown by RF algorithm was equal to 97.26%. But it was soared to 97.36% for the 10 attributes. The performance of RF algorithm is seen much better when there are reducing the number of attributes first and second round of experiments. However, it was in contrast to the 6 attributes in which the J48 algorithms is the best validation performance that was equal to 97.27%. Small changes were found between RF and J48 which is 0.27%. Based on the study, RF algorithm is among the best validation performance for the Science dataset with the highest accuracy results which is 97.36% for 10 attributes. Table 3 shows the used of the 10 attributes that can enhance the performance of RF algorithm.

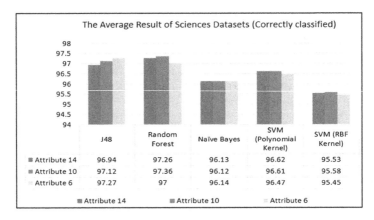

Fig. 1. Average of accuracy measures for sciences datasets for multiple attribute selection

Table 3. Selected attributes for the sciences datasets

Total of attributes	List of attributes
14	CGPA, sponsor, family income, disability, number of dependent, marital status, gender, age, race, state, academic qualification, BI (SPM), study mode, student status (class)
10	CGPA, sponsor, family income, disability, number of dependent, marital status, gender, age, race, student status (class)
6	CGPA, sponsor, family income, disability, number of dependent, student status (class)

The analysis based on the average of accuracy result also was performed on the Engineering dataset. Figure 2 illustrates the average of accuracy results for varying number of attributes that used as the parameter settings. The best validation performance algorithm with the highest classification accuracy had shown by the SVM (Polynomial Kernel) which is equal to 97.52% for varying number of attribute selection. Unfortunately, the different sets of attributes could not lead the SVM (Polynomial Kernel) of algorithms become more accurate. It confirms that the analysis on the performance of these algorithms still remain unchanged even the number of attribute selection has been reduced from first to third round of experiments. Table 4 shows the three different sets of attributes have been used in the predicting model.

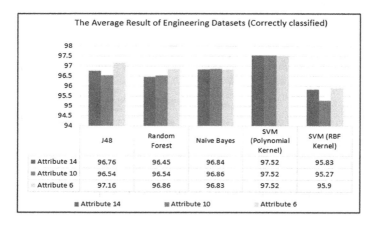

Fig. 2. Average of accuracy measures for engineering datasets for multiple attribute selection

Table 4. Selected attributes for the engineering datasets

Total of attributes	List of attributes
14	CGPA, sponsor, family income, disability, number of dependent marital status, gender, age, race, state, academic qualification, BI (SPM), study mode, student status (class)
10	CGPA, sponsor, study mode, gender, disability, race, marital status, number of dependent, BI (SPM), student status (class)
6	CGPA, sponsor, study mode, gender, disability, student status (class)

Figure 3 shows the comparison on the average of classification results for multiple data mining algorithms on the Business dataset. It can note that for the 14 attributes, the best validation performance has presents by the SVM (Polynomial Kernel) algorithms with results is equal to 98.92%. The experiments also were performed on the 10 attributes, but the performance of the SVM (Polynomial Kernel) algorithms has shown to be inefficient with slightly differences 0.03% between them. However, when the experiment is carried out on the third set of 6 attributes, the best validation performance algorithm with the highest classification accuracy has shown by the NB algorithm was equal to 98.94%. Naïve Bayes algorithm is the classification method that can obtained the best results among the other classifiers have been tested. It shows that when less number of attributes are used, the accuracy of classifiers would be better. Table 5 shows the number of attributes for three experiments using information gain. The third set of 6 attributes is most preferable to choose in which produces the highest classification accuracy that shown by the NB algorithm.

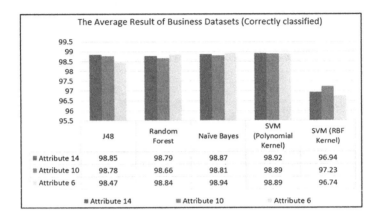

Fig. 3. Average of accuracy measures for business datasets for multiple attribute selection

Table 5. Selected attributes for the business datasets

Total of attributes	List of attributes
14	CGPA, Sponsor, family income, disability, number of dependent marital status, gender, age, race, state, academic qualification, BI (SPM), study mode, student status (class)
10	CGPA, sponsor, state, study mode, academic qualification, BI (SPM), family income, race, age, student status (class)
6	CGPA, sponsor, state, study mode, academic qualification, student status (class)

6 Conclusion and Future Work

In this study, it has presented an analysis on the evaluation measures that can be used in the real world studies in higher education datasets. Therefore, five classifiers have been used and compared based on the accuracy performance measure among these classifiers. The varying attributes selection for the first to third round of experiments have been used in order to determine the best data mining algorithms for each of datasets.

For Science academic program, Random Forest is the best predicting model when it was tested by using with 10 attributes which is 97.36%. The SVM (Polynomial Kernel) is the best predicting model for Engineering academic program. The finding shows the results is 97.52% when it was tested on the first to third sets of attributes. In contrast with the Business academic program, the best predicting model is SVM (Polynomial Kernel) that similar to Engineering academic program dataset. Though Business academic program produced better accuracy result 14 and 10 attributes. In future work, these datasets can be compared and analyzed with other semi-supervised or unsupervised data mining algorithms.

Acknowledgement. The authors would like to thank Research Management Centre and Center of Strategic Planning and Information (CSPI) of Universiti Teknologi MARA in supporting the research.

References

1. Tackling lack of interest in STEM subjects: Malaysia Education Hub (2017). http://www.edumsia.my/article/tackling-lack-of-interest-in-stem-subjects
2. Chen, Y., Johri, A., Rangwala, H.: Running out of STEM: a comparative study across STEM majors of college students at-risk of dropping out early. In: Proceedings of the 8th International Conference on Learning Analytics and Knowledge, pp. 270–279 (2018)
3. Enhancing Academic Productivity and Cost Efficiency, Ministry of Higher Education Malaysia (2017)
4. Taipe, M.A., Mauricio, D.: Predicting university dropout through data mining: a systematic literature. Indian J. Sci. Technol. **12**(4), 1–12 (2019)
5. Adusei-asante, K., Doh, D.: Students' attrition and retention in higher education: a conceptual discussion, pp. 1–10 (2016)
6. Viale Tudela, E.H.: A theoretical approach to the college student drop out. Revista Digital de Investigación En Docencia Universitaria (RIDU) **8**(1), 59–74 (2014)
7. Hoffait, A., Schyns, M.: Early detection of university students with potential difficulties. Decis. Support Syst. **101**, 1–11 (2017)
8. Hegde, V., Prageeth, P.P.: Higher education student dropout prediction and analysis through educational data mining. In: Proceedings of the Second International Conference on Inventive Systems and Control (ICISC 2018), pp. 694–699 (2018)
9. Kang, K., Wang, S.: Analyze and predict student dropout from online programs. In: Proceedings of the 2nd International Conference on Compute and Data Analysis, pp. 6–12 (2018)
10. Martins, L.C.B., Carvalho, R.N., Carvalho, R.S., Victoria, M.C., Holanda, M.: Early prediction of college attrition using data mining. In: Machine Learning and Applications (ICMLA), pp. 1075–1078 (2017)
11. Rahman, N.A.A., Tan, K.L., Lim, C.K.: Supervised and unsupervised learning in data mining for employment prediction of fresh graduate students. J. Telecommun. Electron. Comput. Eng. **9**(2), 155–161 (2017)
12. Sangodiah, A., Beleya, P., Muniandy, M., Heng, L.E., Spr, C.R.: Minimizing student attrition in higher learning institutions in malaysia using support vector machine. J. Theor. Appl. Inf. Technol. **71**(3), 377–385 (2015)

13. Jamil, N.I., Baharuddin, F.N., Maknu, T.S.R.: Factors mining in engaging students learning styles using exploratory factor analysis. Procedia Econ. Finance **31**, 722–729 (2015)
14. Bakhshinategh, B., Zaiane, O.R., Elatia, S., Ipperciel, D.: Educational data mining applications and tasks: a survey of the last 10 years. Educ. Inf. Technol. **23**(1), 537–553 (2017)
15. Yusof, N.N., Mohamed, A., Abdul-Rahman, S.: Reviewing classification approaches in sentiment analysis. In: Berry, M.W., Mohamed, A.H., Wah, Y.B. (eds.) SCDS 2015. CCIS, vol. 545, pp. 43–53. Springer, Singapore (2015). https://doi.org/10.1007/978-981-287-936-3_5
16. Saa, A.A.: Educational data mining & students' performance prediction. Int. J. Adv. Comput. Sci. Appl. **7**(5), 212–220 (2016)
17. Bilal, M., Israr, H., Shahid, M., Khan, A.: Sentiment classification of Roman-Urdu opinions using Naïve Bayesian, Decision Tree and KNN classification techniques. J. King Saud Univ. Comput. Inf. Sci. **28**(3), 330–344 (2016)
18. Ab Jamil, M.Z., Mutalib, S., Abdul-Rahman, S., Aziz, Z.A.: Classification of paddy weed leaf using neuro-fuzzy methods. Malays. J. Comput. (MJoC), **3**(1), 54–66 (2018)
19. Zamani, N.W., Mohd Khairi, S.S.: A comparative study on data mining techniques for rainfall prediction in subang. In: AIP Conference Proceedings 2013, p. 020042 (2018)
20. Wan Yaacob, W.F., Md Nasir, S.A., Wan Yaacob, W.F., Mohd Sobri, N.: Supervised data mining approach for predicting student performance. Indonesian J. Electr. Eng. Comput. Sci. **16**(3), 1584–1592 (2019)
21. Yusoff, M., Jefri, N.J., Kahar, M.S.: Sequential minimal optimization algorithm with support vector machine for mosquito Larvae identification. Adv. Sci. Lett. **23**(5), 4274–4277 (2017)
22. Ibrahim, Z., Kasiran, Z., Isa, D., Sabri, N.: Multi-script text detection and classification from natural scenes. In: Berry, M.W., Hj. Mohamed, A., Yap, B.W. (eds.) SCDS 2016. CCIS, vol. 652, pp. 200–210. Springer, Singapore (2016). https://doi.org/10.1007/978-981-10-2777-2_18
23. Sivakumar, S., Venkataraman, S., Selvaraj, R.: Predictive modeling of student dropout indicators in educational data mining using improved decision tree. Indian J. Sci. Technol. **9**(4), 1–5 (2016)
24. Mittal, V., Anuradha: A real time data mining model to predict academic attrition. Int. J. Res. Sci. Eng. Technol. **4**(7), 46–54 (2017)
25. Frank, E., Hall, M.A., Witten, I.A.: The WEKA Workbench. Online Appendix for "Data Mining: Practical Machine Learning Tools and Techniques, 4th edn. Morgan Kaufmann (2016)
26. Kamaru-Zaman, E.A., Brass, A., Weatherall, J., Rahman, S.A.: Weak classifiers performance measure in handling noisy clinical trial data. In: Berry, M.W., Hj. Mohamed, A., Yap, B.W. (eds.) SCDS 2016. CCIS, vol. 652, pp. 148–157. Springer, Singapore (2016). https://doi.org/10.1007/978-981-10-2777-2_13

The Use of Hybrid Information Retrieve Technique and Bayesian Relevance Feedback Classification on Clinical Dataset

Fatihah Mohd[1]($\boxtimes$), Masita @ Masila Abdul Jalil[1],
Noor Maizura Mohamad Noor[1], Suryani Ismail[1],
and Zainab Abu Bakar[2]

[1] School of Informatics and Applied Mathematics,
Universiti Malaysia Terengganu, 21030 Kuala Terengganu,
Terengganu, Malaysia
mpfatihah@gmail.com
[2] Faculty of Computer and Information Technology,
Al-Madinah International University, Shah Alam, Selangor, Malaysia

Abstract. Retrieval of information related to a subset variable or feature has become the attention of many researchers in data mining fields. The objective of feature selection (FS) is to improve the performance of the prediction. This contributes to providing a better definition of the features, feature structure, feature ranking, feature selection functions, efficient search techniques, and feature validation methods. In this study, a retrieval method that integrates correlation and linear forward selection algorithms to evaluate and generate the subset of clinical features are present. The objective of the research is to find the optimal features of a cancer dataset and to classify the disease into multiple cancer stages: one, two, three, and four. The research methodology is developed based on data mining, knowledge data discovery with four phases: pre-processing, resampling, feature selection, and classification. The proposed Bayesian Relevance Feedback (BRF) for classification is also described to resolve the zero value of posterior probabilities, concentrating on increasing the accuracy in the diagnosis of cancer stages. The experimental works are done on oral cancer dataset by applying WEKA. The analysis on accuracy performance was done on several classification algorithms using 15 optimal features that were chosen by a hybrid features selection method. The result shows that, BRF has outperformed others achieving 97.25% classification accuracy compared to the six classifiers, which are K-Nearest Neighbors Classifier, Multi Class Classifier, Tree-Random, Multilayer Perceptron, Naïve Bayes, and Support Vector Machine.

Keywords: Bayesian Relevance Feedback (BRF) · Classification · Correlation · Data mining · Features selection · Oral cancer diagnosis

1 Introduction

Data mining (DM) is a study of mining or digging deep into raw data to discover patterns and valuable knowledge. DM has six common of function: anomaly detection, association rule learning, clustering, classification, regression, and summarization

© Springer Nature Singapore Pte Ltd. 2019
M. W. Berry et al. (Eds.): SCDS 2019, CCIS 1100, pp. 193–207, 2019.
https://doi.org/10.1007/978-981-15-0399-3_16

[1, 2]. DM has shown its functionalities in numerous fields such as financial [3, 4], education [5, 6], and healthcare [7, 8]. Classification is a main function in data mining and broadly applied in numerous applications. Data mining provides a function to create classification models from training dataset, and apply a model to accurately predict categorical class labels of unclassified datasets [9, 10]. For example, a classification model could be used to support clinicians in diagnostics, therapeutic, or monitoring tasks. It learns from historical patient data and apply knowledge gained to predict a future decision by using machines learning techniques to build models from clinical data and then make inferences on new cases [10].

Before applying a real data for classification, the data is pre-processed to remove outlier, replace missing values, resolve imbalance data and applied a principle data mining on real data [1, 11, 12]. Most data contain more information than is required to build the model, or the less important information. Feature selection (FS), also recognized as variable selection, attribute selection or variable subset selection, is the method of selecting a subset of related features (variables, predictors) for model construction [13]. It can support to generate an accurate predictive model by removing the unrelated, and redundant attributes [13].

Thus, the aim of feature selection covers three reasons: increasing the prediction performance of the predictors or target, making faster and more cost-effective predictors, and giving a better understanding of the fundamental procedure that produced the data [14]. The hybrid FS method was used in this study, to find the optimum feature of the oral cancer dataset (OCDS) to improve the model accuracy.

In this study, the enhancement of Bayesian model classification techniques is also presented. Six classification algorithms, including K-Nearest Neighbors Classifier (KNN/IBK), Multi Class Classifier (MCC), Random Tree (RT), Multilayer Perceptron (MLP), Naïve Bayes (NB), and Support Vector Machine (SVM/SMO) are used in the experimental works for classifying the stage of oral cancer. Oral cancer (OC) is a common disease in outpatient cancer departments. This cancer could be detected by stage one, two, three or four in preparing a suitable treatment and prognosis. The experimental works are done to identify the most performing classification data mining algorithms. The aim of this study is to provide a comprehensive review of different classification techniques in data mining.

The paper is structured into four sections: Sect. 2 defines the material and method used in the study. It covers the research methodology, oral cancer dataset description, feature selection method, used in choosing an optimal feature, the proposed classification model, and metric measurement used to evaluate for all classification algorithms. The experimental works and analysis of finding are discussed in Sect. 3. Lastly, the conclusion of the study is drawn in Sect. 4, regarding to the findings.

2 Material and Method

This section, firstly illustrates a research methodology with four phases: pre-processing, resampling, feature selection, and classification. Then, the oral cancer dataset with 27 attributes used in the experimental works is presented. Next, the enhancement of

Bayesian model with relevance feedback is discussed. Lastly, the metrics used to evaluate the classification algorithm's performance are described in detail.

2.1 Research Methodology

In this section, the research methodology for disease classification is presented. Oral cancer dataset (OCDS) will be first the pre-processed to remove outlier and missing values. Then, resampling method is applied to resolve imbalance issues using the Synthetic Minority Oversampling Technique (SMOTE) [15]. Correlation-Based Filter (CBF) and Linear Feature Selection (LFS) are utilized to find the optimum features for classification model development. The steps of the research methodology are developed based on data mining, knowledge data discovery with four phases: pre-processing, resampling, feature selection, and classification (see Fig. 1).

2.2 Oral Cancer Dataset

The validation of data mining classification algorithms is done by using the dataset of oral cancer, which is reviewed from a record of oral disease, at Hospital Universiti Sains Malaysia (HUSM) in Kelantan [16]. This dataset consists of 27 attributes that are divided into two: 1–26 are independent attribute and 27 is a dependent or response attribute. In oral cancer dataset (OCDS), data is represented as a table of instances. Each instance is defined by a number of measurements, or features, with a label that represents its class. All features in OCDS are in a numeric value. The model of oral cancer attributes is recorded in Table 1. In data mining, the classification task that presented with a specified set of classes, and training objects are labelled with the appropriate class is described as supervised learning [17]. Thus, in this case, the classification on OCDS with a specified cancer stages as a class is referred to as supervised learning.

2.3 Attribute Selection Measures

The model for selected features was generated using a WEKA package being called into a Java program with a 10-fold cross validation. Functions that were used in this study were selected attributes and classified. The FS is employed to discover the most important attributes by eliminating features with poor or no predictive information. For further explanation, to be more related to this process, a term of attribute is assigned as a feature. In order to find the connection between each attribute in OCDS and the output attribute and select only those attributes that have higher connection and remove the attributes with a low connection which value close to zero, a popular technique for selecting the most relevant attributes in OCDS dataset, Correlation Based Feature Selection is applied. Applying correlation, one or multiple attributes depend on another attribute or a cause for another attribute is defined. It also presents the relationship one or multiple attributes with other attributes. In this study, the combination of Correlation Variable Evaluator with ranker and CfsSubset Evaluator with Linear Forward Selection (HCELFS) is proposed as a hybrid FS method. It integrates the benefits of both methods to define the ideal feature subset of the original dataset.

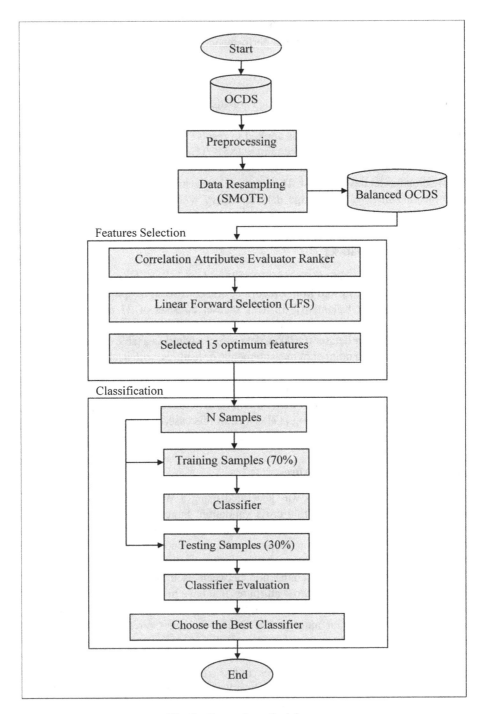

Fig. 1. Research methodology.

Table 1. Oral cancer dataset (OCDS).

	No	Attributes	Description
Demographic	1.	Case Id	Patient_Id
	2.	Age	Patient's age on first time registered
	3.	Sex	Gender of patient (female or male)
	4.	Ethnicity	Patient's ethnicity
Social habit	5.	Smoking	History of smoking
	6.	Quid chewing habit	History quid chewing habit
	7.	Alcohol	History drinking of alcohol
Clinical signs and symptoms	8.	Difficulty in chewing	Difficulty in chewing or swallowing (Dysphagia)
	9.	Ulceration	Painless ulceration more than 14 days
	10.	Neck lump	Existing neck lump
	11.	Loss of appetite	Loss of appetite
	12.	Loss of weight	Loss of weight
	13.	Hoarseness of voice	Hoarseness of voice
	14.	Bleeding	Bleeding
	15.	Burning	Burning sensation in the mouth
	16.	Painful	Painful
	17.	Swelling	Swelling
	18.	Numbness	Numbness
Clinical present	19.	Site	Region of symptom
	20.	Size	Size of symptom
	21.	Lymph node	Lymph node involvement
Histopathology	22.	Histopatological type	Histological type or class
	23.	Differentiation of SCC	Differentiation (Squamous-cell carcinoma (SCC) type)
	24.	Primary Tumour	Primary Tumour
	25.	Regional lymph nodes	Regional lymph nodes
	26.	Distant Metastasis	Distant Metastasis
	Target/Class		
	27.	Stage	Stage of cancer (stage I, II,III, and IV)

Correlation Variable Evaluator with Ranker. This algorithm valuates an attribute by estimating the connection of all attribute to the class. This nominal characterization is carried out based on value as an indicator where the general relationship for nominal variable is accomplished by a weighted average. This algorithm utilized with ranker algorithm so as to rank its traits by using individual assessments.

Correlation Forward Selection Subset Evaluator with Linear Forward Selection. This method estimates the value of a subset of variables by focusing each of the predictive capability of each element into the level of repetition among them. This forward selection considers a limited number of k characteristics. Each set chooses a fixed number *k* of traits, while *k* is expanded in each progression when fixed-width is chosen. The search is used to utilize to start ordering to choose the top *k* variables, or to do a ranking base on similar value but can be used later.

In this hybrid FS model, first the Correlation Variable Evaluator technique filters the related features which results in reducing features subset. Then, CFS Subset Evaluator with Linear Forward Selection (LFS) searches for the most related features producing in optimum feature set applied for classifying the disease.

2.4   Bayesian Relevance Feedback

This study presents also an enhanced Bayesian model as illustrated in Fig. 2, which can adaptively enhance oral cancer diagnosis performance [8]. The main idea of this research is utilized the probability concept and the Bayesian algorithm. The proposed model contains three components: prior, conditional and posterior probability. This study concentrates on posterior calculations to classify stages, according to the same oral cancer cases and symptoms. The process of learning in a relevance feedback is adapted in the posterior computation. The relevant feedback process starts when the system classifies a number of new objects by using existing parameters. Since the classification of the object may be correct or incorrect hence, knowledge experts will be used to monitor newly classified objects to improve classification performance and produce high quality results. At the end of each iteration, the model acquires the appropriate class as generated by the expert, and the corrected object feature combined with their class will be applied in the new iteration [8].

2.5   Classification Model Performance Evaluation Metrics

The experimental works were implemented using Weka 3.8. OCDS including 160 patients' records was utilized for training and testing [8]. The evaluation phase of the proposed model, BRF and another six classifier algorithm to the classification of OCDS is done, which are K-Nearest Neighbors Classifier (KNN/IBK), Multi Class Classifier (MCC), Random Tree (RT), Multilayer Perceptron (MLP), Naïve Bayes (NB), and Support Vector Machine (SVM/SMO). The classifiers participating in the experiments were trained (70%) and tested (30%) using 10 fold cross validation, where each fold has roughly the same class distribution. The performance of classification data mining algorithms was evaluated using following measures.

Accuracy, Sensitivity and Specificity (ASS). This method generally used to describe a clinical diagnostic test, and to validate the performance of the diagnostic test. The true positive rate (TP), false positive rate (FP), true negative rate (TN) and false negative rate (FN) defines the performance data mining algorithm efficiency. In this study, three predictive model performance evaluation metrics were used: accuracy, sensitivity, and specificity [18];

Algorithm Bayesian Relevance Feedback
Input: Dataset, D
Initialize n as the total number of the instances of class A_j in D.

1. Calculate the prior probabilities $P(A_j)$ for each class A_j in D.

$$P(A_j) = \frac{n(A_j)}{\sum_{j=1}^{n} A_j}, j = 1..n$$

2. Calculate the conditional probabilities $P(B_{ij}|A_j)$ for each feature values in D

    i. $P(B_{ij}|A_k) = \dfrac{n(B_{ij})}{n(A_k)}, i = 1...n, j = 1..m, k = 1..x$

3. Define general features and class features in D
4. Assign C for general features and H for supervise features
5. Calculate the posterior probabilities for C and H in D.

$$P(C_i|A_j) = \frac{P(A_j)\prod_{i=1}^{n} P(C_i|A_j)}{\sum_{k=1}^{m} P(A_k)\prod_{i=1}^{n} P(C_i|A_k)}, j = 1...n$$

$$P(H_i|A_j) = \frac{P(A_j)\prod_{i=1}^{n} P(H_i|A_j)}{\sum_{k=1}^{m} P(A_k)\prod_{i=1}^{n} P(H_i|A_k)}, j = 1...n$$

6. Define e,

$$P(e_i|A_j) = \frac{P(C_j) + P(H_j)}{\sum_{j=1}^{4} P(C_j) + P(H_j)}, j = 1..n$$

7. If $\left(\sum e_i \neq 0\right)$ then,
    i.   Add new object to D
    ii.   STOP
8. Else Assign class for new object
9. Procedure: Assign new object (* call Algorithm 2)
    i.   Read value T, N, M for new object
    ii.   Assign class value, stage
    iii.   New object correctly classified
10. Add new object to D
11. Relevance feedback: back to line 1.
Output Differential Probabilities, DP

Fig. 2. Bayesian Relevance Feedback (BRF).

Accuracy. This metric is the precision degree of a classification and is identified from measures of specificity and sensitivity. The best effective predictor would be defined as 100% sensitive and 100% specific. The accuracy also called as the percentage proportion of correctly classified cases to all cases in the set. The higher of the predictive accuracy contribute to the better situation.

$$accuracy = \frac{TP + TN}{(TP + FP + TN + FN)} x\, 100 \tag{1}$$

Sensitivity. This measure is presented as false-negative rate, recall, Type II error, ß error, error of omission, or alternative hypothesis.

$$sensitivity = \frac{TP}{(TP + FN)} x\, 100 \tag{2}$$

Specificity. Measure the capability of a test to be negative when the situation is absent. It is defined as false-positive rate, the precision, Type I error, an error, an error of commission, or null hypothesis, written as:

$$specificity = \frac{TN}{(TN + FP)} x\, 100 \tag{3}$$

Precision, Recall and F-Measure. The precision, recall and F-measure are additional parameters that could help physician determines exactly whether a patient is ill or not. Recall is the same in the application as sensitivity, F-measure is the harmonic mean of both recall and precision, while specificity is the reverse of sensitivity [19, 20]. These metrics are presented as:

$$precision = \frac{TP}{TP + FP} \times 100 \tag{4}$$

$$recall = \frac{TP}{TP + FN} \times 100 \tag{5}$$

$$F\ measure = 2 \times \frac{precision \times recall}{precision + recall} \times 100 \tag{6}$$

n- Fold Cross-Validation. The n-fold cross-validation or rotation estimation is a model applied to estimate how accurately a model will perform in practice. It involves splitting at random a sample data R into complimentary f mutually exclusive subsets (the folds: R1, R2,... Rf) for approximately equal size. In order to reduce variability, the classification model is trained and tested f times. Each time (t{1,2...,f}), it is trained on all but one-fold (Rt) known as training set, and tested on the remaining single fold

(Rt), known as validation or testing set. The cross-validation evaluate of the total accuracy is the average of the n individual accuracy measures denoted by;

$$CA = \frac{1}{f} \sum_{i=1}^{f} Ai \tag{7}$$

Where CA is a cross-validation accuracy, f is the number of folds employed, and A is the accurate measure of each fold. Meanwhile, the cross-validation accuracy based on the random assignment each of the cases into n distinct folds, it is frequently stratified by creating folds that include almost the same proportion of predictor labels as the original dataset to produce findings with lower bias and variance as contrast to regular fold cross validation [21, 22].

Cohen's Kappa (k). Cohen's Kappa (k) is a robust statistic method that have been used for either interrater or intrarater reliability testing [23]. Calculation of Cohen's kappa is formulated as:

$$k = \frac{P_a - P_e}{1 - P_e} \tag{8}$$

Where P_a denotes the actual observed agreement, and P_e denotes chance agreement.

Matthews Correlation Coefficient (MCC). Contrasting the measurement methods described in above section, MCC covers all the cells of the Confusion Matrix in its formula [24],

$$MCC = \frac{TP \times TN - FP \times FN}{\sqrt{(TP + FP) \times (TP + FN) \times (TN + FP) \times (TN + FN)}} \tag{9}$$

Like to Correlation Coefficient, the range of values of MCC lie between -1 to $+1$. A model with a score of $+1$ is a good model and -1 is a bad model. This property is one of the key usefulness of MCC as it leads to easy interpretability.

3 Result and Discussion

3.1 Optimum Features Selected

The feature selection (FS) algorithm began with the OCDS which contain 26 features included predictive feature (Stage) and excluded case_id (1) with balanced OCDS, 160 instances. All string or class feature was standardized and mapped to numerical qualities. Table 2 shows the finding of hybrid FS implemented in this study. The process of hybrid FS started with the Correlation Variable Evaluator with Ranker (FS1). FS1 method resulted to ranking 25 features: 20, 23, 21, 22, 16, 19, 24, 8, 2, 15, 7, 17, 3, 18, 5, 1, 13, 9, 11, 6, 25, 10, 14, 4 and 12. The feature number 12 is removed because of the ranking rate is zero. The finding with 24 features, then subset method eliminates the redundant and unrelated 10 features, covers 7, 5, 1, 13, 11, 6, 25, 10, 14 and 4. The FS algorithm, CfsSubsetEval and Linear Forward Selection (FS2) finished with 15 features: 20, 23, 21, 22, 16, 19, 24, 8, 2, 15, 17, 3, 18, 9, and 24 as optimum features set (sex, ethnicity, ulceration, neck lump, painful, swelling, numbness, site, size, lymph node, histopathology, SCC types, primary tumour, regional lymph nodes and metastasis).

Table 2. Attributes selection with hybrid feature selection technique

FS	FS method	Selected features
FS0	No selected feature	All features (25 features)
FS1	Correlation variable eval ranker	i. Ranked features: 20,23,21,22,16,19,24,8,2,15,7,17,3,18,5,1,13,9,11,6,25,10,14,4,12 (25 features) ii. Remove ranting value = 0 (feature 12)
FS2	CfsSubsetEval LinearForwardSelection	i. Remove irrelevant features = 10 features (7,5,1,13,11,6,25,10,14,4) ii. Optimum features: 15 features (20, 23, 21, 22, 16, 19, 24, 8, 2, 15, 17, 3, 18, 9, and 24)

3.2   Correctly and Incorrectly Instance Classification Performance

Six classifiers were used in the classification which is KNN, MCC, RT, MLP, NB, and SVM. In this research, these six diverse machine learning algorithms were utilized to group the oral cancer growth dataset with 25 features (FS0) and optimum features were selected by the hybrid approach (FS2). Table 3 shows that, hybrid approach with 15 selected features improved the accuracy performance in term of correctly classified for almost classification algorithms (KNN, MCC, RT, NB, and SVM). The best performing was by hybrid approach, FS2 with NB and SVM algorithm at 93.75% correctly classified. Both NB and SVM also contributed the highest percentages of kappa statistic and MCC (91.67%, 91.7%). MLP has the same value of accuracy for both features selection methods with 89.58%. Next score was achieved by MCC and RT with 87.50%. KNN has the lowest performance with remaining 83.33%. Based on the result, the hybrid feature selection approach with 15 features is mainly improving the correctly instance classified, kappa and MCC value for all the classification algorithm.

Table 3. Correctly, incorrectly instance classification, kappa statistic and MCC to build a model among six classifier algorithms with 25 features and optimal selected features (15F) of OCDS

No.	CA	25 Features (FS0)				15 Features (FS2)			
		Corr	Incorr	Kappa	MCC	Corr	Incorr	Kappa	MCC
1	KNN	79.17	20.83	72.22	73.30	83.33	16.67	77.78	78.4
2	MCC	85.42	14.58	80.56	80.40	87.50	12.50	83.33	83.2
3	RT	85.42	14.58	80.56	81.00	87.50	12.50	83.33	84.2
4	MLP	89.58	10.42	86.11	86.20	89.58	10.42	86.11	86.2
5	NB	89.58	10.42	86.11	86.30	93.75	6.25	91.67	91.7
6	SVM	89.58	10.42	86.11	86.00	93.75	6.25	91.67	91.7

Correctly (Corr), Incorrectly (Incorr)

3.3   Accuracy Measurement

The finding in Table 4 is an accuracy measured with criteria's such as TP rate, FP rate, precision, recall and F-measure. The criteria's are described as:

- True Positive (TP) Rate. It denotes the cases that have been correctly classified with respect to a given class.
- False Positive (FP) Rate. It explains the cases that have been falsely or incorrectly classified with respect to a given class.
- Precision: Listed the amount of those cases true to a particular class divided by overall cases classified with respect to that class.
- Recall: It describes the amount of those cases that have been classified by a class divided by the total cases present in the class.
- F-Measure: It is calculated by merging the measure of Recall and Precision.

The finding shows that, hybrid FS approach with 15 selected features improved the accuracy performance of TP rate, FP rate, precision, recall and F-measure. KNN, MCC, RT, NB, and SVM. The highest TP performing was by 15 selected features with NB and SVM algorithm at 0.938. Both NB and SVM contributed also the highest score of precision, recall and F-Measure with the same value (0.939, 0.938, 0.937). MLP has almost the same score of TP, FP, precision, recall and F-Measure for both FS methods. KNN has the lowest performance with remaining 0.833. Based on the result, the hybrid FS method with 15 features successfully improved the TP, FP, Precision, Recall, F-Measure score for almost classifier algorithms.

Table 4. Performance evaluation Comparisons with respect to TP, FP, Precision, Recall, F-Measure for six classifier algorithms with 25 Features and optimal selected features (15F)

No.	CA	25 Features					15 Features				
		TP	FP	Prec	Rec	F	TP	FP	Prec	Rec	F
1	KNN	0.792	0.069	0.817	0.792	0.787	0.833	0.056	0.847	0.833	0.832
2	MCC	0.854	0.049	0.850	0.854	0.848	0.875	0.042	0.872	0.875	0.870
3	RT	0.854	0.049	0.864	0.854	0.855	0.875	0.042	0.896	0.875	0.872
4	MLP	0.896	0.035	0.897	0.896	0.896	0.896	0.035	0.898	0.896	0.895
5	NB	0.896	0.035	0.901	0.896	0.897	0.938	0.021	0.939	0.938	0.937
6	SVM	0.896	0.035	0.894	0.896	0.894	0.938	0.021	0.939	0.938	0.937

Classifier Algorithm (CA), Precision (Prec), Recall (Rec), F-Measure (F)

From confusion matrix (see Table 5), sensitivity, specificity and accuracy of different classifiers have been obtained. Table 6 shows the example of confusion matrix for classification [8].

Table 5. Confusion Matrix

		Actual Class	
		P	N
Predicted Class	P	TP	FP
	N	FN	TN

As example calculation to KNN using confusion matrix:

Table 6. Confusion matrix

		Actual Class	
		P	N
Predicted Class	P	0.833	0.056
	N	0.167	0.944
Total		1	1

$$sensitivity = \frac{0.833}{(0.833 + 0.167)} \times 100 = 83.30$$

Table 7 illustrates the percentage performance of the sensitivity, specificity, accuracy, and error rate of six classification algorithms on testing data by using 25 features and 15 selected features of OCDS. The findings present that NB and SVM give the highest accuracy of 95.85% and lowest error rate (4.15%) compared to other algorithms. Both NB and SVM also showed the highest value of sensitivity and specificity (93.80%, 97.90%). Followed by MLP with the value of accuracy is 93.05%. The performance of KNN on OCDS showed the lowest accuracy with 88.85% and the highest error rate (11.15%).

Table 7. Comparison performance of sensitivity, specificity, and accuracy among six classifier algorithm with 25 features and optimal selected features (15F)

No.	CA	25 Features				15 Features			
		Sens	Spec	Acc	Error	Sens	Spec	Acc	Error
1	KNN	79.20	93.10	86.15	13.85	83.30	94.40	88.85	11.15
2	MCC	85.40	95.10	90.25	9.75	87.50	95.80	91.65	8.35
3	RT	85.40	95.10	90.25	9.75	87.50	95.80	91.65	8.35
4	MLP	89.60	96.50	93.05	6.95	89.60	96.50	93.05	6.95
5	NB	89.60	96.50	93.05	6.95	93.80	97.90	95.85	4.15
6	SVM/SMO	89.60	96.50	93.05	6.95	93.80	97.90	95.85	4.15

Sensitivity (Sens), Specificity (Spec), Accuracy (Acc)

Table 8 and Fig. 3, illustrate the overall performance of seven classification algorithms on testing data (15 features). The finding presents that the Bayesian Relevance Feedback model (BRF) gives the highest accuracy of 97.25% and lowest error rate (2.75%) compared to other algorithms. BRF also showed the highest value of sensitivity and specificity (95.80%, 98.70%). Followed by NB and SVM algorithms with the same value of accuracy (95.85%). The performance of KNN on OCDS showed the lowest accuracy with 88.85%.

Table 8. Overall performance of seven classifiers with optimal selected features (15 features)

No.	PM	KNN	MCC	RT	MLP	NB	SVM	BRF
1	Correctly	83.33	87.50	87.50	89.58	93.75	93.75	95.83
2	Incorrectly	16.67	12.50	12.50	10.42	6.25	6.25	4.17
3	Kappa	77.78	83.33	83.33	86.11	91.67	91.67	94.44
4	MCC	78.4	83.2	84.2	86.2	91.7	91.7	94.54
5	TP Rate	0.833	0.875	0.875	0.896	0.938	0.938	0.958
6	FP Rate	0.056	0.042	0.042	0.035	0.021	0.021	0.013
7	Precision	0.847	0.872	0.896	0.898	0.939	0.939	0.964
8	Recall	0.833	0.875	0.875	0.896	0.938	0.938	0.958
9	F-Measure	0.832	0.870	0.872	0.895	0.937	0.937	0.972
10	Sensitivity	83.30	87.50	87.50	89.60	93.80	93.80	95.80
11	Specificity	94.40	95.80	95.80	96.50	97.90	97.90	98.70
12	Accuracy	88.85	91.65	91.65	93.05	95.85	95.85	97.25
13	Error	11.15	8.35	8.35	6.95	4.15	4.15	2.75

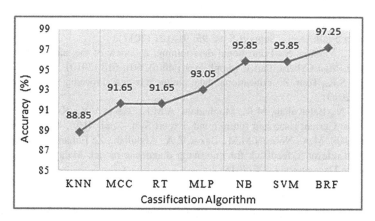

Fig. 3. Accuracy performance of seven classification algorithms.

4   Conclusion

The proposed hybrid FS method, which integrates Correlation Variable Evaluator with ranker and CfsSubset Evaluator with Linear Forward Selection is utilized to choose a significant feature subset from the original OCDS. The validation result of selected features using a hybrid FS approach with seven classifier algorithms, K-Nearest Neighbors Classifier, Multi Class Classifier, Random Tree, Multilayer Perceptron, Naïve Bayes, Support Vector Machine, and Bayesian Relevance Feedback was done in experimental works. The finding presents that, by removing unrelated features and select the optimal features (sex, ethnicity, ulceration, neck lump, painful, swelling, numbness, site, size, lymph node, histopathology, SCC types, primary tumour, regional lymph nodes and metastasis) well contributes higher performance accuracy for all classifier algorithm. The proposed classification algorithm, BRF shows the highest accuracy with 97.25%, compared to the other six classifiers.

Acknowledgement. This study is partially funded by the JKKLA, Universiti Malaysia Terengganu (UMT).

References

1. Han, J., Pei, J., Kamber, M.: Data mining: concepts and techniques, 2nd edn. Elsevier, San Francisco (2011)
2. Liao, S.H., Chu, P.H., Hsiao, P.Y.: Data mining techniques and applications–a decade review from 2000 to 2011. Expert Syst. Appl. **39**(12), 11303–11311 (2012)
3. Ngai, E.W., Hu, Y., Wong, Y.H., Chen, Y., Sun, X.: The application of data mining techniques in financial fraud detection: a classification framework and an academic review of literature. Decis. Support Syst. **50**(3), 559–569 (2011)
4. Carneiro, N., Figueira, G., Costa, M.: A data mining based system for credit-card fraud detection in e-tail. Decis. Support Syst. **95**, 91–101 (2017)
5. Romero, C., Ventura, S.: Educational data mining: a review of the state of the art. IEEE Trans. Syst. Man Cybern. Part C (Appl. Rev.) **40**(6), 601–618 (2010)
6. Mohamad, S.K., Tasir, Z.: Educational data mining: a review. Procedia-Soc. Behav. Sci. **97**, 320–324 (2013)
7. Esfandiari, N., Babavalian, M.R., Moghadam, A.M.E., Tabar, V.K.: Knowledge discovery in medicine: Current issue and future trend. Expert Syst. Appl. **41**(9), 4434–4463 (2014)
8. Mohd, F., Jalil, M.A., Noor, N.M.M., Bakar, Z.A., Abdullah., Z.: Enhancement of Bayesian model with relevance feedback for improving diagnostic model. Malays. J. Comput. Sci. (Spec. Issue December), 1–14 (2018)
9. Dangare, C.S., Apte, S.S.: Improved study of heart disease prediction system using data mining classification techniques. Int. J. Comput. Appl. **47**(10), 44–48 (2012)
10. Kourou, K., Exarchos, T.P., Exarchos, K.P., Karamouzis, M.V., Fotiadis, D.I.: Machine learning applications in cancer prognosis and prediction. Comput. Struct. Biotechnol. J. **13**, 8–17 (2015)
11. Borowska, K., Topczewska, M.: Data preprocessing in the classification of the imbalanced data. Adv. Comput. Sci. Res. **11**, 31–46 (2014)
12. Witten, I.H., Frank, E., Hall, M.A., Pal, C.J.: Data Mining: Practical Machine Learning Tools and Techniques. Morgan Kaufmann, Burlington (2016)

13. Jain, D., Singh, V.: Feature selection and classification systems for chronic disease prediction: a review. Egypt. Inf. J. **19**(3), 179–189 (2018)
14. Guyon, I., Elisseeff, A.: An introduction to variable and feature selection. J. Mach. Learn. Res. **3**(Mar), 1157–1182 (2003)
15. Chawla, N.V., Bowyer, K.W., Hall, L.O., Kegelmeyer, W.P.: SMOTE: synthetic minority over-sampling technique. J. Artif. Intell. Res. **16**(2002), 321–357 (2002)
16. Bakar, Z.A., Mohd, F., Noor, N.M.M., Rajion, Z.A.: Demographic profile of oral cancer patients in East Coast of Peninsular Malaysia. Int. Med. J. **20**(3), 362–364 (2013)
17. Hall, M.A., Correlation-based feature selection for machine learning. University of Waikato, Hamilton, NewZealand (1999)
18. Zhu, W., Zeng, N., Wang, N.: Sensitivity, specificity, accuracy, associated confidence interval and ROC analysis with practical SAS implementations. In: NESUG Proceedings: Health Care and Life Sciences, Baltimore, Maryland, vol. 19, p. 67 (2010)
19. Powers, D.M.: Evaluation: from precision, recall and F-measure to ROC, informedness, markedness and correlation. J. Mach. Learn. Technol. **2**(1), 37–63 (2011)
20. Kautz, T., Eskofier, B.M., Pasluosta, C.F.: Generic performance measure for multiclass-classifiers. Pattern Recogn. **68**, 111–125 (2017)
21. Fushiki, T.: Estimation of prediction error by using K-fold cross-validation. Stat. Comput. **21**(2), 137–146 (2011)
22. Zhang, Y., Yang, Y.: Cross-validation for selecting a model selection procedure. J. Econom. **187**(1), 95–112 (2015)
23. Kraemer, H.C.: Kappa coefficient. Wiley StatsRef: Statistics Reference Online 1–4 (2014)
24. Boughorbel, S., Jarray, F., El-Anbari, M.: Optimal classifier for imbalanced data using Matthews Correlation Coefficient metric. PLoS ONE **12**(6), e0177678 (2017)

An Experience Report on Building a Big Data Analytics Framework Using Cloudera CDH and RapidMiner Radoop with a Cluster of Commodity Computers

Sittiporn Kunnakorntammanop, Netiphong Thepwuttisathaphon, and Supphachai Thaicharoen$^{(\boxtimes)}$

Faculty of Science, Srinakharinwirot University, Bangkok 10110, Thailand
{sittporn.kktn,netiphong.thep,supphachai}@g.swu.ac.th

Abstract. Many real-world data are not only large in volume but also heterogeneous and fast generated. This type of data, known as big data, typically cannot be analyzed by using traditional software tools and techniques. Although an open-source software project, Apache Hadoop, has been successfully developed and used for handling big data, its setup and configuration complexity including its requirement to learn other additional related tools have hindered non-technical researchers and educators from actually entering the area of big data analytics. To support big-data community, this paper describes procedures and experiences gained from building a big data analytics framework, and demonstrates its usage on a popular case study, Twitter sentiment analysis. The framework comprises a cluster of four commodity computers run by Cloudera CDH 6.0.1 and RapidMiner Studio 9.3 with Text Processing, Hive Connector, and Radoop extensions. According to the study results, setting up a big data analytics framework on a cluster of computers does not require advanced computer knowledge but needs meticulous system configurations to satisfy system installation and software integration requirements. Once all setup and configurations are correctly done, data analysis can be readily performed using visual workflow designers provided by RapidMiner. Finally, the framework is further evaluated on a large data set of 185 million records, "TalkingData AdTracking Fraud Detection" data set. The outcome is very satisfied and proves that the framework is easy to use and can practically be deployed for big data analytics.

Keywords: Apache Hadoop · Big data analytics · Cloudera CDH · Computer cluster · RapidMiner Radoop · Sentiment analysis

1 Introduction

Solving many real-world problems involves data that are not only large in volume but also heterogeneous and fast generated. The solutions for these problems cannot be practically implemented using traditional tools and techniques such as relational databases and typical in-memory data analysis tools. A number of commercial and

© Springer Nature Singapore Pte Ltd. 2019
M. W. Berry et al. (Eds.): SCDS 2019, CCIS 1100, pp. 208–222, 2019.
https://doi.org/10.1007/978-981-15-0399-3_17

open-source software tools have been developed for coping with this type of data, and one of them that is stood out is Apache Hadoop.

Apache Hadoop is an open-source software library framework that has been successfully exploited by industries and academia when working with big data. With the framework, tasks of processing large data sets can be reliably and scalably distributed across a cluster of computers using a simple programming model called MapReduce. Apache Hadoop library framework consists of four main modules, Hadoop Common, Hadoop Distributed File System (HDFS), Hadoop YARN, and Hadoop MapReduce. Although Apache Hadoop alone could be sufficiently deployed for handling large data sets, other software tools should also be concertedly applied in order to facilitate the implementation of the solutions for complex problems.

Hadoop ecosystem is a set of open-source software projects that has been productively employed together to provide software solutions for complex big data problems. It consists of Apache Hadoop and a number of related software components such as Pig, Hive, Spark, Oozie, Flume and Sqoop. Some examples of research studies and application domains that utilize Hadoop ecosystem are described as follows. Chennamsetty et al. used Hive as a data warehouse for storing large sets of Electronic Health Records (EHRs) and utilized HiveQL, a query processing language, to retrieve the stored data and generate reports for statistical analysis [3]. Sangeeta presented a study on using FLUME and Hive tools for twitter sentiment analysis [12]. In the study, FLUME was used to retrieve and collect tweets from Twitter, and HIVE Serde was operated to import tweets in JSON format into Hive table. Finally, some statistical analyses were performed through HiveQL query statements. Although these studies presented interesting use cases and applications of Hadoop ecosystem, their analyses were implemented on a single machine. To truly appreciate big data framework capability, analysis tasks should be carried out in a cluster of computers.

The following research studies are selected examples that implemented big data analysis on a cluster of computers. Liu et al. built a research data science platform using 40 industrial computers donated by Yahoo [9]. The purpose of their use of cluster computers rather than cloud-based architecture is to fully understand the Hadoop ecosystem and to use the platform for research studies. The platform is run by Apache Hadoop 2.7.2, HBase 1.1.5, OpenTSDB 2.2.0, and Spark 1.6.1, and evaluated on four test data sets. Logistic regression in Spark ML was selected to execute Wordcount MapReduce program. They found that size of the data sets, the number of CPU cores, and the driver memory contribute to performance. Durby et al. exploited Hadoop MapReduce ecosystem with multi-layered feed forward neural networks for stock market prediction [4]. In their study, a Hadoop cluster of 50 worker nodes run on Ubuntu 1.0.4 is utilized, and Neural Network was configured for parallelism using MapReduce. On the performance evaluation, they found that increasing number of nodes in the cluster can speed up the neural network training time. Even though building a big data analytics platform with Hadoop ecosystem allows implementers to fully understand the underlying principles of big data analytics, it requires installing a number of software tools and configuring them to work together. These complex installation and configuration could frustrate non-technical and inexperienced users on big data to advance into big data analytics domain. An alternative approach is to utilize an enterprise-ready open-source software suite.

Numerous software suites for manipulating big data are available in both commercial and open-source products. Different products have their own advantages and disadvantages. Therefore, which software to choose is dependent on criteria and preferences of the implementers. A research study by Nereu et al. evaluated five big data analytics platforms – Apache Hadoop, Cloudera, Spark, Hortonworks, and HPCC to find out which platform is best fit to small and medium-sized enterprises and non-government organizations [10]. Based on their study, they suggested that Cloudera is better than other platforms for all contexts, specifically when dealing with real-time large data sets. Ivanov et al. reported a performance evaluation on two enterprise big data platforms, DataStax Enterprise (DSE) and Cloudera's Distribution of Hadoop (CDH), using HiBench benchmark suite [7]. From their experimental results, CDH performs better than DSE in almost all test cases with faster execution time, faster read time, and higher throughput on Wordcount and enhanced DFSIO tests. For these reasons, Cloudera CDH is chosen in our study.

After a big data hardware and software platform architecture is established, analyzing data can be realized by either writing computer programs or using software tools. Although writing computer programs is more flexible and more powerful than using software tools, it is generally suitable to only computer and IT professionals. In many cases, simply using software tools can give analyzers enough functionality and analysis power for their tasks. For example, Han *et al.* used Weka for Chinese document clustering [6], Feltrin explored a usage of KNIME for geoscience application domain [5], and Tripathi *et al.* performed a sentiment analysis using RapidMiner [14]. Some comparative studies were conducted to evaluate these tools. Jovic *et al.* compared the following free software tools: RapidMiner, R, Weka, KNIME, Orange, and Scikit-learn, for general data mining projects [8]. They found that RapidMiner, R, Weka, and KNIME contain most of the desired characteristics for a fully-functional data mining platform. Another comparison was conducted by Altahi *et al.* on 19 open-source tools for data mining and knowledge discovery tasks. They found that Weka, Knime, and RapidMiner Studio are the most promising tools for their two evaluation criteria [1]. Since RapidMiner Studio offers visual workflow designer that greatly facilitates data analysis in addition to fully-functional data mining features similar to other tools, it is selected to be used in our study.

Finally, a number of quality criteria should be taken into consideration when building a big data analytics framework. Singh and Reddy presented an investigation of different big data analytics platforms to assess their strengths and weaknesses on the following six performance metrics – scalability, I/O performance, fault tolerance, real-time processing, data size support, and support for iterative tasks [13]. Using a 5-star rating, they provided an overview table summarizing performance scales of different platforms under the study. For example, Peer-to-Peer (TCP/IP) receives 5 stars for scalability and data size supported. Security is another important quality metric for big data analytics framework. Bhathal and Singh presented a study on different types of vulnerabilities of Hadoop framework and proposed some possible solutions to reduce the security risks [2]. They additionally implemented some security attacks to truely understand the security weaknesses of the Hadoop environment. From the results, they discovered that security modules provided by enterprise software suites such as IBM, MapR, Hortonworks, and Cloudera are not sufficient. Security breaches are still possible. That is because securing Hadoop environment not only involves preventing unauthorized access to Hadoop and stored data but is concerned with network security and operating systems security as well. In our study, data size supported, scalability and ease of use are our main focus.

Based on our experiences and to the best of our knowledge, despite the fact that many current real-world data analysis problems are involved with big data, most works, particularly in academia, have been conducted on a small sample data set using a single machine and in-memory software tools. Consequently, the results of the analyses generally lack comprehensiveness and are useful only for ad-hoc studies. This may be because people inexperienced with big data might perceive that building a big data analytics platform requires expensive hardware and software and should be done only by highly-technical professionals. In this paper, we show that this belief is no longer true. We present a big data analytics framework using only a set of commodity computers, an enterprise-grade open-source big data software suite, Cloudera CDH, and an easy-to-use data science tool, RapidMininer Studio. In addition, we demonstrate the usability of the framework on Twitter sentiment analysis and prove the practicality of the framework on a large data set. The experimental result is very satisfactory.

The contribution of this paper is threefold. First of all, it presents detailed descriptions and suggestions on how to build a simple, accessible and affordable big data analytics framework that uses only commodity hardware and open-source software. Secondly, it provides a demonstrating example on how to use the framework to analyze a complex data analysis problem, Twitter sentiment analysis. Finally, it gives performance evaluation on a large data set to confirm the practicality of the framework.

The content in this paper is organized as follows. Section 2 gives a brief overview of technical background. Section 3 provides a description of procedures and methods. Section 4 presents the study results. Finally, Sect. 5 concludes the study.

2 Technical Background

This section briefly describes concepts and technologies for understanding the methodology in this paper. The content is divided into three parts: Cloudera CDH, RapidMiner Radoop, and sentiment analysis.

2.1 Cloudera's Distribution of Hadoop (CDH)

Cloudera's Distribution of Hadoop (CDH) is a 100% open source enterprise-grade big data analytics platform distribution provided by Cloudera. It integrates Apache Hadoop Core with a number of key open-source Apache projects such as Accumulo, Flume, HBase, Hive, Hue, Impala, Kafka, Pig, Sentry, Spark and Sqoop. Cloudera CDH allows enterprises to perform end-to-end big data workflows right out of the box. Moreover, it provides Cloudera Manager (CM), a Web-based management tool for managing CDH clusters, helping the installation process, and containing functionality for cluster configuration, resource allocation, and real-time monitoring. Cloudera also offers a virtual machine version, Cloudera CDH QuickStart, for interested users to be familiar with the platform. Users can install it in a single machine for their own exploration and experimentation. Cloudera CDH Quickstart is available in three favors, VirtualBox, VMWare and Docker.

In this study, to really experience a real-world big data analytics platform as possible, rather than using Cloudera QuickStart virtual machine, a complete CDH distribution was installed in a cluster of commodity computers. Moreover, instead of utilizing the latest CDH 6.2.0, CDH 6.0.1 was chosen for our framework because all of its software component versions are supported by the current version of RapidMiner Radoop 9.3. Unsupported versions of some CDH components can cause connection problems between CDH and RapidMiner Radoop. Cloudera CDH 6.0.1 is available at https://www.cloudera.com/downloads/cdh/6-0-1.html.

2.2 RapidMiner Radoop

Radoop is a plug-in extension of RapidMiner Studio introduced by Prekopcsak *et al.* [11]. It provides code-free operators for using Hadoop, Hive, and Spark to carry out a number of data analysis and data mining tasks such as Naive Bayes, Linear and Logistic Regressions, Decision Tree, Random Forest, Support Vector Machine, K-Means Clustering, Scoring, and Validation. In addition, it offers a visual workflow designer and data processing operators similar to typical in-memory RapidMiner operators but can be run in-parallel in Hadoop clusters. Such data processing operators are data access, data blending and data cleansing. **Hadoop Nest** operator is the main operator in Radoop. To run processes synchronously in a Hadoop cluster, all operators must be executed inside **Radoop Nest** operator.

In this paper, since Radoop does not yet provide operators for text processing, text pre-processing tasks were carried out in a local computer using typical in-memory RapidMiner Studio operators. However, predictive modeling and evaluation tasks for sentiment analysis were performed in a Hadoop cluster within **Hadoop Nest** operator.

2.3 Sentiment Analysis

Sentiment analysis is an automatic process for determining the sentiment polarities of people opinions whether they are positive, negative, or neutral. These opinions usually are presented in a textual format. Sentiment analysis has been applied to many application domains such as product/movie review, customer services, crime mitigation, and stock prediction.

Sentiment analysis approach can be divided into three main categories: (i) lexical-based method, (ii) supervised-learning method, and (iii) hybrid method. Lexical-based sentiment analysis relies on extracting and mapping words to a sentiment category (positive, negative, or neutral) and then uses the mapping results to compute sentiment scores. Subsequently, a threshold is applied to the total score calculated from all words in the sentence to determine the polarity. In contrast, supervised-learning method learns a classification model from labeled data of text representation and utilizes the model to predict the polarity of the whole textual sentence. Finally, the hybrid method exploits both lexical and supervised-learning approaches.

In this paper, a machine-learning approach, Logistic Regression, was used for predicting sentiments of Twitter data.

3 Methodology

In this section, detailed descriptions of setting and configuring a Cloudera CDH cluster, connecting Cloudera CDH to RapidMiner Radoop, and Twitter sentiment analysis are presented.

3.1 Build a Cloudera Cluster

In this study, a cluster of four computers is constructed. It consists of one TP-Link Gigabit router and four desktop computers. The router was configured to assign fixed private IP addresses to these computers as 192.168.0.2, 192.168.0.3, 192.168.0.4, and 192.168.0.5, respectively. CentOS 7 was then installed in all computers.

Due to compatibility requirements, rather than installing the latest Cloudera CDH 6.2.0, a previous version, Cloudera CDH 6.0.1, was selected. The reason is that the highest version of Spark supported by RapidMiner Radoop 9.3 is Spark 2.2 which is available in Cloudera CDH 6.0.1. Based on our experiences, version incompatibility can cause a connection problem between Cloudera CDH and RapidMiner Radoop later during the data analysis process.

A computer with 16 GB of RAM was chosen as a name node and the remaining three computers with 8 GB, 8 GB, and 12 GB of RAM as data nodes. It is recommended to use a computer with the largest size of RAM to act as a name node because it generally performs many roles. Based on our experiences, with similar CPU speed and number of cores, the larger the RAM size, the smoother an analysis process is running.

Before proceeding to install Cloudera CDH, it is highly necessary to configure the system of all computers in the cluster to meet all the requirements. Some examples of Linux commands used in our study are given below. Note that lines preceded by a ">" are corresponding to Linux commands, lines preceded by "#" are comments, and those lines without ">" or "#" are content of a file open from their preceding command.

- Setting up time zone
    >timedatectl list-timezones | grep Asia
    >timedatectl set-timezone Asia/Bangkok
- Configuring and synchronizing NTP services
    >sudo yum install ntp
    >sudo nano/etc./ntp.conf
    ### Comment out existing public servers such as server 0.centos.pool.ntp.org burst
    ### and add the following three servers instead.
    ### The server names suitable to a specific country in Asia can be found at
    ### https://www.pool.ntp.org/zone/@
    server 1.th.pool.ntp.org iburst
    server 3.asia.pool.ntp.org iburst
    server 1.asia.pool.ntp.org iburst
    >sudo systemctl start ntpd
    >sudo systemctl enable ntpd

>*sudo ntpdate-u 1.th.pool.ntp.org*
>*sudo hwclock–systohc*
- Disabling SELinux
  >*sudo nano/etc./selinux/config*
  ### Change the line: SELINUX = enforcing to SELINUX = permissive or
  ### SELINUX = disabled
- Disabling Firewall
  >*sudo systemctl disable firewalld*
  >*sudo systemctl stop firewalld*
- Disabling IPv6
  >*sudo nano/etc./sysctl.conf*
  net.ipv6.conf.all.disable_ipv6 = 1
  net.ipv6.conf.default.disable_ipv6 = 1
- Configuring host names (change the host name to match names of individual computers in the cluster)
  >*sudo nano/etc./sysconfig/network*
  HOSTNAME = master.xxx.xx.xx
  >*sudo nano/etc./hosts*
  192.168.0.2master.xxx.xx.xxmaster
  192.168.0.3node1.xxx.xx.xx node1
  192.168.0.4node2.xxx.xx.xx node2
  192.168.0.5node3.xxx.xx.xx node3
  127.0.0.1 localhost
  ::1

The first step for installing Cloudera CDH is to download and install Cloudera Manager in a computer established as a name node (or master host), which can be done by the following three commands.

>*wget https://archive.cloudera.com/cm6/6.0.1/cloudera-manager-installer.bin*
>*chmod u + x cloudera-manager-installer.bin*
>*sudo./cloudera-manager-installer.bin*

With Cloudera Manager successfully installed, the next step is to log into Cloudera Manager and install Cloudera CDH. This step can be carried out as follows: (i) open a Web browser, (ii) point to Cloudera Manager log-in page such as http://yourhost-name:7180/cmf/login, and (iii) log into the system. The default username and password are admin/admin. After logging in, the installation process begins. Based on our available hardware resources, Cloudera Express version was adopted, and according to the scope of our data analysis tasks, Data Engineering services (consisting of HDFS, YARN, ZooKeeper, Oozie, Spark, Hive, and Hue) were installed. During the installation process, Cloudera CDH services can also be distributed to be installed in computer data nodes by specifying their host names. Some memory configurations such as VM Swappiness and memory allocation may need to be performed during and after the installation for performance optimization and for fixing some warnings.

The final step for building a Cloudera cluster is to install RapidMiner Studio and three extensions – Text Processing, Hive Connector, and Radoop, in the computer name node.

3.2 Connect Cloudera CDH with RapidMiner Radoop

After completing a Cloudera CDH cluster setup, a connection between Cluster CDH and RapidMiner Radoop can be established by using "Import from Cluster Manager" option under "Manage Radoop Connections" menu of RapidMiner Studio. With this approach, almost all parameters and values are already configured except Spark version. In this paper, Spark version 2.2 was selected which is the highest version supported by the latest version RapidMiner Radoop 9.3 at the time of writing.

Finally, to be certain that all software services can be effortlessly working together, a series of tests should be performed on individual Radoop settings and ended with a complete full/integration test.

3.3 Analyze Tweet Sentiments on the Topic of Global Warming and Climate Change

In this study, sentiment analysis problem was chosen for demonstrating the usability of the proposed framework because it is more complex than a typical classification problem. Sentiment analysis is a text classification problem whose tasks can be divided into two steps: (i) text pre-processing and (ii) predictive modeling and evaluation. Text pre-processing is a process of cleansing and transforming raw text into a representation that is suitable to further analysis such as an input to a machine-learning algorithm. Two commonly known text representation formats to date are vector space model and word embedding where the first format was used in this paper. Predictive modeling and evaluation is a process of building a predictive model from one set of data and evaluating the model performance using another data set. The first data set is called a training set and the later is a test set.

Since the current version of RapidMiner Radoop does not yet provide operators for handling text data, text pre-processing tasks were carried out locally in computer name node. All other tasks, predictive modeling and evaluation, were implemented and executed concurrently in the Hadoop cluster.

Data. Two data sets were used in this study.

- The first data set was downloaded from https://www.figure-eight.com/data-for-everyone/, which is a publicly-available sentiment analysis data set on the topic of global warming and climate change. This data set contains 5,679 tweets with three types of polarity: Yes, No, and N/A ('Yes' means that the owner of a tweet believes that global warming and climate change are really happening, 'No' means the opposite, and 'N/A' is corresponding to a missing label). This data set was used for demonstrating processes of building a predictive model and evaluating the model performance using RapidMiner Radoop operators on textual data.
- The second data set was directly retrieved from Twitter using **Search Twitter** operator in RapidMiner Studio. The search keyword is "Global Warming and

Climate Change-filter:retweets AND-filter:replies", where filters are added to the keyword search to eliminate duplicate tweets. This data set was used for illustrating how tweets can be retrieved and how a new text representation data set can be constructed using a set of features/words defined from another data set.

Text Pre-processing. Text pre-processing process is divided into two parts. The first part is building a training set and the second part is building a test set. As described previously, these two parts are run locally in the computer name node using typical in-memory operators of RapidMiner Studio.

As shown in Fig. 1, started from the top-left operator, a twitter data set in CSV format is read into RapidMiner workspace in a tabular format, followed by selecting only relevant attributes, filtering out undesired rows of data, setting one column as a class label, and sampling a subset of data (500 rows of "Y" and 500 rows of "N"). When text data are imported into RapidMiner, they are automatically converted to nominal data. As a result, before feeding them into **Process Document from Data** operator, they need to be converted back to text.

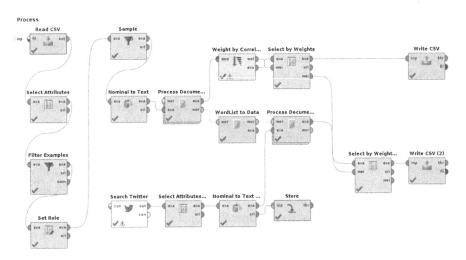

Fig. 1. An overview of text processing visual flow diagram.

Inside **Process Document from Data** operator, a series of text pre-processing tasks was performed as follows: tokenization, case transformation, English stopword removal, user-defined stopword removal, and word stemming (shown in Fig. 2). The outputs from this operator are a vector space model with TF*IDF weighting scheme and a list of words as a dictionary. A statistical correlation was used for weighting importance of individual words and the top 150 correlation words were selected as features.

Fig. 2. Text processing visual flow diagram for the tweet data set.

Figure 3 shows a sequence of text pre-processing tasks inside another **Process Document from Data** operator for the retrieved tweets.

Fig. 3. Text processing visual flow diagram for the retrieved tweets.

Text representations produced from the downloaded data set and retrieved data set were saved into two CSV files for the next step (see **Write CSV** operators in **Fig.** 1).

Predictive Modeling and Evaluation. Predictive modeling and performance evaluation processes are illustrated in Figs. 4, 5 and 6. Since some data files are saved into a local computer, these processes contain a combination of some in-memory operators and in-cluster operators. Note that all in-cluster operators must be put or executed inside **Radoop Nest** operator.

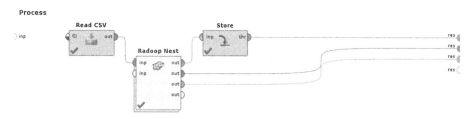

Fig. 4. An overview of building and evaluating a predictive model in Hadoop cluster.

In Fig. 4, a text representation data set is read from a local data file created in the previous step and fed into **Radoop Nest** operator. The output from **Radoop Nest** operator is a predictive model which is saved into a data file and will be used for making predictions on the retrieved tweet data set.

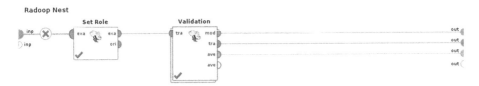

Fig. 5. Inside Hadoop Nest operator, a training data set is split for building evaluating a predictive model using Split Validation operator.

In Fig. 5, a sentiment column is set as a class label and the data set is split into training set and test set using **Split Validation** or **Validation** operator. Its three outputs are a predictive model, training data set (optional), and the classification performance.

Fig. 6. Inside Split Validation operator, a logistic regression model is built and evaluated.

Figure 6 shows a sequence of operators for building and evaluating a predictive model inside **Split Validation** operator.

After having a predictive model, the model can then be used to predict the polarities of tweets retrieved from Twitter as shown in Figs. 7 and 8. The **Retrieve** operator retrieves a predictive model saved in RapidMiner workspace and **Read CSV** operator reads text representation data set of the retrieved tweets created in the text pre-processing step. **Apply Model** operator applies the predictive model to predict polarities of the retrieved tweets. In this case, the prediction results must be manually verified by the researchers.

Fig. 7. Predictive model and pre-processed tweets are fed into **Radoop Nest** operator for predictions.

Fig. 8. An **Apply Model** operator is used for making predictions on the retrieved tweets.

3.4   Evaluate the Framework on a Large Data Set

To evaluate the practicality of the framework for real-world big data analysis, an additional experiment was conducted on a large data set, "TalkingData AdTracking Fraud Detection". The data set (training.csv), was downloaded from Kaggle. It consists of eight attributes, 185 million rows of data with the size of 7.54 GB. For this size of data, processing it using traditional software tools on a typical computer with 16 GB of RAM can cause the system to freeze. For the presented framework, the data set was used for building a predictive model and evaluated the model performance successfully. The whole process took approximately one and a half hour to complete. Figures 9 and 10 display the underlying processes for building and evaluating a predictive model using a large data set.

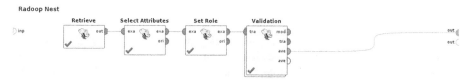

Fig. 9. Retrieve a large data set previously stored in Hive and feed it into Radoop Nest.

Fig. 10. Build a predictive model using Logistic regression and evaluate its performance.

4   Experimental Results

A Cloudera cluster with its hardware and software specifications resulted from our study is illustrated by a UML deployment diagram in Fig. 11.

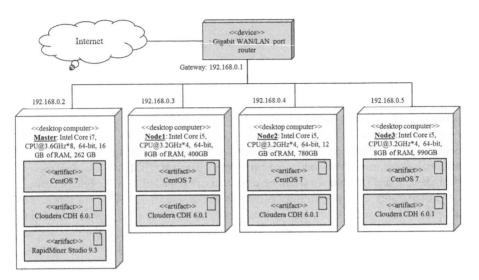

Fig. 11. A cluster of four commodity computers with hardware and software specifications where the master computer acts as a name node and node1, node2 and node3 computers function as data nodes.

As shown in Fig. 11, the cluster consists of one master host and three worker hosts. The master host is set as a name node and worker hosts as data nodes. Since Rapid-Miner Studio is installed in the master host, all in-memery data analysis tasks will be performed on this computer. Therefore, the master host is set to have the largest RAM size among all computers in the cluster.

Data in Table 1 is the Twitter data analysis results generated by RapidMiner Studio as a classification performance matrix in terms of accuracy, precision, and recall. It is given for only confirming the usability of the framework.

Table 1. Classification performance.

	True Y	True N	Class precision
Pred. Y	131	22	85.62%
Pred. N	24	125	83.89%
Class recall	84.52%	85.03%	

(Accuracy = 84.77%)

5 Conclusions

In this paper, a big data analytics framework was constructed, a case study on twitter sentiment analysis was presented, and an experiment of data analysis using a large data set was performed. The framework is built on a cluster of four commodity computers in which one computer is set as a name node and the remaining ones are set as data nodes.

Cloudera CDH 6.0.1 and RapidMiner Studio 9.3 with Text Processing, Hive Connector, and Radoop extensions are installed on the name node computer.

Based on the study results, a big data analytics framework can be constructed by using only commodity computers and open-source software applications. Building a successful framework requires careful software configurations and detailed examination of supporting software versions between Cloudera CDH and RapidMiner Radoop. In our study, instead of selecting the latest version, an earlier version, Cloudera CDH 6.0.1, is used because it contains Spark 2.2, the highest version supported by RapidMiner Radoop at the time of writing. In addition, based on our experiences in this study, size of RAM is one of key factors for efficiently conducting a data analysis. As a consequence, it is very important to check Cloudera hardware requirements at an early stage of a project. For the name node computer, we recommend using RAM whose size is larger than that specified in the Cloudera requirements because it is also used for running RapidMiner Studio/Radoop processes.

Cloudera CDH software suite allows Hadoop ecosystem to be readily set up on a cluster computers. Visual workflow designer in RapidMiner Studio makes big data analysis tasks become easier for people with no programming experiences. Therefore, we hope that the methodology and framework presented in this paper can be used both as a starting point of learning for researchers, educators, or professionals in any domain who are interested in the area of big data analysis and as a research and big data analysis tools for experienced users.

For future work, the framework could be used to collaborate with sensor devices from Internet of Things system for real-time big data analysis.

References

1. Altalhi, A.H., Luna, J.M., Vallejo, M.A., Ventura, S.: Evaluation and comparison of open source software suites for data mining and knowledge discovery: open source software suites for data mining and knowledge discovery. Wiley Interdiscip. Rev. Data Min. Knowl. Discov. **7**(3), e1204 (2017)
2. Bhathal, G.S., Singh, A.: Big data: Hadoop framework vulnerabilities, security issues and attacks. Array **1**, 100002 (2019)
3. Chennamsetty, H., Chalasani, S., Riley, D.: Predictive analytics on electronic health records (EHRs) using Hadoop and Hive. In: 2015 IEEE International Conference on Electrical, Computer and Communication Technologies (ICECCT), pp. 1–5. IEEE, Coimbatore (March 2015)
4. Dubey, A.K., Jain, V., Mittal, A.P.: Stock market prediction using Hadoop MapReduce ecosystem, p. 6 (2015)
5. Feltrin, L.: KNIME an open source solution for predictive analytics in the geosciences [software and data sets]. IEEE Geosci. Remote Sens. Mag. **3**(4), 28–38 (2015)
6. Han, P., Wang, D.B., Zhao, Q.G.: The research on Chinese document clustering based on WEKA. In: 2011 International Conference on Machine Learning and Cybernetics, pp. 1953–1957. IEEE, Guilin (July 2011)
7. Ivanov, T., Niemann, R., Izberovic, S., Rosselli, M., Tolle, K., Zicari, R.V.: Performance evaluation of enterprise big data platforms with HiBench. In: 2015 IEEE Truscom/BigDataSE/ISPA, pp. 120–127. IEEE, Helsinki (August 2015)

8. Jovic, A., Brkic, K., Bogunovic, N.: An overview of free software tools for general data mining. In: 2014 37th International Convention on Information and Communication Technology, Electronics and Microelectronics (MIPRO), pp. 1112–1117. IEEE, Opatija (May 2014)
9. Liu, F.C., Shen, F., Chau, D.H., Bright, N., Belgin, M.: Building a research data science platform from industrial machines. In: 2016 IEEE International Conference on Big Data (Big Data), pp. 2270–2275. IEEE, Washington DC (December 2016)
10. Nereu, J., Almeida, A., Bernardino, J.: Big data analytics: a preliminary study of open source platforms. In: Proceedings of the 12th International Conference on Software Technologies, pp. 435–440. SCITEPRESS - Science and Technology Publications, Madrid (2017)
11. Prekopcsák, Z., Makrai, G., Henk, T., Gáspár-Papanek, C.: Radoop: analyzing big data with RapidMiner and Hadoop. In: RCOMM 2011: RapidMiner Community Meeting and Conference, p. 13. Rapid-I (June 2011)
12. Sangeeta: Twitter data analysis using FLUME & HIVE on Hadoop framework. Spec. Issue Int. J. Recent Adv. Eng. Technol. 4(2), 119–123 (2016)
13. Singh, D., Reddy, C.K.: A survey on platforms for big data analytics. J. Big Data 2(1), 8 (2015)
14. Tripathi, P., Vishwakarma, S.K., Lala, A.: Sentiment analysis of english tweets using rapid miner. In: 2015 International Conference on Computational Intelligence and Communication Networks (CICN), pp. 668–672. IEEE, Jabalpur (December 2015)

Multi-stage Clustering Algorithm for Energy Optimization in Wireless Sensor Networks

Israel Edem Agbehadji[1], Richard C. Millham[1(✉)],
Simon James Fong[2], Jason J. Jung[3], Khac-Hoai Nam Bui[4],
and Abdultaofeek Abayomi[1]

[1] ICT and Society Research Group, Department of Information Technology,
Durban University of Technology, Durban, South Africa
richardml@dut.ac.za
[2] ICT and Society Research Group,
Department of Computer and Information Science,
University of Macau/Durban University of Technology, Taipa, Macau SAR
[3] Chung-Ang University, Seoul, South Korea
[4] Korea Institute of Science and Technology Information, Daejeon, South Korea

Abstract. Clustering technique is one of the approach to optimize energy consumption, balance load and increase lifetime of networks in wireless sensor network (WSN). In this paper, a novel multi-stage clustering algorithm is proposed for heterogeneous energy environment. The proposed multi-stage approach combines the behaviour of a bird and the distributed energy efficient model. The behaviour of the bird is expressed in the form of mathematical expression and then translated into an algorithm. The algorithm is then combined with the distributed energy efficient model to ensure efficient energy optimization. The proposed multi-stage clustering algorithm (referred to as DEEC-KSA) is evaluated through simulation and compared with benchmarked clustering algorithms. The result of simulation showed that the performance of DEEC-KSA is efficient among the comparative clustering algorithms for energy optimization in terms of stability period, network lifetime and network throughput. Additionally, the proposed DEEC-KSA has the optimal network running time (in seconds) to send higher number of packets to base station successfully.

Keywords: Edge computing · Load balancing · Wireless sensor network ·
Clustering algorithm · Kestrel-based search algorithm ·
Heterogeneous environment · Internet of Things (IoT) analytics

1 Introduction

The era of internet of things (IoT) enhances data sharing among connected objects and people [1]. IoT is an ecosystem consisting of sensor devices and equipment connected together to form a network for data transmission or reception [2]. Basically, network facilitates communication among people and "things" and encourage intelligent collaboration. The intelligence relates to how quick a decision is made on a network to increase performance and lifetime of a network. Network lifetime is the probability of a

© Springer Nature Singapore Pte Ltd. 2019
M. W. Berry et al. (Eds.): SCDS 2019, CCIS 1100, pp. 223–238, 2019.
https://doi.org/10.1007/978-981-15-0399-3_18

network to continuously be available taking into consideration the network load. Network intelligence is crucial in edge computing particularly when many devices with different energy requirements can be connected at any time, and as the network scales-up, the lifetime of network could be maintained.

This paper proposes an intelligent energy efficient model for wireless sensor networks (WSN) based on the behaviour and characteristics of a bird to develop a clustering technique. The rest of this paper is organized as follows: Sect. 2, related work. Section 3, proposed model. Section 4, conclusion.

2 Related Work

A classical energy consumption model guides the design of low level energy consumption devices for transmitting and receiving of messages. This section will review previous works on semantic framework, load balance and bio-inspired techniques to load balance, edge computing and wireless sensor networks.

2.1 Semantic Framework for Energy Management

Fundamentally, semantic framework apply semantic web technology to support web based decision making for energy optimization [3]. This framework provides a standard model for energy optimization because it enable different energy related data source such as energy production needs, energy prices, weather data and end-users' behaviour, to be collected and integrated. Data integration supports interoperability of heterogeneous systems which are implemented through logic based rules.

2.2 Load Balance

Load balance is a way to control energy usage and rate of usage. Basically, load balancing do not seek equal distribution of load on entire network of nodes but on how to balance load on single node of a network based on status of a network. A method to achieve load balance is clustering. In clustering, several devices (e.g., IoT devices, home appliances, sensor-enabled devices) are connected to form a cluster. Each cluster has a cluster head that connects another cluster head to form a network of clusters. Generally, clustering techniques are applied to both homogeneous (that is, devices have same initial energy) and heterogeneous (that is, when devices do not have same initial energy) networks. In respect of optimality, optimal number of cluster refers to the number of clusters that can help reduce energy dissipation.

Although benchmarked clustering algorithms such as Low Energy Adaptive Clustering Hierarchy (LEACH), Power-Efficient Gathering in Sensor Information Systems (PEGASIS), Hybrid energy-efficient distributed clustering (HEED) are applied to homogeneous network, the author of [4] indicates that it is a challenge to devise an energy-efficient clustering algorithm for heterogeneous network because of the complication in network operation and the energy configuration of connected devices (that is, heterogeneity of initial energy).

2.3 Bio-inspired Techniques

Bio-inspired techniques have also played a role in clustering. In a sense that, it helps to find an optimal way to build a cluster by taking into considerations the energy makeup of devices and the distance among clusters with their respective nodes. In order to create a dynamic load balance, bio-inspired techniques explore and exploit different search areas to find an optimal way to form a cluster. In view of this, bio-inspired techniques are applied in clustering model for energy efficiency and dynamic load balance. Example of bio-inspired techniques are namely Particle Swarm Optimization (PSO), Genetic Algorithms (GA) etc. The advantage of a bio-inspired technique is the ability to jump out of any local search that might not lead to an ideal cluster formation. In respect of this advantage, several energy clustering models are integrated with bio-inspired techniques for clustering so as to solve load balancing problems. Example of such integration includes Energy-aware Clustering for WSNs using the PSO Algorithm; Cluster-based WSN Routing using the ABC Algorithm; Whale Optimization for clustering and energy optimization for Wireless Sensor Network etc.

2.4 Edge Computing

Edge computing performs real-time analytics to find the best way to send data to the cloud, instead of performing analytics directly on the cloud. Generally, the advantage of edge computing is the efficient distributed computing such that when a sensor node dies or goes off quickly because of workload in collecting, aggregating and sending data to a BS, different sensor node is selected as cluster head.

Figure 1 depicts the edge analytics architecture for edge devices (e.g., sensor enabled) to perform analytics in real time or in nearby locations (e.g., ceilings).

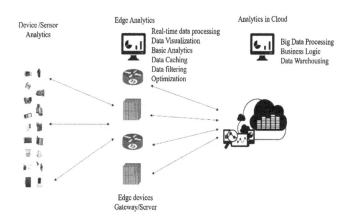

Fig. 1. Edge analytics architecture

Figure 1 shows the edge analytics architecture which consists of three analytics layers namely device/sensor layer, edge analytics layer and analytics in cloud layer.

Firstly, the edge analytics layer is the layer that supports Fog computing. Basically, Fog computing is a computing framework that liaises between device/sensor layer and analytics in cloud layer. Mostly, in Fog computing, data analytics is delegated to edge devices/gateway rather than being delegated to a central cloud server.

Secondly, analytics in cloud layer supports further data processing and storage particularly through the use of the internet. Thirdly, the device/sensor layer houses all data collection devices such as smart appliances, low powered devices that are connected to a single location mostly referred to as sink node or Base Station (BS). It is significant to note that fog computing devices may not necessarily be at the edge of a network, but rather it may reside close to the edge of the network. In essence, fog computing and edge computing are both close to the IoT end-devices, but the edge computing devices are often more closer. In many related works, fog computing and edge computing have been used interchangeably.

In edge computing, the application of bio-inspired search method provides an optimal way to dynamically adjust energy consumption in real-time without being stuck in the local search.

2.5   Wireless Sensor Networks

Wireless sensor networks (WSN) consist of sensor nodes equipped with the capability to sense, compute and communicate with other sensor nodes. Routing algorithms are required when a node is unable to send data to base stations. In WSNs, data is transmitted in real-time and clustering algorithm continuously update their network status. Each iteration to update the network is referred as a round. Sensor devices have different parts that consumes energy namely the micro controller processing, radio transmission and receiving, transient energy, sensor sensing, sensor logging and actuation. Additionally, the location of sensor nodes in a cluster, which leads to having different transmit distance to a cluster head also consume energy. In WSN, energy consumption in data transmission often reduces and this affect overall performance of a network, stability of network and efficiency of information transmission. The description of WSN is depicted in Fig. 2.

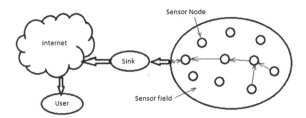

Fig. 2. Structure of wireless sensor network

The sink node shown (see Fig. 2) represents the edge device gateway of the edge analytics architecture (see Fig. 1).

3 Proposed Model

The paper proposes a novel clustering scheme exploiting Kestrel-based search algorithm to optimize energy consumption in heterogeneous environments. The approach to optimize energy consumption is based on the behaviour and characteristics of birds/animals mostly referred to as bio-inspired approach.

3.1 Proposed Clustering Algorithm Based on Kestrel-Based Search Algorithm

The Kestrel-based search algorithm (KSA) is based on characteristics namely random encircling, trail evaporation based on half-life period, position and velocity of kestrels [5]. Although the algorithm has been applied to different problem domain such as missing value estimation [5], association rule mining [6], feature selection in classification [7, 8] etc. In this study, we applied Kestrel-based search algorithm to clustering in the case of heterogeneous energy requirements. Basically, Kestrel achieves this optimization by changing position, velocity and trail evaporation, and ensures randomness. The KSA starts by initialing a set of random kestrels at the set-up phase of first round/iteration to determine the energy requirements of heterogeneous devices and find an optimal parameter. Devices are said to be heterogeneous because each has different energy requirement. The position of KSA is expressed as follows:

$$x_{i+1}^k = x_i^k + \beta_o e^{-\gamma r^2}(x_j - x_i) + f_i^k \tag{1}$$

Where x_{i+1}^k is the current best position of kestrel that represents candidate solution; x_i^k is the previous position of Kestrel based on random encircling formulation [5]; $\beta_o e^{-\gamma r^2}$ is the attractiveness which indicates the light reflected from a trail, where the variable β_o represents initial attractiveness, r represents distance measurements expressed using Minkowski distance [5], γ represents variation of light intensity between [0, 1]; x_j represents a Kestrel with a better position; f_i^k is the frequency of bobbing as expressed by [5]. The random encircling formulation is expressed by:

$$\vec{x}(t+1) = \vec{x_p}(t) - \vec{A} * \vec{D} \tag{2}$$

Thus:

$$\vec{A} = 2 * \vec{z} * \vec{r2} - \vec{z} \tag{3}$$

$$\vec{D} = \left| \vec{C} * \vec{x_p}(t) = \vec{x}(t) \right| \tag{4}$$

$$\vec{C} = 2 * \vec{r1} \tag{5}$$

Where $\vec{A}$ is coefficient vector, $\vec{D}$ is the encircling value obtained, $\vec{x_p}(t)$ is the position vector of the prey, $\vec{x}(t+1)$ represents the previous position of Kestrels. Where $\vec{C}$ is the coefficient vector, $\vec{x}(t)$ indicates the position vector of a Kestrel, r1 and r2 are random numbers generated between 0 and 1, and as Kestrels shift the centre of encircling it maximise the chances of locating its prey hence the constant value of 2. $\vec{z}$ represents a parameter to control the active mode with $\vec{z}_{hi}$ as the parameter for flight mode and $\vec{z}_{low}$ as the parameter for perched mode, which linearly decreases from 2 (high active mode value) to 0 (low active mode value) respectively during the iteration process. This is expressed as:

$$\vec{z} = \vec{z}_{hi} - (\vec{z}_{hi} - \vec{z}_{low}) \frac{itr}{Max\_itr} \qquad (6)$$

Where itr is the current iteration, Max_itr is the total number of iterations which are performed during the search. Other Kestrels that are involved in the search update their position according to the best position of the leading Kestrel.

Finally, the velocity of Kestrel is updated using the expression:

$$v_{t+1}^k = v_t^k + x_t^k \qquad (7)$$

Where v_{t+1}^k is the current best velocity, v_t^k represents the initial velocity, whilst x_t^k represents the current best position of a Kestrel.

3.2   Trail Evaporation

In meta-heuristic algorithms, ant use trails both to trace the path to a food source and to prevent themselves from getting stuck in a single food source. Thus, ants, using these trails, can search many food sources in a search space. As ants continue to search, trails are drawn and pheromones are deposited on a trail. This pheromone help ants to communicate with each other about the location of food sources. Therefore, other ants continuously follow this path and also deposit substances for the trail to remain fresh. Similar to ants, Kestrels use trails in search of food sources. However, these trails are rather deposited by preys which provides an indication to Kestrels on availability of food sources. The assumption is that the substances deposited by a prey is similar to pheromone deposited on ants' pheromone trail. Additionally, when the source of food depletes, Kestrels no longer follow this path that leads to the location of a prey. Consequently, pheromone trail begins to diminish with time at an exponential rate causing trails to become old. This diminishment denotes the unstable nature of the trail substances which can be theoretically stated as: if there are N unstable nodes (that is, different energy requirements), then the rate at which the "substance" decays with time t is expressed by:

$$\frac{dN}{dt} = -\gamma N \qquad (8)$$

Thus, decay rate (γ) with time (t) is simplified as:

$$\gamma_t = \gamma_o e^{-\varphi t} \qquad (9)$$

Where γ_o represents initial value, t is the time of decay. The decay constant φ which shows how long it takes for a "substance" to decay is re-expressed as:

$$\varphi = \frac{ln\,0.5}{-t_{\frac{1}{2}}} \tag{10}$$

Where φ is the decay constant and $-t_{\frac{1}{2}}$ is the half-life period. If the value of decay constant is greater than 1, then the trail is considered as new else the trail is considered as old, which is expressed by:

$$if\ \varphi\ \rightarrow\ \begin{cases} \varphi > 1, & trail\ is\ new \\ 0, & otherwise \end{cases} \tag{11}$$

Where φ is the decay constant.

In WSNs, the heterogeneity of devices on the edge of networks, makes it impossible for all nodes to go off at the same time. Similarly, it could be said that each node has its own half-life. In this regard, the decay rate of nodes on the network can be determined by applying the half-life formulation. Moreover, as nodes are heterogeneous, a degree of randomness is introduced which is accounted for by the decay process.

1. Heterogeneous network model for energy optimization

The energy model finds energy dissipated when sensor node transfer or receive data on network. In this paper, DEEC model is adopted and integrated with the behaviour and characteristics of kestrel formulation (see Sect. A). In the transfer of data, some amount of energy is dissipated and to optimize this energy the radio energy dissipation model [9] shown (see Fig. 3) is applied.

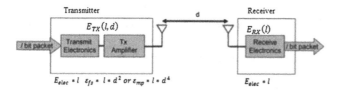

Fig. 3. First-order radio model

In this model, the energy required by the transmit amplifier $E_{TX}(l, d)$ to transmit an *l-bit* message over a distance d between a transmitter $E_{TX}(l, d)$ and receiver $E_{RX}(l)$ is expressed by:

$$E_{TX}(l, d) = \begin{cases} l * E_{elec} + l * \varepsilon_{fs} * d^2 & if\ d \le d_o \\ l * E_{elec} + l * \varepsilon_{mp} * d^4 & if\ d \ge d_o \end{cases} \tag{12}$$

Where d_o is the threshold transmission distance between the transmitter and receiver which is expressed as $d_o = \sqrt{\frac{\varepsilon_{fs}}{\varepsilon_{mp}}}$. E_{elec} is the energy consumption in electronics for sending or receiving a bit, ε_{fs} and ε_{mp} represents amplifier parameters for free state and two ray model respectively. d^2 and d^4 refers to short and long distance transmissions respectively. When an l-bit packet is received, the energy required by receiver $E_{RX}(l)$ is expressed by:

$$E_{RX}(l) = l * E_{elec} \tag{13}$$

In this section, we described the energy model for multi-level heterogeneous network. In multi-level heterogeneous model, energy of sensor nodes are randomly distributed in a size of $M \times M$ meters. In this context, the total initial energy for all sensor node is expressed by:

$$E_{total} = \sum_{i=1}^{n} E_o(1 + a_i) = E_o\left(\left(n + \sum_{i=1}^{n} a_i\right)\right) \tag{14}$$

Where E_o represents the initial energy of a node, in this context a node has initial energy $E_o(1 + a_i)$ which is a_i times more than the lower/initial bound E_o.

When a cluster is formed, each cluster head dissipates energy in receiving a signal from nodes, then aggregates the signals and transmit the aggregate signal to the BS which is far from nodes. Thus, a cluster-head should have enough energy to reach a BS. The energy of cluster is expressed as.

$$E_{CH} = l * \frac{n}{k} * (E_{elec} + E_{DA}) + l * \varepsilon_{mp} * d^4_{toBS} \tag{15}$$

Where E_{CH} represents energy of a cluster head, n is the number of sensor nodes, k is the number of clusters, E_{elec} is transmitter electronics, E_{DA} is energy for aggregating data, l is the data packet, ε_{mp} transmitter amplifier in long distance d^4_{toBS} to a base station. Similarly, the energy dissipated by non-cluster heads E_{n-CH} is expressed by:

$$E_{n-CH} = l * E_{elec} + l * \varepsilon_{fs} * d^2_{toCH} \tag{16}$$

Where l is the data packet, E_{elec} is transmitter electronics, ε_{fs} represents transmitter amplifier for free state in short distance d^2_{toCH} to cluster-head. The distance both short and long distances are expressed [4] by:

$$d^4_{toBS} = \frac{M}{\sqrt{2\pi k}}, \quad d^2_{toCH} = 0.765\frac{M}{2} \tag{17}$$

Where M is the size of sensor field, k is the number of clusters, d^2_{toCH} is short distance to cluster-head, d^4_{toBS} is the long distance to base station. Thus, total energy E_{total} dissipated by a cluster is expressed by:

$$E_{Cluster} = E_{CH}\left(\frac{n}{k} - 1\right)E_{n-CH} \approx E_{CH} + \frac{n}{k}E_{n-CH} \tag{18}$$

$$E_{total} = k * E_{Cluster} \tag{19}$$

Where $E_{Cluster}$ is the energy dissipated by a cluster among cluster members, E_{CH} represents energy of cluster head, n is the number of sensor nodes, k is the number of clusters, E_{n-CH} is energy dissipated by non-cluster. Additionally, each non-cluster head sends l-bits message to the cluster head in a round, therefore total energy dissipated in the network during a round is expressed by:

$$E_{round} = l * \left(2 * n * E_{elec} + n * E_{DA} + k * \varepsilon_{mp} * d_{toBS}^4 + n * \varepsilon_{fs} * d_{toCH}^2\right) \tag{20}$$

Where E_{round} represents energy a round, l is the data packet, n is the number of sensor node, E_{elec} is transmitter electronics, E_{DA} is energy for aggregating data, ε_{mp} transmitter amplifier in long distance d_{toBS}^4 to a base station, ε_{fs} is transmitter amplifier for free state in short distance d_{toCH}^2 to cluster-head, and k is the number of clusters. The BS and CH are always located at a distance, and due to randomization, the distance to send a data packet is always computed in each round/iteration. In view of this, the threshold transmission distance d_o and d are compared to find the distance to send packets to BS and CH. In respect of clusters, the optimal number of clusters k_{opt} which replaces k is expressed as:

$$k_{opt} = \frac{\sqrt{n}}{\sqrt{2\pi}} * \sqrt{\frac{\varepsilon_{fs}}{\varepsilon_{mp}}} * \frac{M}{d_{toCH}^2} \tag{21}$$

Where n is the number of nodes, ε_{mp} transmitter amplifier, ε_{fs} represents transmitter amplifier for free state in short distance d_{toCH}^2 to cluster-head, M is the size of sensor fields.

It is important to select an optimal cluster-head, therefore a probability threshold $T(s_i)$ is applied to determine the optimal cluster head in a round. If the probability is less than a threshold $T(s_i)$ value, the node is selected as cluster head for that round. The $T(s_i)$ is expressed by:

$$T(s_i) = \begin{cases} \dfrac{p_i}{1-p_i\left(r \bmod \frac{1}{p_i}\right)} & if \ s_i \in G \\ 0 & otherwise \end{cases} \tag{22}$$

Where p_i is user set probability for a cluster head, r represents the current round, G is the set of nodes represents the reference value heads in the previous $\frac{1}{p_i}$ rounds. Thus,

$$p_i = \frac{p_{opt}n(1+a)E_i(r)}{\left(n + \sum_{i=1}^n a_i\right)\bar{E}(r)} \tag{23}$$

Where p_{opt} represents the reference value of the average probability of p_i, n is the number of nodes, $E_i(r)$ is the residual energy, $\bar{E}(r)$ is estimated energy that serves as standard reference energy for each node. This reference energy indicates that each node has its own energy in each round to keep the network alive, and this introduces some heterogeneity on the network. Moreover, in a heterogeneous network, it is important to ensure that there is enough energy for data transmission. In view of this, both initial energy and residual energy level of nodes are used to select cluster-heads at each round. Since nodes have different energy requirements, the network identifies the best node base on the average energy $\bar{E}(r)$ at a round of the network which is computed by:

$$\bar{E}(r) = \frac{1}{n} * E_{total}\left(1 - \frac{r}{R}\right).$$
(24)

Where R is the total network lifetime. The assumption for considering network lifetime is that, should all the nodes die simultaneously, R is the total of rounds from the time the network begins to time the node dies. Furthermore, network lifetime can be categorized into two periods such as the stable and unstable periods. Whereas, the stable period refers to the period from the beginning of transmission to the period the first node dies; and the unstable period refers to a period from the death of first node till the death of last node on the network. Therefore, the energy consumed by the network in each round is denoted by E_{round} and therefore R is expressed by:

$$R = \frac{E_{total}}{E_{round}}$$
(25)

Where E_{total} is the total energy.

2. Objective function
The strategy for objective function is to select maximum of the minimum energy on the network so as not to select a weak node in the lifetime of the network. This objective function is expressed by:

$$objective\ function = \alpha * f_1 + (1 + \alpha) * f_2$$
(26)

$$f_1 = \max_{1 \in [1,K]}\left(\sum_{s_i \in C_k} \frac{d(sensor_i, CH_k)}{|C_k|}\right)$$
(27)

$$f_2 = \frac{\sum_{i=1}^{N} E(sensor_i)}{E(CH_k)}$$
(28)

Where α is user-defined parameter between 0 and 1, N is the number of sensors; $|C_k|$ is the number of sensors that belong to a cluster C_k; f_1 is the maximum average distance of sensors and their cluster heads; f_2 is the ratio of total initial energy of all sensors with the total current energy of cluster head in a round.

In respect of the DEEC, the fitness function is expressed by equation $T(s_i)$, similar to [10]. It is ideal to have same fitness function for each proposed and comparative algorithms. However, in this study divergent fitness functions was applied.

3.3 Parameter Setting for Simulation of Network

The parameters for Kestrel-based search algorithm (KSA) are $zmin = 0.2$ (that is, parameter for perched mode), $zmax = 0.8$ (that is, parameter for flight mode) [5]. Following the network parameter settings by Jadhav and Shankar [11], the network parameter for our energy model is set. Initially, in the energy model there are 1000 nodes, the transmitter electronics is set as 5 nJ/bit, the initial energy is between 0.5 J and 0.6 J, and other parameters are define in Table 1. *Popt* represents the optimum probability which is set to 0.1. The network parameter settings is summarized in Table 1.

The proposed multi-stage DEEC-KSA model consists of the following process; determination of clusters using random encircling, evaluation of fitness of each encircled position, and determination of energy consumption. There are different optimal parameter for KSA whereas the comparative algorithms used optimal parameter (p_{opt}) defined in Table 1.

The performance of network is evaluated in terms of death of the first node and remaining active nodes, network lifetime, network throughput in terms of packets sent to BS. Death of the first node is the rounds of iteration for nodes to die. In WSNs, performance tends to decline with the nodes' death. Normally, the network is in a stable period before the first node dies.

The death of the first node shows that the network is in unstable period and its performance starts to decline. On the other hands, last node dead is the number of remaining active nodes on a network. Therefore, stability of network is categorized into stable and unstable periods. The stable period is a period from the beginning of transmission to the period the first node dies; and unstable period is period from death of first node till the death of last node on the network. Network lifetime is the number of surviving nodes on the network. Network throughput is define as the number of data packets successfully received at BS. In other words, it is expressed as the number of packets sent to BS minus number of packets dropped [12, 13].

Table 1. Parameter settings

Parameter	Values
Number of nodes	1000
Transmitter electronics E_{elec}	5 nJ/bit
Initial energy E_o	0.5 J–0.6 J
Data aggregation E_{DA}	5 nJ/bit/message
Transmitter Amplifier ε_{mp} if $d \geq d_o$	0.0013 pJ/bit/m^4
Transmitter Amplifier ε_{fs} if $d \leq d_o$	10 pJ/bit/m^2
Data packet size l bit	4000
Network size	100 m × 100 m
p_{opt}	0.1

3.4   Simulation Results

In this section, we present the simulation result of the proposed DEEC-KSA algorithm and compare with existing clustering algorithms namely Distributed Energy-Efficient Clustering (DEEC), Developed Distributed Energy-Efficient Clustering (DDEEC), and extended version of Distributed Energy-Efficient Clustering (E_DEEC) with normal, advance and super node classification.

The clustering algorithm is evaluated in terms of stability of network on first node death (FND) and number of remaining active nodes at the final round (LND), network lifetime, and network throughput in terms of packets sent to BS.

1. Comparison of first node death
In WSNs, network is in a stable period before first node dies. When first node dies then network performance tends to decline which results in unstable period. Analysis results are presented in Tables 2, 3, 4 and 5 for heterogeneous initial energy values between 0.5 J and 0.6 J.

Table 2 shows the performance of allive nodes for the proposed and comparative algorithms for 1000 nodes with initial energy of 0.5 J. Based on the result presented in Table 2, it shows performance of first node death for DEEC-KSA, DEEC, DDEEC and E_DEEC as 1147, 1014, 685 and 421 respectively. The proposed DEEC-KSA ensures the probability of selecting a cluster head is lower at 0.4. The first node's death in DEEC, DDEEC and E_DEEC is completely different from DEEC-KSA. Although, the comparative algorithms ensure that the probability of selecting a cluster head is 0.1 and avoid waste of energy, it was unable to extend the death of first node after several rounds of iteration. DEEC-KSA has the advantage of extending the allive of first node's death. The proposed DEEC-KSA, to a large extent, has the best performance in terms of first node's death in the simulation involving different levels of initial energy and number of nodes as shown on Tables 3, 4 and 5.

Table 2. Allive Nodes during Network Lifetime for 1000 nodes using 0.5 J

Algorithms	Popt	FND	Tenth node dead (tenth_dead)	LND	Packets to BS	Time (s)
DEEC-KSA	0.44	1147	1615	2920	2155711	191.34
DEEC	0.1	1014	1238	3744	297055	2547.65
DDEEC	0.1	685	1096	3423	136403	1414.83
E_DEEC	0.1	421	864	128	1931194	697.27

Table 3. Allive Nodes during Network Lifetime for 1000 nodes (initial energy of 0.6 J)

Algorithms	Popt	FND	Tenth node dead (tenth_dead)	LND	Packets to BS	Time (s)
DEEC-KSA	0.22	1215	2319	4579	2760260	235.83
DEEC	0.1	1132	1426	4397	396400	2858.04
DDEEC	0.1	913	1354	4121	180331	1858.37
E_DEEC	0.1	763	1076	185	2265855	667.07

Table 4. Allive Nodes during Network Lifetime for 2000 nodes (initial energy of 0.5 J)

Algorithms	Popt	FND	Tenth node dead (tenth_dead)	LND	Packets to BS	Time (s)
DEEC-KSA	0.001	1428	1643	3018	4473702	387.27
DEEC	0.1	905	1231	3778	515169	10554.42
DDEEC	0.1	506	1017	3642	228926	37424.51
E_DEEC	0.1	325	870	278	3839531	2258.57

Table 5. Allive Nodes during Network Lifetime for 2000 nodes (initial energy of 0.6 J)

Algorithms	Popt	FND	Tenth node dead (tenth_dead)	LND	Packets to BS	Time (s)
DEEC-KSA	0.36	375	1230	2504	2691844	361.7
DEEC	0.1	1075	1438	4754	657618	14532.6
DDEEC	0.1	743	1276	4382	270626	5765.4
E_DEEC	0.1	374	1084	406	4559852	2678.0

Based on result presented in Table 2, it shows the last node death represent the number of remaining active nodes at the final round. DEEC, DDEEC, DEEC-KSA and E-DEEC has 3744, 3423, 2920 and 128 number of active nodes respectively at the end rounds of iteration. Thus, DEEC-KSA is third in the number of nodes retained after last node's death. To a large extent, DEEC-KSA is third in retaining nodes on the network as shown on Tables 3, 4, and 5.

2. Comparison of network lifetime performance

Figure 4 shows the graphical display of surviving/allive nodes in each round of iteration for different initial energy level. There are different number of nodes and corresponding network lifetime variation curves per rounds. When the network begins running, it is observed that DEEC-KSA has 1000 nodes in the 1500th round, while the comparative algorithms have their number of nodes reduce earlier. In comparison of network lifetime performance, it is observed that DEEC-KSA is optimal as compared with the three comparative algorithms where DDEEC network lifetime performance is second, DEEC is third while E_DEEC is fourth, as evident in Figs. 5, 6 and 7.

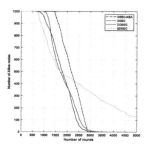

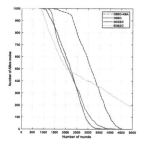

Fig. 4. Allive nodes per round for 1000 nodes with 0.5 J

Fig. 5. Allive nodes per round for 1000 nodes (Initial energy of 0.6 J)

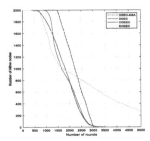

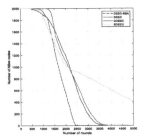

Fig. 6. Allive nodes per round for 2000 nodes (Initial energy of 0.5 J)

Fig. 7. Allive nodes per round for 2000 nodes (Initial energy of 0.6 J)

3. Comparison of the Network Throughput

In WSN, network throughput is fundamental to evaluate efficiency of algorithms. It refers to number of data packets in the network successfully sent at BS. As cluster member node sends information in the form of packets to CH, the CH fuses the information it sensed and finally sends to BS as packets. Simulation is performed using 1000 nodes with heterogeneous initial energy between 0.5 J and 0.6 J, and simulation result is presented in Figs. 8, 9, 10 and 11. Based on Fig. 8, the number of packets sent successfully to BS using DEEC-KSA, E_DEEC, DEEC and DDEEC are 2.2×10^5, 1.9×10^5, 0.3×10^5 and 0.2×10^5, respectively with respect to the number of rounds. The result shows that DEEC-KSA has the highest network throughput in all cases of simulation results (Figs. 9, 10 and 11).

In respect of time, the simulation results in Tables 2, 3, 4 and 5 indicates that when the network begins running the proposed DEEC-KSA has the least time (in terms of network running time) to send successful packets to BS, E_DEEC is second, DEEC is third and DDEEC is fourth in terms of time to send successful packets. This suggests that the proposed DEEC-KSA is the efficient clustering algorithm since it spent less time to send higher number of packets to BS. It is ideal that efficient clustering algorithms spend less time to send higher number of packets thereby reducing the energy consumption in WSNs.

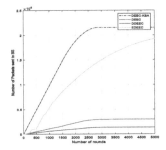

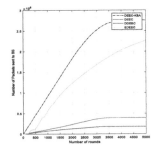

Fig. 8. Packets sent to BS on 1000 nodes per round with initial energy of 0.5 J

Fig. 9. Packet sent to BS on 1000 nodes (Initial energy of 0.6 J)

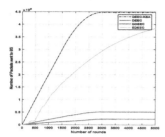

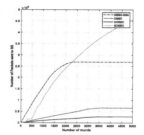

Fig. 10. Packet sent to BS on 2000 nodes (Initial energy of 0.5 J)

Fig. 11. Packet sent to BS on 2000 nodes (Initial energy of 0.6 J)

Based on the simulation result obtained and subsequently presented in Tables and Figures, once a node runs out of its energy, it is considered to be dead and it can no longer transmit or receive any data. Thus simulation ends when all the nodes in the network run out of their energy. High energy efficiency means low energy consumption and long stability period. From the simulation results, it is evident that network lifetime (in round) increases in DEEC-KSA with different initial energy between 0.5 J and 0.6 J.

The stability of algorithms for dynamic environment was considered by [14] in terms of estimation error and its accuracy. The authors' [15] considered stability of algorithms in terms of number of nodes in the network lifetime variation curves. Stability of the proposed DEEC-KSA and comparative algorithms were considered in terms of the number of nodes in each round. In this regard, when the network begins running the number of nodes in the algorithms were the same with different number of rounds. In most cases, the proposed DEEC-KSA goes through several rounds and has the highest number of nodes with limited network running time as compared with the comparative algorithms used in this experiment.

4   Conclusion

This paper presented a multi-staged approach in the form of algorithm which is based on a bio-inspired approach and DEEC approach for clustering in WSNs. The simulation result shows that the proposed DEEC-KSA performed optimally in comparison with the existing benchmarked clustering algorithms for heterogeneous environments. Additionally, the proposed DEEC-KSA has the highest network throughput spending limited time, and it has best network stability than the comparative algorithms considered. The proposed multi-stage DEEC-KSA algorithm is efficient in terms of stability period, network lifetime performance and network throughput in terms of packets successfully sent to BS relative. Additionally, the proposed multi-stage DEEC-KSA has optimal network running time to send successfully the packets to BS. Base of these, the proposed DEEC-KSA intelligently optimizes energy therefore, it can be concluded that our proposed multi-stage DEEC-KSA provides energy efficient clustering algorithm that ensures longer stability period for WSNs. Future work should focus on the use of different methods to analyse the stability of the proposed algorithm.

Acknowledgement. The authors are thankful to the research supported grant by both the National Research Foundation of South Africa with grant number 117799 and the National Research Foundation of Korea (NRF) grant funded by the Korea government (MSIP) (NRF-2018K1A3A1A09078981).

References

1. Siow, E., Tiropanis, T., Hall, W.: Analytics for the Internet of Things: a survey. ACM Comput. Surv. 1–35 (2018)
2. Ristl, A.: The Internet of Things: IoT analytics from the edge to core to cloud. DellEMC, p. 45 (2017)
3. Sicilia, A., et al.: A semantic decision support system to optimize the energy use of public buildings. In: CIB W78 Conference 2015 (2015)
4. Qing, L., Zhu, Q., Wang, M.: Design of a distributed energy-efficient clustering algorithm for heterogeneous wireless sensor networks. Comput. Commun. **29**, 2230–2237 (2006)
5. Agbehadji, I.E., et al.: Bioinspired Computational Approach to Missing Value Estimation. Math. Prob. Eng. **2018**, 16 (2018)
6. Agbehadji, I.E., Millham, R.C., Fong, S.: Kestrel-based search algorithm for association rule mining and classification of frequently changed items. In: IEEE International Conference on Computational Intelligence and Communication Networks, Dehadrun. IEEE (2016)
7. Agbehadji, I.E., et al., Kestrel-based Search Algorithm (KSA) and Long Short Term Memory (LSTM) network for feature selection in classification of high-dimensional bioinformatics datasets. In: Federation Conference of Computer Science and Information Systems (FedCSIS), Poznan, pp. 15–20 (2018)
8. Agbehadji, I.E., et al.: Integration of Kestrel-based search algorithm with Artificial Neural Network (ANN) for feature subset selection. Int. J. Bio-Inspired Comput. 12 (2019)
9. Liu, J.-L., Ravishankar, C.V.: LEACH-GA: genetic algorithm-based energy-efficient adaptive clustering protocol for wireless sensor networks. Int. J. Mach. Learn. Comput. **1** (1), 79–85 (2011)
10. Ari, A.A.A.: Bio-inspired solutions for optimal management in wireless sensor networks. In: Artificial Intelligence [cs.AI]. Université Paris-Saclay, p. 139 (2016)
11. Jadhav, A.R., Shankar, T.: Whale optimization based energy-efficient cluster head selection algorithm for wireless sensor networks. In: Neural and Evolutionary Computing, p. 22 (2017)
12. Behzad, M., Ge, Y.: Performance optimization in wireless sensor networks: a novel collaborative compressed sensing approach. In: International Conference on Advanced Information Networking and Applications, pp. 749–756. IEEE Computer Society (2017)
13. Liaqat, M., et al.: Distance-based and low energy adaptive clustering protocol for wireless sensor networks (2016)
14. Chen, L.: Algorithm design and analysis in wireless networks, in Data Structures and Algorithms. Université Paris-Sud: Laboratoire de Recherche en Informatique (UMR 8623) Université Paris-Sud, p. 163 (2017)
15. Towfic, Z.J., Sayed, A.H.: Stability and performance limits of adaptive primal-dual networks, pp. 1–16 (2015)

Computational and Artificial
Intelligence

Case Based Reasoning for Diagnosing Types of Mental Disorders and Their Treatments

Sri Mulyana[⊠], Sri Hartati, and Retantyo Wardoyo

Department of Computer Science and Electronics,
Faculty of Mathematics and Natural Sciences, Universitas Gadjah Mada,
Yogyakarta, Indonesia
{smulyana, shartati, rw}@ugm.ac.id

Abstract. Mental health is still a serious problem in Indonesia. Basic health research data in 2013 recorded the prevalence of severe mental disorders in Indonesia reaching 1.7 per mil. On the other hand, the availability of mental health services and experts in the field is not yet adequate and not evenly distributed. Therefore, developing a system to help diagnosing the types of mental disorders and their treatment can be an alternative to overcome these inequalities. Case-Based Reasoning is one of the reasoning methods in expert systems. On Case-Based Reasoning, a base case is required containing cases with sotions that have been achieved. To find a solution to a given new problem, the system will look for cases on the case base that have the highest level of similarity. In this research, a case-based reasoning system has been developed for dagnosing the types of schizophrenic disorders and mood disorders along with their treatment. The case base was constructed based on medical records of mental disorder patients, which were obtained from the collaboration with a Mental Hospital in Yogyakarta. In addition, it also considers Guidelines for Classification and Diagnosis of Mental Disorders in Indonesia III, which contains 10 categories of mental disorders.

Keywords: Case-based reasoning · Case base · Mental disorders · Medical records

1 Introduction

Mental health is still a serious problem in Indonesia. The 2013 Basic Health Research Data noted the prevalence of severe mental disorders in Indonesia reached 1.7 per mil. This means that 1–2 people out of 1,000 residents in Indonesia experience severe mental disorders. This is exacerbated by the lack of mental health services and facilities in various regions of Indonesia so that many sufferers of mental health disorders have not been handled properly.

The number of people with mental disorders that have not been handled medically, due to the lack of professional mental health workers in Indonesia. This certainly further hampers efforts to prevent and deal with community mental health problems. The number of professional mental health workers in Indonesia is still unable to meet the minimum quota set by the World Health Organization or WHO. At present

© Springer Nature Singapore Pte Ltd. 2019
M. W. Berry et al. (Eds.): SCDS 2019, CCIS 1100, pp. 241–251, 2019.
https://doi.org/10.1007/978-981-15-0399-3_19

Indonesia with a population of around 250 million new people has about 451 clinical psychologists (0.15 per 100,000 inhabitants), 773 psychiatrists (0.32 per 100 thousand inhabitants), and soul nurses 6500 people (2 per 100,000 residents). Even though WHO sets the standard for the number of psychologists and psychiatrists with a population of 1: 30000 people, or 3 people per 100,000 residents. This condition is exacerbated by the uneven distribution and only concentrated in big cities. Therefore, developing a system to help diagnose the types of mental disorders and their management can be used as an alternative to overcome these inequalities.

Case-Based Reasoning (CBR) has become a successful technique for knowledge-based systems in many domains. Case-Based Reasoning (CBR) means using previous experience in similar cases to solve new problems. The basic idea of Case-Based reasoning is the assumption that similar problems have similar solutions.

Case-Based Reasoning (CBR) consists of four main steps, namely [1]:

(1) *Retrieve*: that is retrieving the same problem. In this step a search or calculation process is carried out from cases that have the highest level of similarity.
(2) *Reuse*: that is reusing information and knowledge in the case to overcome new problems. In this step a solution of a similar case is sought in the previous conditions for new problems.
(3) *Revise*: that is reviewing the solution given. In this step a solution is found for a similar case in the previous conditions for the problem that occurred later.
(4) *Retain*: that is exploring the part of previous experience to be used in solving the next problem.

The relationship between these steps can be presented in Fig. 1.

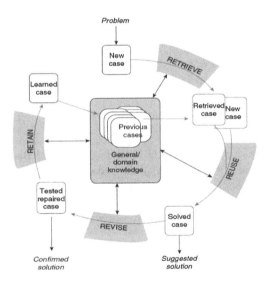

Fig. 1. Case-based reasoning cycles [1]

2 Literature Review

The case-based reasoning approach has developed very rapidly [2]. Some researchers who began the research in the field of Case-Based reasoning (CBR) were Schank with dynamic memory [3] which focused on the influence of knowledge, learning and memory, Carbonell in the field of analogy [4], Kolodner [2] and Rissland [5] who works in the field of formal reasoning. After that, the development of the CBR continued with various studies by Kolodner and his students [6–8], and research by Hammond and others in the case-based planning field [9–11], and research by Ashley and Rissland with the HYPO system for legal reasoning [12].

An important field of application in solving CBR-based problems is for diagnosing. In this area of diagnosing, CBR-based problem solving is often used to adjust old cases with new problems. CASEY [13] has been known as a case-based system for diagnosing the health problems of a patient by matching the diagnosis of the pessary that was previously known. The case-based system for other diagnosing is PROTOS [14].

CBR research continues to develop, and is increasingly attractive to researchers. After describing how the development of the begining of CBR research, the following are some CBR studies in the last 3 years.

The application of the CBR method has been developed in various applied fields, including CBR to diagnose infectious diseases. In the study, a case had attributes such as temperature, symptoms of dizziness (intensity and area), coughing, intensity of bowel movements, vomiting, and urination. All these attributes have numerical values from 1 to 4 to express the level of intensity, namely never, low, medium and high.

Ongoing research in recent years shows that CBR applications in the medical domain are developing quite well. Chakraborty has developed a CBR system to detect swine flu disease called SFDA (Swine Flu Diagnostic Assistant) [15]. Pant and Joshi have also developed a CBR system in the field of neurology, namely the NDS (Neurology Diagnosis System) [16]. In 2013 a CBR-based system was also developed to campaign for warning to stop smoking. The system contains early warnings about the dangers of smoking sent via cell phone [17]. The development of a system to detect mood disturbances (mood) with case-based reasoning has also been developed [18] and has also developed a case-based system for the diagnosis of heart disease [19]. In the future, CBR systems may be able to provide more services in the medical field and will be more integrated into the clinical environment.

3 Research Methodology

3.1 Research Data Collection

In this study, the data used was obtained based on medical records of mental patients, which were collected from a hospital in Yogyakarta according to the mechanism and applicable Operational and Procedure Standard (SOP). In accordance with the regulations, the statement states that patient data is confidential, because the data collected does not include the patient's name. The process of data collection is done by copying the contents of the patient's medical record through google form through the link: http://ugm.id/formpsik.

3.2 Implementation Procedure

The procedure for implementing this study begins with building a case base. The collection of patient medical record data that has been obtained is a source of reference in conducting this research. The case base preparation stage is shown in Fig. 2.

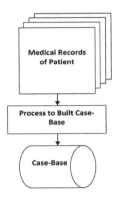

Fig. 2. Stages for case base preparation

Then a case-based reasoning system was developed to help diagnosing the types of schizophrenic mental disorders and mood disorders and their treatment. The series of case-based reasoning processes is shown in Fig. 3.

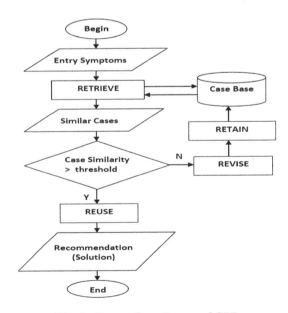

Fig. 3. Process flow diagram of CBR

Based on the flow diagram, the first step taken is to provide input in the form of features related to the new problem. Then the system will carry out the RETRIEVE process, which is looking for the type of mental disorder and its treatment from the case base by first calculating the level of similarity between cases. A case that has the greatest level of similarity, and its value is greater or equal to the specified threshold value, will be recommended as a solution to new problems. This process is known as REUSE. If there are no cases that have the required level of similarity, then it is possible to do the REVISE process, which is to revise together with experts, so that the new problem can be made a case. With the RETAIN process, it can be saved to the case base, as a new case.

4 Research Result and Discussion

As previously stated, the case for this research was obtained from medical records from a hospital in Yogyakarta. Based on these medical records and after normalization, a case is represented by a **.json** data structure that contains the following features: **id, age, systole, diastole, symptoms, axis-I, axis-II, axis-III, axis-IV, V-axis, treatment**. In this study there were 18 actual data that were used as the basis of the case, some examples of which were given in Table 1.

Table 1. Some examples of case base

Id	8187	7730	81670	93887	78784
Age	52	51	43	40	36
Systole	120	130	112	107	130
Diastole	80	90	68	77	80
Symptoms	G1, G26, G22, G21	G22	G10, G16, G18, G22, G17, G14, G23, G11	G15, G16, G22, G21	G10, G14, G15, G7, G0, G22, G21
AXIS I	F 25.0	F20.3	F20.3	DD Psikotik	skiz paranoid
AXIS II	nan	skizoid	DBN	Skizoid	skizoid
AXIS III	There is no diagnosis	There is no diagnosis	There is no diagnosis	There is no diagnosis	Oedem R Thorax ante
AXIS IV	nan	Disconnect the drug	Not clear	Friend problem	Not clear
AXIS V	bad	bad	nan	bad	bad
Treatment	Risperidone, THP, Clozapine, if you don't want it to be injected with half-ampoule lodomer, diazepam 1 ampoule	Haloperidol, THP, CpZ	CLOZAPIN 25 1-0-1, HDL 5 1-0-1, THP 2 1-0-1	injLodomer1 amp IM, risperidon 2 mg 1-0-1, thp 2 1-0-1	cpz 100 0-0-1 risperidon 2 1/2-1/2-/ 1/2

In the stored symptoms use codes, where each code indicates the symptoms as shown in Table 2. For example if in a case there is 'G0', then it means that there is a symptom of 'talking want to die' in that case.

Table 2. Symptom codes

G0	Speak want to date	G9	Request goods	G18	Disconnected from medicine
G1	Confused	G10	Rampage	G19	Socialization
G2	Rejected by people around	G11	Threatening	G20	Hard to tell
G3	Emotional	G12	Annoying other people	G21	Difficult to eat
G4	Things and understandings are denied	G13	Shutting him/herself	G22	Insomnia
G5	Relapsing	G14	Hurtinghim/herself	G23	Screaming
G6	Seizuring	G15	Hurting other people	G24	Don't want activity
G7	Wandering around	G16	Damaging goods	G25	Don't want to take a shower
G8	Angry	G17	Walking back and forth	G26	Miscommunication

Implementation of the system is divided into 3, namely: data processing, similarity calculation, and interface. The system is built using the Python programming language. In the data processing section, semi-manual case bases are formed from the raw data in the .xls format. The similarity calculation process is carried out by reading the .json data one by one with the name 'diagnosis' function, where it is repeated to read the old case on line 13. The code for reading the data, is shown in Fig. 4.

```
1   def diagnosa(gejala_baru, threshold=0, jsondir='./data'):
2       """
3       menghitung similarity gejala baru dengan
4       semua file gejala lama di direktori jsondir
5
6       :param gejala_baru: list - data gejala baru pasien
7       :param jsondir: string - direktori folder data json pasien
8       :return similaritylist: list - list dari tuple nama file dan
9                               hasil perhitungan similarity-nya dengan
10                              gejala_baru
11      """
12      hasil = []
13      for filename in os.listdir(jsondir) :
14          with open(os.path.join(jsondir,filename)) as jsonfile:
15              kasus = transform_json_object(json.load(jsonfile))
16              try:
17                  # [gejala] == [set(gejala), usia, sistol, diastol]
18                  gejala = kasus[1:5]
19                  data = [item for n, item in enumerate(kasus) if n not in range(1,5)]
20                  similarity = similarity_multi(gejala_baru, gejala)
21                  data.insert(0, similarity)
22                  if similarity > threshold:
23                      # [sim+data] == [similarity, id, [aksis], tatalaksana]
24                      hasil.append(data)
25              except :
26                  pass
27      hasil.sort(key=lambda x: x[0], reverse=True)
28      return hasil
```

Fig. 4. Code for data reading

The "diagnose" function also makes comparisons between new problems with cases on a case base to calculate the level of similarity and at the same time doing the sorting process of the similarity value, and then displayed.

The process of calculating the level of similarity is shown in Fig. 5 which starts from line 20 by comparing symptoms of new problems with symptoms on case-based. The process will call the 'similarity_multi' function. There are 2 types of calculation of the level of similarity, namely symptoms in the form of symbolic data and data having numerical value. For symptoms, the calculation on line 18 uses the Jackard method by utilizing the advantage of data set type in the Python programming language. The method is stated in Eq. 1.

$$Sim(X, Y) = \frac{|X \cap Y|}{|X \cup Y|} \tag{1}$$

The calculation of the level of similarity of numerical data, on the line 21–23 use a formula as stated in Eq. 2.

$$Sim(a, b) = 1 - \frac{|b - a|}{B - A} \quad a, b \in [A, B] \; dan \; A < B \tag{2}$$

with A and B are the lower and upper limits of the interval.

The symptom feature is a factor that has a higher influence to determine the level of similarity compared to other features (age, systole, diastole), so we can use this weight. Calculation of the similarity level of both numerical data and its weight is carried out on lines 26–29, by generating similarity level values in the range 0–1.

```
1    def similarity_multi(baru, lama):
2        """
3        Menghitung similarity dari kasus baru dan 1 kasus lama.
4        Masing masing kasus berupa list yang terdiri dari:
5        [set(gejala), usia, sistol, diastol]
6        Perhitungan similarity total menggunakan bobot dimana kemiripan gejala
7        lebih determinan daripada yang lain.
8
9        :param baru: list
10       :param lama: list
11       :return: float
12       """
13
14       bobot = [6, 1, 1, 1]
15       sim = []
16
17       # menghitung similarity antar gejala
18       sim.append(len(baru[0].intersection(baru[0]))/len(baru[0].union(lama[0])))
19
20       # menghitung similarity dari 2 angka,
21       sim.append(max(0, (1 - (abs(baru[1]-lama[1]) / (max_val[0]-min_val[0])))))  # usia
22       sim.append(max(0, (1 - (abs(baru[2]-lama[2]) / (max_val[1]-min_val[1])))))  # sistol
23       sim.append(max(0, (1 - (abs(baru[3]-lama[3]) / (max_val[2]-min_val[2])))))  # diastol
24
25       # perhitungan similarity berbobot
26       similarity = 0
27       for i, s in enumerate(sim):
28           similarity += bobot[i] * s
29       similarity /= sum(bobot)
30
31       return similarity
```

Fig. 5. Program code for calculating level of similarity

The development of interfaces on this system uses *a tkinter library* which is divided into 3 windows. The first window is the main window seen in Fig. 6 where input symptoms, age, systole and diastole from a case are written. Limitation of the similarity or *threshold* of the results displayed is also given here with the number 50% as the *default* number. If the 'diagnose' button is pressed, the similarity search process runs and will release the result window shown in Fig. 7. In this window automatically the

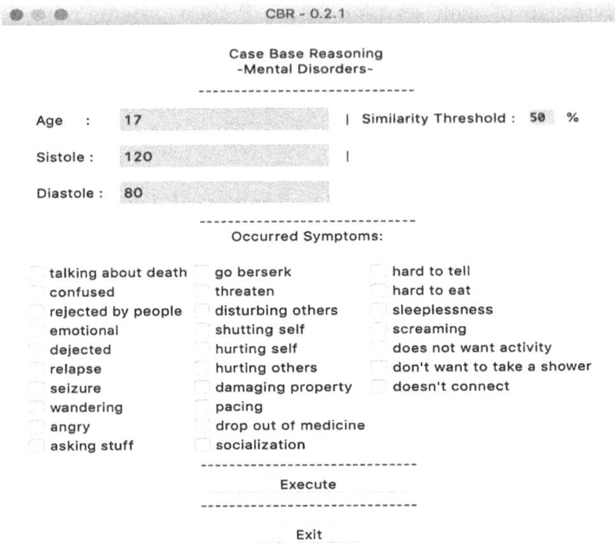

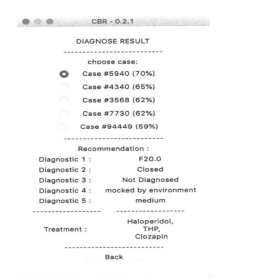

Fig. 6. Main window display

Fig. 7. Window display of process result

highest similarity cases are selected and indicated by axes and corresponding treatment. In this window there is also the ability to view other available case data that have similarities above the threshold. If, there are no cases that have a similarity more than the threshold, then the notification window will come out as in Fig. 8.

Fig. 8. Window display of not found result

4.1   Testing Result

The following is an example of implementation, for example there is a new problem, namely: A 41-year-old patient, with a blood pressure of 120/80. In the patient, the symptoms of anger and rampage appear. Input of the patient's condition is shown in Fig. 9.

```
●  ●  ●                          CBR - 0.2.1

                          Case Base Reasoning
                           -Mental Disorders-

                    --------------------------------

       Age    :    41              |  Similarity Threshold :  70   %

      Sistole :    120             |

      Diastole :   80

                    --------------------------------
                           Occurred Symptoms:

     talking about death ☑ go berserk          hard to tell
     confused              threaten            hard to eat
     rejected by people    disturbing others   sleeplessness
     emotional             shutting self       screaming
     dejected              hurting self        does not want activity
     relapse               hurting others      don't want to take a shower
     seizure               damaging property   doesn't connect
     wandering             pacing
   ☑ angry                 drop out of medicine
     asking stuff          socialization

                    --------------------------------
                              Execute
                    --------------------------------

                              Exit
```

Fig. 9. Window display of symptom input

After processing, the results of a similarity level of 99% are obtained with the case of 5940. Therefore the case can be recommended to provide solutions to the new problems with the axis and treatment as shown in Fig. 10.

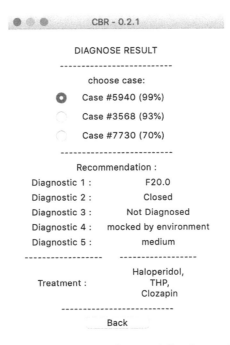

Fig. 10. Window display of types of disorders produced

5 Conclusion

Based on the results of experiments on case-based reasoning systems to help diagnose the types of schizophrenic disorders and disorders of mood and treatment, it can be summarized as follows:

1. A case-based reasoning system has been implemented to help diagnose the types of schizophrenic disorders and mood disorders and their management
2. The system has been able to select and determine cases that have the highest level of similarity above the specified threshold, where the solution to the case can be recommended as a solution to new problems.

References

1. Aamold, A., dan Plaza, E.: Case-based reasoning: foundation issues, methodological variation and system approach. AI Commun. **7**(1), 39–59 (1994)
2. Kolodner, J.: Reconstructive memory, a computer model. Cogn. Sci. **7**, 281–328 (1983)
3. Schank, R.C.: Dynamic Memory: A Theory of Learning in Computers and People. Cambridge University Press, Cambridge (1982)
4. Carbonel, J., Knoblock, C.A., Minton, S.: Prodigy: an integrated architecture for planning and learning. In: Van Lehn, K. (ed.) Architecture for Intelligence, the Twenty Second Carnegie Mellon Symposium on Cognition. Erlbaum Publ., Mahwah (1991)
5. Rissland, E.L.: Examples in legal reasoning: legal hypotheticals. In: Proceedings IJCAI 1983, Karlsruhe (1983)
6. Kolodner, J., Simpson, R.L., Sycara, K.: A process model of case-based reasoning in problem solving. In: IJCAI-1985 (1985)
7. Kolodner, J.: Capitalizing on failure through case-based inference. In: Proceedings Ninth Annual Conference of the Cognitive Science Society. Erlbaum, New Jersey (1987)
8. Sycara, E.P., Resolving adversarial conflicts: an approach to integrating case-based and analytic methods. Ph. D. thesis. Georgia Tech (1987)
9. Hammond, K.: CHEF: a model of case-based planning. In: Proceedings of AAAI-1986 (1986)
10. Hammond, K.: Explaining and repairing plans that fail. In: IJCAI-1987, pp. 109–114 (1987)
11. Collins, G.: Plan creation: using strategies as blueprints. Ph. D. thesis. Yale University (1987)
12. Ashley, K.D., Rissland, E.L.: Compare and contrast: a test experience. In: Proceeding AAAI 1987 (1987)
13. Koton, P.: Using experience in learning and problem solving. Ph. D. thesis. Computer Science Dept. MIT (1988)
14. Bareiss, R., Poter, W., Weir, C.C.: Protos: an exemplar-based learning apprentice. Int. J. Man-Mach. Stud. **29**, 549–561 (1989)
15. Chakraborty, B., Srinivas, I., Sood, P., Nabhi, V., Ghosh, D.: Case based reasoning methodology for diagnosis of swine flu. In: IEEE GCC Conference and Exhibition, 19–22 Februari 2011, Dubai, United Arab Emirates, pp. 132–135 (2011)
16. Pant, S., dan Joshi, S.R.: Case-based reasoning in neurological domain. IEEE 978-1-4673-2590-5/12 (2012)
17. Ghorai, K., Saha, S., Bakshi, A., Mahanti, A., Ray, P.: An mhealth recommender for smoking cessation using case based reasoning. In: 46th Hawaii International Conference on System Sciences, pp. 2695–2704 (2013)
18. Mulyana, S., Hartati, S., Wardoyo, R., Winarko, E.: Case-based reasoning with input text processing to diagnose mood [affective] disorders. Int. J. Adv. Res. Artif. Intell. (IJARAI) **4**(9), 1–5 (2015)
19. Wahyudi, E., dan Hartati, S.: Case-based reasoning for diagnosis of heart disease. Indonesian J. Comput. Cybern. Syst. (IJCCS) **11**(1), 1–10 (2017)

Study of Score Fusion and Quality Weighting in the Bio-Secure DS2 Database

Saliha Artabaz[1] and Layth Sliman[2(⊠)]

[1] Laboratoire de Méthodes de Conception de Systèmes LMCS,
Ecole Nationale Supérieure d'Informatique ESI, Oued-Smar, Alger, Algeria
s_artabaz@esi.dz
[2] Ecole d'Ingénieur Généraliste en Informatique et Technologies
du Numérique Efrei, 30–32 Avenue de la République, Villejuif, France
layth.sliman@efrei.fr

Abstract. A uni-biometric system suffers from unbalanced accuracy because of image quality, features extraction weakness, matching algorithm and limited degrees of freedom. This can be overcome by using multiple evidences of the same identity (Multi-biometrics fusion). In a previous work, we proposed new fusion functions based on arithmetic operators and search the best ones using Genetic Programming on the XM2VTS score database. The objective function is based on the Half Total Error Rate (HTER) (a threshold dependent metrics), from the Expected Performance Curve (EPC), of fused matching scores. In this paper, we select ten functions from the generated ones and apply them on matching scores of different biometric systems, which are provided by the bio-secure database. This database provide 24 streams that we use to generate 1000 multi-biometric combinations that we, then, use to conduct our comparative study. Since the result of fusion can be biased and requires a good quality assessment to evaluate the degree of reliability of a processed scheme, we use quality weights on the proposed functions and we compare the results with existing approaches. The proposed quality weights help to reduce the Equal Error Rate (EER a threshold-independent metric) since the obtained matching scores are results of different fusions of instances, sensors and evidences. The EER range is optimized along the tested functions. To confirm that our proposed functions give better score results than the existing functions based on arithmetic rules, we perform multiple statistical significance tests to check the reliability of our experimentation.

Keywords: Multi-biometrics · Fusion · Quality weights · Genetic Programming · Optimized search

1 Introduction

Multi-biometrics address several traditional biometric systems drawbacks. Mainly, they aim at reducing system errors. In fact, experiments show that combining different evidences enhance accuracy [1–6].

Multi-biometrics fusion considers different levels and evidences such as image, feature, score, decision and rank level. The most used level in the literature is the score

© Springer Nature Singapore Pte Ltd. 2019
M. W. Berry et al. (Eds.): SCDS 2019, CCIS 1100, pp. 252–261, 2019.
https://doi.org/10.1007/978-981-15-0399-3_20

level [16, 17]. At this level, combined scores are easier to fuse and provide rich information at the same time [4]. Furthermore, the ease of accessing and accuracy that out-perform other levels make it the best level [17]. At this level, fusion considers matching scores: the result of comparing between different evidences, instances, provided by different sensors and processed with different algorithms. Hence, to identify the best system, the evaluation must take into account all these parameters. In fact, studying different systems is necessary to conclude whether the used strategy for fusion improve baseline systems accuracy or not. In this paper, we propose a comparative study between fusion of several biometric systems. To do that, we generate different combination of biometric systems from a score database and apply our generated fusion functions from a previous work [7].

As the quality is one of the main factors affecting the overall performance of biometric systems, using quality measurement can lead the fusion process to reach better results. In this paper, we are interested in the score fusion with quality weighting. We use an optimized generation using GP in a previous work [7] to get several fusion functions. In addition, we aim to optimize the score distribution by integrating the template-query quality as weights. We perform our experiments on the Biosecure score database [13]. The proposed approach outperforms the baseline uni-biometric systems. We conduct a comparative study between the best-computed fusion functions. In addition, we give a complete view of different multi-biometric systems to test the most reliable ones according to the Equal Error Rate.

This paper is organized as follows: First, in Sect. 2, we introduce the studied field.

Section 3 illustrates the used weighted fusion functions and introduce database build on to conduct experiments. After that, we give experimental results in Sect. 4 with comparative study between proposed fusion functions and studied multi-biometric systems. Finally, we conclude and list some perspectives of our work.

2 Multi-biometrics and Score Level Fusion

Multi-biometrics is a merged field that addresses unimodal biometric system weaknesses. Researchers who studied different fusion methods to assess their effectiveness, mostly affirm that Multi-biometrics improve the accuracy of baseline systems. The fusion is needed to enhance baseline systems accuracy or to face non-universality of all used modalities. To optimize Multi-biometrics fusion, many challenges must be handled. This includes multiple data source incompatibility, matchers' scores normalization, as well as the noise that affects system performances and can result in false positive or negative authentication. Fusion at the score level is the most used fusion [1] due to its low fusion complexity that outperforms other levels. However, data is facing the same challenges cited before (data source incompatibility, matchers' scores normalization and noise). Consequently, the quality measurement [14] is becoming inherent in biometric systems as it allows to predict biometric system performances. We can define sample quality [14] as "scalar quantity that is monotonically related to the performance of biometric matchers". The effectiveness of a sample quality evaluation and the different ways to provide this scalar can be found in [14].

Many recent research works [8–12, 15] are interested in the sample quality to get a well-adjusted fusion function, these works use the computed scalars for each sample used as weights. Other works consider the quality as a measure of the performance of a biometric system. The provided scalar from data source is used to select an adaptive solution according to sensed biometric signals, which can vary for each authentication [14]. The challenge in that case is to reach the best performances, as the signal quality is variable. In other case, the simple way is to consider the signal quality as a scalar that quantify relative signal degradation under noise. So, this scalar may be used as weight in a fusion process or contribute as an indicator in image enhancement, which is supposed to improve accuracy. *Theofanos* [17] studies the impact of image reconstruction on false acceptance and proposes a signal quality measurement to optimize it without increasing the false acceptance. In addition, the essential issue is to deal with unsupervised environment and different constraints to provide adapted systems that match with security requirements [8]. Therefore, the system quality measure can be seen as is a degree of trust that provides the reliability of the system and ensures its interoperability with regard to processed data heterogeneity [15]. All cited methods use weighting to control and take a decision depending on the data quality. Other proposals discussed using system reliability indicator [14] to estimate matcher weights.

Based on a single system, we can neither prove nor deny the effectiveness of the proposed fusion or weighting impact on system accuracy. In fact, getting excellent accuracy with a fusion strategy on a specific combination does not imply the effectiveness of this strategy. The question that we can ask is: *'given a predefined fusion strategy, in which cases the fusion strategy works better than initial fused systems'*. As well, we can look for the best combination and fusion strategy to meet security requirements. In addition, the challenge is to prove that the used fusion strategy is robust even employed for different multi-biometric systems

3   Materials and Methods

Choosing fusion method in score level is very crucial to enhance provided system performances. The rule-based function is the most relevant to be used thanks to its simplicity. In this paper, we provide experimental results of different fusion functions that we compute with GP in a previous work [7]. In our comparative study, we give an analysis of different multi-biometric systems. We test different combinations of biometric scores and compare between them. Our functions are constructed using Genetic Programming 'GP' that is based on tree structure where each node represents an operation. Using GP, we explore the search space of different trees that represent the fusion functions by applying mutation and crossover operations. The two operations apply modifications on nodes of the tree. The crossover modifies nodes of a selected tree using nodes' contents of another one. The mutation modifies randomly the nodes using other operations.

We can see in Fig. 1 the enhancement of the average HTER with GP simulation using crossover to evolve the population and roulette selection to get the best list. The HTER average is reduced to a range of [0.5%]. The graph oscillations show the progress of the GP in Fig. 1(a), which does not converge systematically.

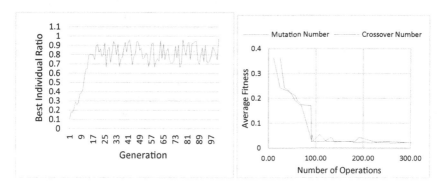

Fig. 1. Impact of mutation and crossover number on fitness average [7].

We select a number of trees obtained from the simulation. We use them to fuse different biometric systems of the Bio-secure database of scores. We use a user-specific weighting in order to improve baseline systems accuracy. The weight gives the sample fidelity to the claimed ID. We use the following formula to compute weight:

$$\left(\frac{1}{2} * \left((x_t - \bar{x})^2 + (x_q - \bar{x})^2\right)^{1/2}\right)^{-1} \tag{1}$$

where x_t is the template quality (claimed id) and x_q is the query quality (true id).

An insignificant value means that the two measurements are distinct and inversely. To normalize the quality, we use the standard deviation between max and min value computed from the development set.

4   Experiments and Discussion

We choose the Biosecure score database [13] as it is the only score database that offers a quality evaluation of its sets. We select a subset of functions that provide low HTER error in XM2VTS database. We take different functions along processed generations. Experiments are done according to these steps:

1. Generate 1000 different configurations to select 8 scores from 24 scores of the database. This allows testing different combinations of biometric systems in order to select the best ones. The scores must be filtered to get subset that contains sufficient number of data;
2. Test the baseline systems of these configurations;
3. Test each function for:
   (a) fusion of the baseline scores;
   (b) fusion using the weights scalars on normalized scores.
4. Compare between functions and analyze statistically the improvement of EER comparing to usual operators used for fusion.

Here is the list of functions used for our experimental study. These functions were selected according to their accuracy and number of fused scores.

Table 1. List of used functions for fusion.

Identifier	Function
Fct1	avg (S1, avg (min (min (S2, S3) + S4 * S5, S6), min (S7, S8)))
Fct2	S1 + avg (min (avg (S2 + S3, min (S4, S5)), S6), min (S7, S8))
Fct3	S1 + avg (max (avg (S2 + S3, avg (S4, S5)), S6), min (S7, S8))
Fct4	S1 + avg (min ((min (S2, S3)) + S4 + S5, S6], min(S7, S8))
Fct5	avg (max (max (S1, S2), (S3 + ((S4 * S5) - S6))), avg (S7, S8))
Fct6	max (avg (S1, S2), S3) + avg (S4, S5) + min (S6, S7) + S8
Fct7	avg (S1, S2) * S3 + avg (S4, S5) + min (S6, S7) + S8
Fct8	S1 + avg (max (avg (S2 + S3, S4 + S5), S6), min(S7, S8))
Fct9	avg (avg (min (S1, min (max (S2, S3), S4)), S5), min (avg (S6, S7), S8))
Fct10	avg (avg (S1, avg (S2, S3)), min (S4, S5) + S6 + S7 + S8);

We achieve these experiments using the Biosecure protocol with the selected functions cited in Table 1 above. This protocol uses datasets of two sessions with different impostors.

Figure 2 compares between some statistics of the tested functions upon the evaluation set. We rank functions according to the variance of the used statistics between the two sessions to evaluate quality of each function. Then, we use the sum of these ranks to get the best ones. We can observe that the best functions are respectively Function 3, 8, 5 and 2 whose average of ranks does not exceed rank 5. The function 3 applied gives small Standard Deviation on a limited range of EER values (max EER = 33%). To verify our results, we take, as an example, selected input scores. In Fig. 3, we see that function 2 is more relevant in this case since it gives the lowest variance with EER = 0.85% on session 2 of the dataset.

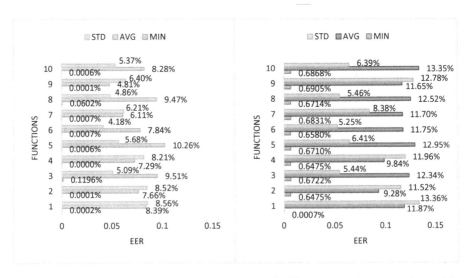

Fig. 2. MIN, AVG and STD equal error rate of the studied functions on the two sessions of the evaluation set.

For instance, in Fig. 3, which shows the multi-biometric system minimizing the EER, we can see that function 1 gives the lowest EER and the best functions EER is near the computed average since the standard deviation do not exceed 10%. The best combination in this case is function 1, 2 and 4 if we consider the EER and disparity between the two sessions of the evaluation set at the same time.

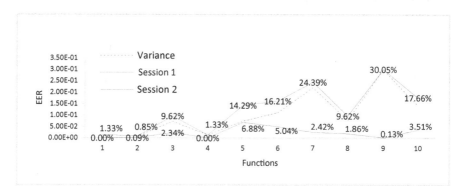

Fig. 3. Results of the multi-biometric system with the lowest EER on the two sessions of the evaluation set.

The example illustrated in Fig. 4 shows that score fusion in these cases allows to reduce errors caused by divergent scores. Indeed, functions 2, 5 and 8 give already good results and optimize Area Under Curve. For instance, the fused scores represent respectively: Face (CANON), Face (CANON with Flash), Iris, two fingerprints taken with different devices as described in the database. The other scores are filtered for dummies values included in the tested database.

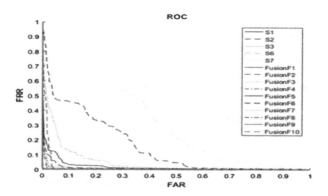

Fig. 4. ROC curves (EER = 0.8% for Function 2) of all used functions compared to a subset of fused scores and baseline systems.

To analyze the effectiveness of our results, we perform a significance test to compare between applying fusion of normalized data and one of the proposed weighted fusion function that give the best statistics. We perform a t-test on two samples of equal averages and unequal variances for the same sample size. The obtained probability (=0.027) is under the tail probability fixed to 0.05. The obtained t value 1.9607 is found to be less than the standard t value 2.2126. Therefore, statistically, we conclude that our approach is significantly better than the fusion applied with normalization.

In order to show our results, we present the whole combined systems in Fig. 5. The figure shows the ratio of multi-biometric systems depending on the number of functions that give an EER less than the referenced EER. The ratio of multi-biometric systems decreases gradually according to the number of functions with an EER under a pre-defined value. For example, five functions reduce the EER of 21% of multi-biometric systems below 5%. For 26% of the systems, we get a function that reduce the EER below 0.1% (only 3% of these systems get an insignificant error (less than 0.001%) for only one function).

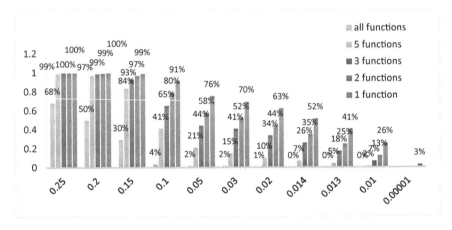

Fig. 5. Cumulative ratio of multi-biometric systems depending on the number of functions verifying the corresponding EER.

From the above analysis, we can observe that the improvement is guaranteed comparing to initial operators (Sum, Product, Min, Max) since the EER provided by our functions is less than the one achieved by the best operator for 88.6% of multi-biometric systems. To prove that at least one of the proposed functions outperforms the usual fusion operators, we must use t-test to verify whether the difference between the outputs is significant. To do so, as a first assessment, we use a paired test to analyze average difference in one direction without taking into consideration the variance. As a result, a t-value equal to 1.63 is obtained which is less than the critical t-value (equal to 1.64). Hence, we can conclude that the hypothesis of significant improvement can be assumed (i.e. P-value <0.1). Consequently, the proposed fusion is significantly better than usual fusion operators with 99% confidence interval. As a second assessment, we use a paired test to analyze average difference assuming that the two variances are different. The test is almost successful (P-value = 0.05 < 0.1 and t-value = 1.58 < 1.64).

In our study, we reach an EER under 0.013% and we get two multibiometric systems that optimize the EER to the range of [0.12%, 1.30%] from baseline scores with EER in the range of [1%, 99.55%] (see Fig. 6). This means that we reach a range improvement of 98%. As a result, we can conclude that our functions outperform the results obtained in [16] using RS-ADA, on the same database, for fusion that reaches an EER equal to 1.98%.

Table 2. Multi-biometric systems details.

	Face	Fingerprints with the same sensor for template and query	Fingerprints with different sensors for template and query
1	Webcam (low resolution) LDA-based face verifier	Thermal: right/left thumb, right index, Optical: left thumb, right index NIST fingerprint system	Left index, left middle finger NIST fingerprint system
2	Webcam (low resolution) LDA-based face verifier	Thermal: right thumb, right index, Optical: right/left thumb, right index	Left index, left middle finger

Table 2 gives details of the fused evidences in these resulting systems (see Fig. 6). The selected multi-biometric systems use face, multiple instances of fingerprint and multiple captures using different sensors. Furthermore, the fusion is done on scores comparing between fingerprint queries and templates taken with different sensors.

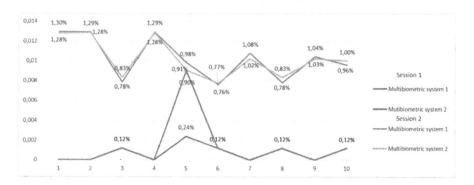

Fig. 6. Equal Error Rate for the best two multibiometric systems (EER less than 1.3% for all functions).

5 Conclusion

In this paper, we study some generated functions based on primitive fusion rule. These are the result of applying Genetic Programming proposed in a previous work to get the best rules combination using the XM2VTS. Afterward, we apply weighting on the generated functions and then we perform experiments on the Biosecure score database to compare between different combinations of the provided scores and find the best solution for fixed range of EER. The significance test confirm the improvement comparing to usual fusion operators. The provided functions can be tested with multi-algorithm biometric systems or using other databases in order expand the study and validate the results. We aim, later, to study fusion of features and classifiers using the proposed functions.

References

1. AlMahafzah, H., AlRawashdeh, M.Z.: Performance of multimodal biometric systems at score level fusion. In: Zeng, Q.-A. (ed.) Wireless Communications, Networking and Applications. LNEE, vol. 348, pp. 903–913. Springer, New Delhi (2016). https://doi.org/10.1007/978-81-322-2580-5_82
2. Kadam, A., Ghadi, M., Chavan, A., Jawale, P.: Multimodal biometric fusion. Int. J. Eng. Sci. Comput. 7(5), 12554–12558 (2017)
3. Anzar, S.M., Sathidevi, P.S.: Optimization of integration weights for a multibiometric system with score level fusion. In: Proceedings of the Second International Conference on Advances in Computing and Information Technology (ACITY) July 13–15, 2012, Chennai, India – vol. 2. Springer, Heidelberg, pp. 833–842 (2013). https://doi.org/10.1007/978-3-642-31552-7_85
4. Conti, V., Militello, C., Sorbello, F., Vitabile, S.: A frequency-based approach for features fusion in fingerprint and iris multimodal biometric identification systems. IEEE Trans. Syst. Man Cybern. Part C (Appl. Rev.) 40, 384–395 (2010)
5. Damer, N., Opel, A.: Multi-biometric score-level fusion and the integration of the neighbors distance ratio. In: Campilho, A., Kamel, M. (eds.) ICIAR 2014. LNCS, vol. 8815, pp. 85–93. Springer, Cham (2014). https://doi.org/10.1007/978-3-319-11755-3_10
6. Eskandari, M., Toygar, Ö.: Score level fusion for face-iris multimodal biometric system. In: 2013 Proceedings of the 28th International Symposium on Computer and Information Sciences, pp. 199–208. Springer, Cham. https://doi.org/10.1007/978-3-319-01604-7_20
7. Artabaz, S., Sliman, L., Benatchba, K., Dellys, H.N., Koudil, M.: Score level fusion scheme in hybrid multibiometric system. In: Badioze Zaman, H., et al. (eds.) IVIC 2015. LNCS, vol. 9429, pp. 166–177. Springer, Cham (2015). https://doi.org/10.1007/978-3-319-25939-0_15
8. Alonso-Fernandez, F., Fierrez, J., Bigun, J.: Quality Measures in Biometric Systems. Encyclopedia of Biometrics, pp. 1287–1297. Springer, US (2015). https://doi.org/10.1007/978-3-642-27733-7
9. Alonso-Fernandez, F., Fierrez, J., Ramos, D., Gonzalez-Rodriguez, J.: Quality-based conditional processing in multi-biometrics: application to sensor interoperability. IEEE Trans. Syst. Man Cybern. - Part A Syst. Hum. 40, 1168–1179 (2010)
10. Mohammed Anzar, S.T., Sathidevi, P.S.: On combining multi-normalization and ancillary measures for the optimal score level fusion of fingerprint and voice biometrics. EURASIP J. Adv. Signal Process. 1, 1–17 (2014)

11. Zhang, D., Lu, G., Zhang, L.: Finger-knuckle-print verification with score level adaptive binary fusion. Advanced Biometrics, pp. 151–174. Springer, Cham (2018). https://doi.org/10.1007/978-3-319-61545-5_8

12. Poh, N., Bourlai, T., Kittler, J.: A multimodal biometric test bed for quality-dependent, cost-sensitive and client-specific score-level fusion algorithms. Pattern Recogn. J. **43**(3), 1094–1105 (2009)

13. Grother, P., Tabassi, E.: Performance of biometric quality measures. IEEE Trans. Pattern Anal. Mach. Intell. **29**, 531–543 (2007)

14. Kabir, W., Ahmad, M.O., Swamy, M.N.S.: Score reliability based weighting technique for score-level fusion in multi-biometric systems. In: 2016 IEEE Winter Conference on Applications of Computer Vision (WACV), pp. 1–7 (2016)

15. Lip, C.C., Ramli, D.A.: Comparative study on feature, score and decision level fusion schemes for robust multibiometric systems. Frontiers in Computer Education. Springer, Heidelberg, pp. 941–948 (2012). https://doi.org/10.1007/978-3-642-27552-4_123

16. Lumini, A., Loris, N.: Overview of the combination of biometric matchers. Inf. Fusion **33**, 71–85 (2017). Elsevier

17. Theofanos, M.: Biometrics systematic uncertainty and the user. In: Proceeding of Biometrics: Theory Applications and Systems (BTAS 2007), IEEE, pp. 1–6 (2007)

Arabic Phonemes Recognition
Using Convolutional Neural Network

Irwan Mazlin, Zan Azma Nasruddin$^{(\boxtimes)}$, Wan Adilah Wan Adnan,
and Fariza Hanis Abdul Razak

Faculty of Computer and Mathematical Sciences, Universiti Teknologi MARA,
40450 Shah Alam, Selangor Darul Ehsan, Malaysia
irwanmazlin@gmail.com,
{zanaz,adilah,fariza}@tmsk.uitm.edu.my

Abstract. This paper focuses on a machine learning that learn the correct pronunciation Arabic phonemes. In this study, the researchers develop using convolutional neural network as feature extraction in order to enhance the performance of the model and Multi layer perceptron as the classifier to classify classes. Different parameters of CNN model are used in order to investigate the best parameter for the recognition purpose. The dataset have been recorded from experts using smartphone which consist of 880 recorded audios to train the model (210 for each class). The researchers have experimented the models to measure the accuracy and the cross entropy in the training process.

Keywords: Convolutional Neural Network · Arabic phonemes recognition · Signal processing · Speech recognition

1 Introduction

In this era of technology, many Muslims use technology as their alternative to seek the information about the religion especially in Islam. There are some applications that have been produced by the developer for users with the ability for them to read Quran by using a smartphone. In fact, most of the Muslims are able to read Quran but they do not notice they make mistakes when read the content of the Quran. They do not read it with correct pronunciation and sometimes they do not even know how to pronounce the phonemes [1]. There are some researchers suggest that people also need to recite the Quran with tajwid. Basically, tajwid is the correct pronunciation when pronounce every Arabic phonemes [2].

There are many applications that are able for user to read the content of the Quran but they do not able to get the feedback from the application in order for them to improve their pronunciation in which they are not able to do two-way communication when using the application [3]. Nowadays, many applications are able to control the device functionality by using voice only. Google developer has developed Google assistance that is able for the user to search something by using their voice only without put any word using the keyboard on the screen [4]. The signal processing can be used to analyze and modify the signal to improve the performance of the system. It is also

© Springer Nature Singapore Pte Ltd. 2019
M. W. Berry et al. (Eds.): SCDS 2019, CCIS 1100, pp. 262–271, 2019.
https://doi.org/10.1007/978-981-15-0399-3_21

can be assumed as the process of converting the signal from analog to digital before the system trained with the data provided by the researchers [5].

In this paper, the researchers are developing a system that is able to recognize the separation of Arabic phonemes. But, the system learns the data that contains basic pronunciation of it. The separation of the Arabic phonemes will be used as an input data. There are 29 of the separation of the words used in the Quran. The researchers using the Convolutional Neural Network for the feature extraction and the classification. This algorithm is a popular technique that used in an image processing. The Convolutional Neural Network also shows better result for the acoustic modelling in a small and large vocabulary task [6]. Related studies were done including in [7] and [8].

Based on this paper [9], the CNN does not have the limitation on its constraint which are the parameters and multiplies, the researcher has introduced this algorithm architecture only strides the filter in frequency in order the counter the limitation of the constraint issue. The researcher has proved that this algorithm is an effective algorithm for the signal processing without involving the multiple convolutional blocks. The detail about the architecture of convolutional neural network is explained in Sect. 2, and the elaboration of the evaluation of the model is in Sect. 3 using accuracy, precision and recall.

2 Convolutional Neural Network Architecture

Convolutional neural network is usually used in image recognition. However, it also can be used in speech recognition to recognize any phonemes that has been spoken by the user. The training phases are only occurred on the server side to train any data that have been collected from the expert. Hence, CNN can be applied in many machine learning domains such as image and speech recognition. The CNN's algorithm is only able to translate, rotate and scale due to its abilities high level semantics [4]. In this paper, the users will recognize the Arabic phonemes. CNN algorithms also are used to improve keyword spotting [10]. For example, google team have developed a device that is able to search anything in the search engine by using voice. In this paper, the researchers are producing Arabic phonemes recognition using this algorithm to produce a model. In fact, CNN are more useful compare with the DNN algorithm. This is because the DNN algorithm ignore input topology something that shows in any form of order without effecting the network and the model of translation are not designed in explicit designed [4]. Furthermore, it is proved that CNN algorithms are very suitable in processing the image. However, some of the researchers have made a shock that CNN algorithm also can be applied in signal form using spectrogram as an input. But it needs to translate the two dimensions, one of them is time (Conference and Processing, 2014). Figure 1 shows the illustrations of the Convolutional Network Architecture which consists of the Input Layer, Convolutional Layer, Rectified Linear Unit, Sub-Sampling and the Full Connected layer.

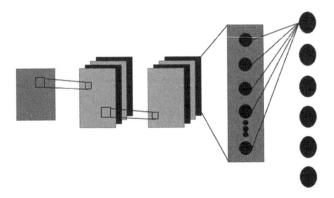

Fig. 1. Convolutional Neural Network architecture

There are three phases involved to train the model which are data processing, feature extraction and classification. The details are explained in Sects. 3, 4 and 5.

2.1 Data Processing

The spectrogram of the recorded audio needs to its attribute which are the duration and frequency. Originally, the spectrogram consists many values of frequency and the size of the spectrogram is not constant. So the attribute is changed to 16 kilohertz and 60 s the duration of each data. This is the step the researchers should do to prepare the input. After finish change the attribute of the samples, the sample is used as an input to the CNN model to extract the feature and train the sample to classify the classes. The Figs. 2 and 3 show the samples of spectrogram before and after editing the time and frequency of the sample. In this study, the size of spectrogram image is [98 × 40].

Fig. 2. Spectrogram before change its attribute

Fig. 3. Spectrogram after change its attribute

2.2 Feature Extraction

Typically, the system will receive the input signal W (t x f) where t and f are time and frequency. Then, the researchers declare the window of time to stride over the for feature extraction purpose. The size of the filter must be t $\leq$ m and f $\leq$ r. It is important the number of feature maps in the convolutional layer is equals to the number of filter used during training. After finish extract the feature in the convolutional layer, the researchers use the useful feature in the pooling layer to help to adapt the variety of the in the time-frequency and other characteristic of the recorded sound. The purpose of doing pooling the feature is to reduce the time-frequency space in the convolutional layer. Most of CNN algorithms work fine in the image recognition because they have the low convolutional layer and higher in full-connected layer for classification [11].

Convolution Layer
The convolutional layer is the layer that will do the linear operation which means do the Adding and Multiplication. The size of feature maps in convolutional layer depends on the filter that we used. In this study, in the convolutional layer, the researchers used 64 number of filters, the size of filter is [2 × 2] and the number pixel striding is [1 × 1]. The number of feature maps can be produced by using this formula $(m - q+1)$. $(n - r+1)v$ where w and v is the value of filter striding. Figure 4 is the sample of feature maps in convolutional layer. Figure 4 shows the sample of feature maps generated by the filter after stride the filter over the input image.

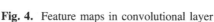

Fig. 4. Feature maps in convolutional layer

Rectified Linear Units (ReLU) Layer
In this step, the researchers use ReLU as the activation function during feature extraction. It will replace the negative pixel value into zero where it will reduce the white color in the feature maps. The main purpose of the ReLU is to present non-linearity in the Convolutional Network. After finish doing the ReLU function, it will produce Rectified feature maps. ReLU can be performed by using this formula *output* = max (0, *input*). In this study, the researchers apply this activation after

performing convolution of every layer. Figure 5 is the feature maps in rectified linear unit. Feature maps can be generated by striding the filter over the input image.

Fig. 5. Feature maps in rectified linear unit

Down-Sampling Layer

The down-sampling layer is the layer that is takes the highest pixel value between matrix. In this study, the researchers use 16 number of filters, the size of filter is [3 × 3] and the number pixel striding is [2 × 2]. The number of feature maps can be produced by using this formula (−+1). (−+1).. Where m and q is time in spectrogram and filter, n and r is frequency in spectrogram and filter, and ms is the pixel striding.

Full Connected Layer

Figure 6 shows the architecture of the Full Connected Layer. Basically, full connected layer uses a Multi-Layer Perceptron by using an activation function of softmax in the output layer.

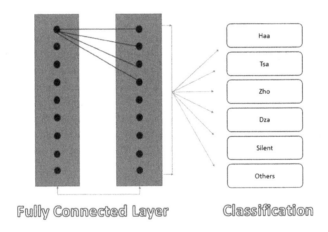

Fig. 6. Full Connected Layer

It can contain more than one layer in the full connected layer but in this study, the researchers use 2 hidden layers with 120 nodes in the first hidden layer and 100 nodes in the second hidden layer and it has 6 nodes of classes in the output layer which are Silence, other, Haa, Dza, Zaa, and Zho. It is important to combine both Convolutional and down sampling layer compared to take the pixel value from one of these layers separately. The nodes in the full connected layer will be connected to every neuron on the second layer. Softmax is summed up the output probabilities from the full connected layer and it must be equal to 1. For example, the model detects Haa class (0.85, 0.15, 0.00, 0.00, 0.00, 0.00), then, the summation of the other class should be 1.

3 Experiment Setup

The researchers have tested different parameter settings of model to train the model. Table 1 shows the parameter settings of the model. The size of the feature maps can be controlled by tuning the parameter of the model which are the depth and the filter striding. The depth is the number of filter use during striding the filter over the image. Stride is how many pixel of the filter will jump during striding the image. There are 80% data is used for training and 10% is used for testing and another 10% for validation. The model will to be tested with the parameter that has been set based on the Table 1. By using the right parameter, the highest accuracy could be produced.

Table 1. Parameter setting of the model

Parameter/ model	Stride 1st convolved	Stride 2nd convolved	Stride 1st down-convolved	Stride 2nd down-convolved	Depth 1	Depth 2
A	[1 × 1]	[2 × 2]	[2 × 2]	[3 × 3]	64	16
B	[1 × 1]	[2 × 2]	[2 × 2]	[3 × 3]	64	16
C	[2 × 2]	[3 × 3]	[4 × 4]	[5 × 5]	64	16
D	[2 × 2]	[3 × 3]	[3 × 3]	[4 × 4]	64	16

4 Evaluation and Discussion

Figures 7, 8, 9, 10, 11, 12, 13 and 14 show line graphs produced by training the model A, B, C, and D using CNN architecture based on the parameter settings that have been explained from previous section. There are two types of lines graph. The first one is the red line graph represents the cross entropy of the model when train the data. The lower of the red line graph, the lower the error rates that made by the model. For the second one is the blue line graph. It is called the validation of the training. Same goes to the accuracy line graph which is the accuracy of the training prediction.

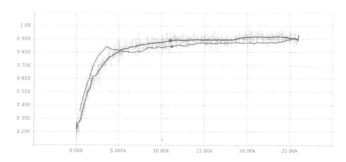

Fig. 7. Accuracy model A (Color figure online)

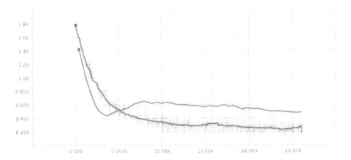

Fig. 8. Cross entropy model A (Color figure online)

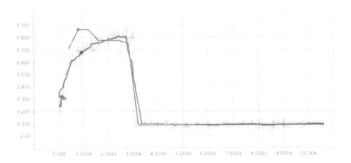

Fig. 9. Accuracy model B (Color figure online)

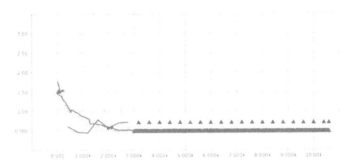

Fig. 10. Cross entropy model B (Color figure online)

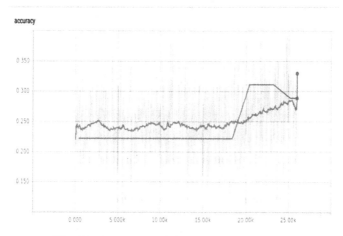

Fig. 11. Accuracy model C (Color figure online)

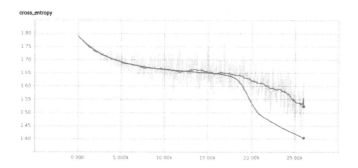

Fig. 12. Cross entropy model C (Color figure online)

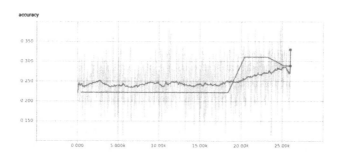

Fig. 13. Accuracy model D (Color figure online)

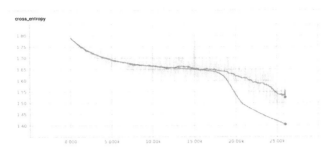

Fig. 14. Cross entropy model D (Color figure online)

It can be made a hypothesis of the higher the red line graph, the higher the accuracy of the model can be predicted. If this red line graph goes downward, it means the model cannot be used because the model is unable to recognize the Arabic phonemes. For the second line graph represents the validation of the model during training process. It same goes the previous picture which is the validation of the model. Models A shows the highest accuracy with 94.5% and lowest entropy with probability of 0.38. The model A produce the highest accuracy and the lowest probability of cross entropy compared to the others model. This is because this model might cause overfitting or did not suitable with the parameter for training based on the data. The model A is suitable used for recognition purpose. Table 2 summarized the accuracy of each model.

Table 2. Accuracy of the model

Parameter/model	A	B	C	D
Learning rate	0.001	0.005	0.001	0.001
Accuracy	94.5%	8.7%	34.7%	39.2%

5   Conclusion and Future Works

In this paper, the study presented the new model for Arabic phonemes recognition by using parameter setting for training. The model A shows the highest accuracy and lowest probability of cross entropy compared to others. However, the model A takes long time to train the model because the usage of learning rate is small. The smaller the number of striding and learning rate, the longest of the duration of the training cycle. It is because the value of striding the filter and the value of learning rate of the model A is smaller compared to others model. The total duration training for the model is around 7 to 8 h. Model B takes 4 h to train the model, model C and D takes 2 h to train the model. The development of the application in this study have been shared in GitHub: (IRWANMARYAM, 2018). The researchers are using keras framework as a tool to develop the model. This research still requires some improvement in order to reduce time taken for training and produce the highest accuracy.

References

1. Wahidah, A., Suriazalmi, M., Niza, M.: Makhraj recognition using speech processing. In: 7th International Conference on Computing and Convergence Technology (ICCCT) 2012, pp. 689–693 (2012)
2. Leaman, O.: The Qur'an: An Encyclopedia. Routledge, Abingdon (2006). T. Qur, T. Qur, A. Ency-
3. Arshad, N.W., Aziz, S.N.A., Hamid, R., Karim, R.A., Naim, F., Zakaria, N.F.: Speech processing for makhraj recognition. In: International Conference on Electrical, Control and Computer Engineering 2011, pp. 323–327 (2011)
4. Sainath, T.N., Parada, C.: Convolutional neural networks for small-footprint keyword spotting. In: 16th Annual Conference of the International Speech Communication Association. INTERSPEECH 2015, pp. 1478–1482 (2015)
5. Wahidah, A., et al.: Makhraj recognition for Al-Quran recitation using MFCC. Int. J. Intell. Inf. Process. 4(2), 45–53 (2013)
6. Abdel-hamid, O., Jiang, H., Penn, G.: Applying convolutional neural networks concepts to hybrid NN-HMM model for speech recognition. In: Department of Computer Science and Engineering, York University, Toronto, Canada, pp. 4277–4280 (2012)
7. Khahriri, F.A, Ibrahim, Z., Rashidin, R., Ismail, N., Ahmad, A.: Malay dialect translator for android. In: Language Invention, Innovation and Design Exposition (LIID2017), UiTM Shah Alam (2017)
8. Mazlin, I., Nasruddin, Z.A., Hamzah, P., Abdul Aziz, M.: Musafir application development using mobile application development life cycle. In: 3rd International Conference on Innovation in Computer Science and Engineering 2019 (iCiCSE) (2019)
9. Tóth, L.: Combining time-and frequency-domain convolution in convolutional neural network-based phone recognition. In: 2014 IEEE International Conference on Acoustics, Speech and Signal Processing (ICASSP), pp. 190–194. IEEE (2015)
10. Chen, G., Parada, C., Heigold, G.: Small-footprint keyword spotting using deep neural networks. In: Acoustics, Speech and Signal Processing, no. i, pp. 1–5, 2014
11. Chao, Q., Xiao-Guang, G., Da-Qing, C.: On distributed deep network for processing large-scale sets of complex data. In: 2016 8th International Conference on Intelligent Human-Machine Systems and Cybernetics, pp. 395–399 (2016)

Forecasting the Search Trend of Muslim Clothing in Indonesia on Google Trends Data Using ARIMAX and Neural Network

Novri Suhermi[1,2], Suhartono[1(✉)], Regita Putri Permata[1],
and Santi Puteri Rahayu[1]

[1] Department of Statistics, Institut Teknologi Sepuluh Nopember,
Surabaya 60119, Indonesia
suhartono@statistika.its.ac.id
[2] Department of Mathematics and Statistics, Lancaster University,
Lancaster LA1 4YF, UK

Abstract. The trend of muslim fashion has significantly raised the search trend for the brands of hijab and sarong in Indonesia. The aim of this study is to forecast the search trend for hijab and sarong based on google trends data. The Hijab brands include Rabbani, Zoya, Dian Pelangi, Elzatta, and Shafira, while the sarong brands include Gajah Duduk, Wadimor, Atlas, Mango, and Sapphire. We apply several forecasting methods such as Holt-Winters' Exponential Smoothing, ARIMA, ARIMAX, FFNN and ERNN. The data contains calendar variation effect due to the Eid al-Fitr days use different calendar system. The results show that FFNN yields the most accurate forecast on 6 out of 10 brands. The forecast results for year 2019 period show that the search trend for Atlas brand is predicted to be the highest of all sarong brands. On the contrary, all the hijab brands' trend search will decrease in this period.

Keywords: ARIMA · ARIMAX · ERNN · FFNN · Forecasting · Google trends

1 Introduction

Indonesia is the largest Muslim-majority country in the world with 87% of Indonesian population or approximately 258 million people identifying themselves as Muslim [1]. It gives a big opportunity for business on Muslim clothes. With a large market opportunity, many Muslim fashion companies have developed their businesses specifically in the field of hijab and sarong production. This is inseparable from the role of designers and entrepreneurs as producers who make hijab (headscarves and veils) as industrial commodities. Several examples of hijab brands that are well-known in Indonesia are Elzatta, Shafira, Dian Pelangi, Zoya, and Rabbani. The growth of Muslim fashion in Indonesia not only happens on hijab products, but also sarongs that are famous among every male muslim where they are most likely to wear sarong for doing prayer. Several examples of famous sarong brands are Atlas, Wadimor, Mangga, Gajah Duduk, and Shappire.

© Springer Nature Singapore Pte Ltd. 2019
M. W. Berry et al. (Eds.): SCDS 2019, CCIS 1100, pp. 272–286, 2019.
https://doi.org/10.1007/978-981-15-0399-3_22

The search terms in Google are recorded and collected in a website called is Google trends [2]. Google Trends was launched in 2006 and has been used as business research to find out the search patterns of specific terms on Google. The search volume on Google Trends reached 2 Trillion in 2016 [3]. Application of Google Trends data has been carried out by a number of researchers, one of them used Google Trends data to forecast current economic activity, which resulted in a promising results where Google Trends data has closely similar patterns to the sales pattern in several sectors such as retail sales data, automotive goods sales data, house sales data, and tourism data [4]. A study on the calendar variation effect in forecasting the sales of Muslim clothes in Indonesia which is influenced by Eid al-Fitr days compared Time Series Regression-ARIMAX model, SARIMA, and Neural Network. The results showed that Time Series Regression-ARIMAX model which included the calendar variation effect out performed forecast among the other models [5].

In this study, we apply several statistical and machine learning methods to forecast the consumer interest on hijab and sarong based on Google Trends data which contained calendar variation effect where hijab and sarong sales significantly increase in Ramadan month. There are 5 methods that will be compared, namely the Holt-Winters' Exponential Smoothing model, the ARIMA model, the ARIMAX model, the Feed Forward Neural Network (FFNN) model, and the Elman-Recurrent Neural Network (ERNN) model. ERNN is a development method of Neural Network commonly referred to as an Autoregressive Moving Average Neural Network (ARMA-NN) where it also involves lag of error variables obtained from the difference of actual values and predicted values [6]. The purpose of this study is to obtain the best model for forecasting search trend for hijab and sarong brands based on Google Trends data. We use out-of-sample criteria for forecast evaluation.

2 Time Series Models for Forecasting

Time series analysis aims to find the pattern of historical time series data and extrapolate the pattern in the past which is used to forecast the future. Several time series models are given in the following.

2.1 Exponential Smoothing

Holt-Winters' Exponential Smoothing (HWES) is a forecasting method which is specialized in data with seasonal pattern [7]. Exponential Smoothing is a procedure that repeats calculations continuously using the latest data based on the exponential calculation of the average smoothing of past data, which is given as follows [8].

$$\hat{Y}_{t+h} = (L_t + hT_t)S_{t-s+h}, \tag{1}$$

where L_t denotes the smoothing value, Y_t is the actual value at. time t, S_t is the seasonal components, h is the forecast horizon, and T_t is the trend component.

2.2   ARIMA

ARIMA (p, d, q) is a nonstationary time series model derived from ARMA (p, q) such that it requires d times differencing to be stationary. The general form of ARIMA (p, d, q) is given in the following [9].

$$\phi_p(B)(1 - B)^d Y_t = \theta_0 + \theta_q(B)a_t. \tag{2}$$

Multiplicative ARIMA model with s seasonal period is denoted by ARIMA $(p, d, q)(P, D, Q)^s$. The general form of seasonal multiplicative ARIMA model is given as follows.

$$\Phi_P(B^s)\phi_p(B)(1 - B)^d(1 - B^s)^D \dot{Y}_t = \theta_q(B)\Theta_Q(B^s)a_t, \tag{3}$$

where $\phi_p(B)$ denotes AR operator, $\Phi_P(B^s)$ denotes seasonal AR operator, $\theta_q(B)$ denotes MA operator, $\Theta_Q(B^s)$ denotes seasonal MA operator, and a_t white noise residual with zero mean and constant variance σ_a^2.

2.3   ARIMAX

ARIMAX model is a development of ARIMA model which includes exogeneous variable. In this study, we add dummy variables including trend, monthly seasonal, and calendar variation effects in the model [10]. The general form of ARIMAX model is shown in the following.

$$Y_t = \delta_0 + \delta_1 V_{1,t} + \ldots + \delta_g V_{g,t} + \frac{\theta_q(B)\Theta_Q(B^S)}{\phi_p(B)\Phi_P(B^S)(1 - B)^d(1 - B^S)^D} a_t, \tag{4}$$

where $V_{g,t}$ is the g-th dummy and δ is the coefficient of dummy variable.

2.4   Nonlinearity Test

Nonlinearity test is conducted in order to test if the data contains nonlinear pattern. White test is one of nonlinearity test which is developed based from Neural Network model. White test uses F statistic to test the significance of nonlinear components which are given as follows [11].

$$Y_t = \delta_0 + \delta_1 Y_{t-1} + \ldots + \delta_p Y_{t-p} + \lambda f(Y_{t-1}, Y_{t-2}, \ldots, Y_{t-p}) + a_t, \tag{5}$$

where λ denotes the nonlinear parameter.

2.5   Feed Forward Neural Network

Feed Forward Neural Network (FFNN) is one of popular nonlinear model which is widely used for time series forecasting. FFNN consists of input layer, hidden layer, and output layer. Each layer contains elements called neurons. Each neuron will receive information only from the neurons in the previous layer [12]. FFNN model for univariate time series data with p inputs, q hidden neurons, and 1 single output, denoted by FFNN(p, q), can be expressed as follows.

$$\hat{Y}_{(t)} = f^o \left[\sum_{j=1}^{q} \left[w_j^o f^h \left(\sum_{i=1}^{p} w_{ji}^h X_{i(t)} + b_j^h \right) + b^o \right] \right], \tag{6}$$

where w_{ji}^h are the weights that connect input layer to hidden layer, w_j^o are the weights that connect hidden layer to output layer. $f(.)$ is called the activation function. b^o and b_j^h are the biases. $X_{i(t)}$ are the input values and $\hat{Y}_{(t)}$ are the predicted output values. In this study, we use tangent hyperbolic (tanh) function as the activation function, which is given as follows.

$$f(x) = \frac{\sinh(x)}{\cosh(x)} = \frac{e^x - e^{-x}}{e^x + e^{-x}} \tag{7}$$

2.6   Elman Recurrent Neural Network

Elman Recurrent Neural Network (ERNN) is one type of neural network where the neurons connect back to other neurons and the information flow is multi-directional such that the activation of neurons can flow around in a loop [13]. This type of neural network model has a sense of time and memory of previous network states which enables it to learn sequences which vary over time. In the context of time series data, the general form of ERNN model with p inputs, q hidden neurons, and 1 single output, denoted by ERNN(p, q), is given as follows.

$$\hat{Y}_{(t)} = f^o \left[\sum_{j=1}^{q} w_j^o a_{j(t)} + b^o \right], \tag{8}$$

where $a_{j(t)} = f^h \left[\sum_{i=1}^{p} \left(w_{ji}^h X_{i(t)} + w_{ji}^c a_{i(t-1)} \right) + b_j \right]$ for t = 2, 3, ... We initialize the value of $a_{j(t)} = f^h \left[\sum_{i=1}^{p} \left(w_{ji}^h X_{i(t)} \right) + b_j \right]$ for t = 1. w_{ji}^c are the weights that connect context layer to hidden layer. The context layer is what differs the ERNN to FFNN. The remaining notations are equivalent to FFNN model. The architectures of FFNN and ERNN are shown in Fig. 1.

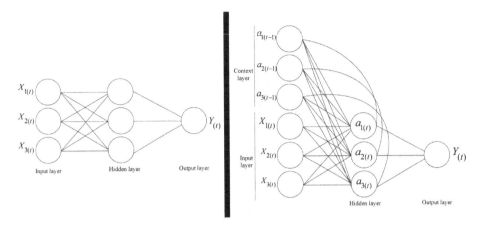

Fig. 1. (Example) the neural architecture of FFNN (left) and ERNN (right)

2.7   Model Selection

We use Root Mean Square Error of Prediction (RMSEP) and Symmetric Mean Absolute Percentage Error of Prediction(sMAPEP) as the criteria for model selection. The formula of RMSEP and sMAPEP are given respectively as follows [9]:

$$RMSEP = \sqrt{\frac{\sum_{l=1}^{L}(Y_{n+l} - \hat{Y}_n(l))^2}{L}}, \tag{9}$$

$$sMAPEP = \frac{1}{L}\sum_{i=1}^{L}\frac{|Y_{n+l} - \hat{Y}_n(l)|}{(|Y_{n+l}| + |\hat{Y}_n(l)|)/2}, \tag{10}$$

where Y_{n+l} denotes the actual value, $\hat{Y}_n(l)$ denotes the forecast value, and L denotes forecast horizon.

3   Dataset and Methods

3.1   Dataset

In this study, we use Google Trends data that is obtained from www.trends.google.com. The dataset contains monthly trend search of five hijab brand and five sarong brands from period January 2011 to September 2018. We split the data into training and testing sets, where the training set contains data from January 2011 to December 2016, and the rest data becomes testing set. List of variables used in this study is given in Table 1.

Table 1. List of variables

Hijab search trend (%)		Sarong search trend (%)	
Variable	Search keyword	Variable	Search keyword
$Y_{1,t}$	"hijab rabbani"	$Y_{6,t}$	"sarung gajah duduk"
$Y_{2,t}$	"hijab dian pelangi"	$Y_{7,t}$	"sarung wadimor"
$Y_{3,t}$	"hijab zoya"	$Y_{8,t}$	"sarung atlas"
$Y_{4,t}$	"hijab elzatta"	$Y_{9,t}$	"sarung mangga"
$Y_{5,t}$	"hijab shafira"	$Y_{10,t}$	"sarung sapphire"

* sarung is Indonesian word for sarong

We use several dummy variables which are given as follows.

1. Trend dummy variable with $t = 1, 2, \ldots, n$
2. Monthly seasonal dummy variables $M_1, M_2, M_3, .., M_{12}$ which represent the months from January to December, respectively.
3. $V_{i,t} = 1$ for Eid al-Fitr days occurred on i-th week and t-th month where $i = 1, 2, 3, 4$ and $t = 1, 2, 3, \ldots, 12$.
4. $V_{i,t-1} = 1$ for Eid al-Fitr days occurred before i-th week and on t-th month where $i = 1, 2, 3, 4$ and $t = 1, 2, 3, \ldots, 12$.

3.2 Methods

Based on the models we present in the previous section, the analysis of modeling in this study is given in the following.

1. Descriptive analysis to identify the characteristics of search pattern for hijab product and sarong product on Google.
2. Perform modeling and forecasting
   a. Modeling the training set and Forecasting the testing set using Exponential-Smoothing.
   b. Modeling the training set and Forecasting the testing set using ARIMA and ARIMAX models based on Box-Jenkins procedure.
   c. Modeling the training set and Forecasting the testing set using FFNN and ERNN with input variables based on the lag variables as well as the trend, seasonal, and calendar variation components from ARIMAX model. The data is then normalized.
   d. Comparing the forecast accuracy of Exponential Smoothing, ARIMA, ARIMAX, FFNN, dan ERNN using RMSEP, dan sMAPE to find the best model.
3. Forecasting the search trend of hijab and sarong products for 15 months ahead by using the best model chosen.

4   Results

4.1   Descriptive Statistics

Firstly, we do descriptive analysis for hijab brands as shown in Table 2. It is shown that brand Dian Pelangi in average has the highest score of search trend, compared to other brands.

Table 2. Descriptive statistics for search trend of hijab product

Variable	Mean	St.Dev	MIN	MAX
$Y_{1,t}$	3.699	2.746	0	13
$Y_{2,t}$	9.600	11.160	0	62
$Y_{3,t}$	19.540	18.970	1	100
$Y_{4,t}$	6.892	7.861	0	55
$Y_{5,t}$	0.731	0.809	0	6

Table 2 also shows that brand Dian Pelangi is most favorite brand based on Google search with score 100 on July 2014, which was Eid al-Fitr. The second most favorite brand is Zoya with the highest score of 62%. A survey from Marketing and Frontier Consulting Group, top brand award organizer stated that the top 3 most favorite brand in year 2016–2018 are Zoya, Rabbani and Elzatta, which are shown in Table 3.

Table 3. Top Brand Award for Hijab product

Brand	TBI 2016	TBI 2017	TBI 2018
Zoya	44.80%	39.40%	24.90%
Rabbani	21.30%	26.80%	24.50%
Elzatta	12.60%	13.00%	19.80%

Hijab Dian Pelangi and Hijab Zoya are the products that are most often searched for by consumers but because of the brand loyalty which is a measure of customer relationship to a brand related to price and quality [14]. Descriptive analysis of sarong brands is shown in Table 4.

Table 4. Descriptive statistics for search trend of sarong product

Variable	Mean	St.Dev	Min.	Max.
$Y_{6,t}$	5.860	6.507	0	35
$Y_{7,t}$	8.590	15.620	0	100
$Y_{8,t}$	6.226	8.091	0	36
$Y_{9,t}$	2.409	4.049	0	25
$Y_{10,t}$	1.011	1.426	0	8

Table 4 shows that the most favorite brand is Wadimor. The second most favorite brand is Atlas. Meanwhile, the least favorite brand is Sapphire. The score of Sapphire brand is very small compared to other brands. The time series plots of search trends of each brand are given in Figs. 2 and 3.

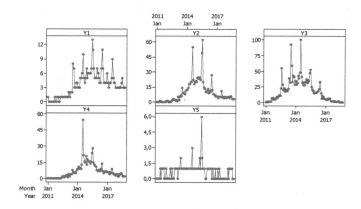

Fig. 2. Time series plots of search trend for hijab brand

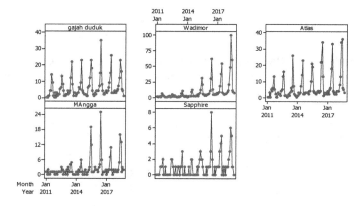

Fig. 3. Time series plots of search trend for sarong brand

Figures 2 and 3 show that the search for interest in hijab and high sarong brands in certain months shows the effect of the effects of calendar variations on the month of Eid al-Fitr. Eid Al-Fitr 2011,2012 and 2013 occured in August. Eid al-Fitr 2014 and 2016 occured in July. Eid al-Fitr 2017 and 2018 occured in June. With the different dates on each Eid al-Fitr, the descriptive analysis of the data is divided into three time intervals on the Eid al-Fitr, i.e. the beginning of the month from 1 to 10, the middle of the month 11 to 20, and the end months 21 to 31. The pattern has a downward trend in interest in Hijab search because of consumer loyalty and competition in marketing or there are new models of hijab brands that are more in demand by the middle and lower classes.

4.2   Forecast Results

Exponential Smoothing. Forecasting using Holt-Winter's Exponential Smoothing (HWES) is a forecast by taking into account the seasonal occurrence of each data. HWES modeling using level, trend, and seasonal parameters is presented in Table 5. Table 5 shows that the sMAPEP of search trend forecast for rabbani brand is 0.214 ($Y_{1,t}$), the lowest among the other brands. Meanwhile the least accurate forecast based on sMAPEP is for atlas brand ($Y_{8,t}$).

Table 5. Estimated parameters of HWES Model

Data	Level	Trend	Seasonal	RMSEP	sMAPEP
$Y_{1,t}$	0.255	0.0220	0.0001	0.870	0.214
$Y_{2,t}$	0.593	0.0185	0.0001	2.192	0.302
$Y_{3,t}$	0.425	0.0150	0.2970	1.505	1.237
$Y_{4,t}$	0.358	0.0001	0.0001	5.224	0.628
$Y_{5,t}$	0.139	0.0113	0.0001	0.920	1.932
$Y_{6,t}$	0.007	0.0070	0.0002	8.038	0.619
$Y_{7,t}$	0.036	0.0001	0.9630	22.732	0.370
$Y_{8,t}$	0.012	0.0001	0.0001	11.134	4.811
$Y_{9,t}$	0.066	0.0001	0.0002	5.003	0.587
$Y_{10,t}$	0.004	0.0001	0.0001	1.551	0.682

ARIMA Modeling. ARIMA modeling using the Box-Jenkins procedure consists of model identification, parameter estimation, diagnostic checking, and forecasting data. ARIMA models along with forecast accuracy are shown in Table 6.

Table 6. Forecast evaluation of ARIMA

Data	ARIMA model	RMSEP	sMAPEP
$Y_{1,t}$	(0, 1, [2, 12])	0.5483	0.1000
$Y_{2,t}$	([1, 12], 1, 0)	1.1380	0.2310
$Y_{3,t}$	(0, 1[11, 12, 13])	1.0933	1.4133
$Y_{4,t}$	(0, 1, 1)	1.4760	0.3940
$Y_{5,t}$	([1, 2, 9],1, 0)	0.8220	1.9620
$Y_{6,t}$	$(1, 1, 0)^{12}$	7.0092	0.3210
$Y_{7,t}$	$([1, 11], 0, 0)(0, 1, 0)^{12}$	23.4900	0.3540
$Y_{8,t}$	$([21], 0, 0)(0, 1, 0)^{12}$	3.2590	0.3930
$Y_{9,t}$	$([3, 9], 0, 0)(1, 1, 0)^{12}$	3.5180	0.6220
$Y_{10,t}$	$(0, 1, 1)^{12}$	2.7750	1.3845

From the modeling results in Table 6, the ARIMA forecast is more accurate than the HWES forecast, with 6 out of 10 models. The atlas brand forecast accuracy significantly increases from 4.811 to 0.393, based on sMAPEP score. It implies that ARIMA model, which is more complex than HWES model, is able to significantly improve the forecast accuracy.

ARIMAX Modeling. Modeling the search trend data on Hijab and sarong with ARIMAX is done by modeling data using time series regression with trend dummy variables, monthly dummy variables and calendar variation dummy variables. This analysis step is carried out for all products. Table 7 shows the best ARIMAX model obtained for each brand along with the accuracy of testing sets.

FFNN Modeling. In this study, we use single hidden layer and we experiment the number of hidden neuron from 1 to 5 units. A single hidden layer neural network is flexible and capable to converge way faster for linear target functions than multiple hidden layer networks [15]. The network input is chosen based on the lag variables in the ARIMAX model as well as the trend dummy, monthly dummy and calendar variation dummy [16, 17]. FFNN model on search trend for Gajah Duduk brand uses the lag variable Y_{t-13}, which is based on ARIMAX model order. Optimum hidden neurons chosen are 2 units with 10 times replication. The results for Gajah Duduk brand is shown in Fig. 4.

Table 7. Forecast evaluation of ARIMAX

Data	ARIMA model	RMSEP	sMAPEP
$Y_{1,t}$	([11, 17], 0, 0)	1.901	0.2950
$Y_{2,t}$	([1, 2, 11, 23], 1, 0)	22.843	1.2590
$Y_{3,t}$	([1, 35], 0, 0)	22.340	1.1360
$Y_{4,t}$	([24], 0, 0)	5.382	1.0430
$Y_{5,t}$	([12, 48], 0, 0)	1.028	1.6910
$Y_{6,t}$	([13], 0, 0)	14.736	0.7450
$Y_{7,t}$	([1, 8], 1, 0)	30.130	0.8870
$Y_{8,t}$	([13, 23], 0, 0)	6.832	0.4211
$Y_{9,t}$	([1, 10, 12], 0, 0)	2.705	0.7350
$Y_{10,t}$	([3, 1, 0])	0.718	0.4508

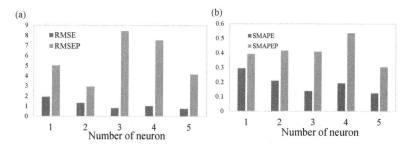

Fig. 4. Forecast accuracy comparison for Gajah Duduk brand with different number of hidden neuron based on (a) RMSE and RMSEP, (b) sMAPE and sMAPEP

Figure 4 shows that the more number of neurons used to form the FFNN architecture, the values of RMSE and SMAPE, which are the values of accuracy in training data, decrease as the number of neurons in the hidden layer increases as shown in blue bars. However, different values of RMSEP and sMAPEP indicate that the up and down values along with the increase in the number of hidden neurons. It means that the optimum selection of hidden neurons is needed in order to obtain prediction with a minimum error value. The forecast results of testing set is shown in Table 8.

ERNN Modeling. ERNN modeling uses input from the ARIMAX model in the form of lag and dummy calendar variations. On brand Gajah Duduk, we also use the input Y_{t-13}, based on the ARIMAX model. The ERNN optimum neuron obtained is 4 units. Hence, the best ERNN model is ERNN (9, 4).

The values of the RMSE and sMAPE criteria indicate that every addition of the hidden neuron does not always improve forecast accuracy. Then, it is necessary to experiment the hidden neurons in order to obtain the optimum hidden neurons. The results of ERNN forecast for each brand is shown in Table 9.

Table 8. Forecast evaluation of FFNN

Variable	Input	Number of neuron	RMSEP	sMAPEP
$Y_{1,t}$	$Y_{1,t-11}$, $Y_{1,t-17}$	4	0.6610	0.157
$Y_{2,t}$	$Y_{2,t-1}$, $Y_{2,t-2}$, $Y_{2,t-11}$, $Y_{2,t-12}$, $Y_{2,t-23}$, $Y_{2,t-24}$	4	1.1810	0.245
$Y_{3,t}$	$Y_{3,t-1}$, $Y_{3,t-35}$	3	0.4670	0.247
$Y_{4,t}$	$Y_{4,t-24}$	4	2.0150	0.512
$Y_{5,t}$	$Y_{5,t-12}$	3	0.5100	1.076
$Y_{6,t}$	$Y_{6,t-13}$	2	2.9430	0.416
$Y_{7,t}$	$Y_{7,t-1}$, $Y_{7,t-2}$, $Y_{7,t-8}$, $Y_{7,t-9}$	2	14.8690	0.258
$Y_{8,t}$	$Y_{8,t-13}$, $Y_{7,t-23}$	4	1.5980	0.198
$Y_{9,t}$	$Y_{9,t-1}$, $Y_{9,t-10}$, $Y_{9,t-12}$	3	1.6313	0.375
$Y_{10,t}$	$Y_{10,t-1}$, $Y_{10,t-2}$, $Y_{10,t-3}$	4	2.0770	0.883

Table 9. Forecast evaluation of ERNN

Variable	Input	Neuron	RMSEP	sMAPE
$Y_{1,t}$	$Y_{1,t-11}$, $Y_{1,t-17}$,	5	2.043	0.310
$Y_{2,t}$	$Y_{2,t-1}$, $Y_{2,t-2}$, $Y_{2,t-11}$, $Y_{2,t-12}$, $Y_{2,t-23}$, $Y_{2,t-24}$,	3	1.614	0.300
$Y_{3,t}$	$Y_{3,t-1}$, $Y_{3,t-35}$	2	7.107	1.151
$Y_{4,t}$	$Y_{4,t-24}$	5	5.174	1.070
$Y_{5,t}$	$Y_{5,t-12}$	4	0.401	1.092
$Y_{6,t}$	$Y_{6,t-13}$	4	3.477	0.498
$Y_{7,t}$	$Y_{7,t-1}$, $Y_{7,t-2}$, $Y_{7,t-8}$, $Y_{7,t-9}$	4	34.102	0.685
$Y_{8,t}$	$Y_{8,t-13}$, $Y_{7,t-23}$	4	7.243	0.520
$Y_{9,t}$	$Y_{9,t-1}$, $Y_{9,t-10}$, $Y_{9,t-12}$	4	4.725	0.615
$Y_{10,t}$	$Y_{10,t-1}$, $Y_{10,t-2}$, $Y_{10,t-3}$	2	1.826	0.689

4.3 Models Comparison

After obtaining the best model for each method on each brand, then we compare the performance between the 5 models. We then choose the best models which are selected based on the smallest forecast error in testing set. The frequency of the model winning for each brand of hijab and sarong is shown in Fig. 5.

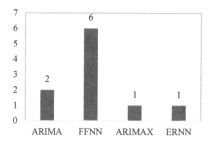

Fig. 5. Frequencyof winning model in each brand

Figure 5 shows that FFNN is the best model among the others, where it is successfully out performs among other models, with 6 out of 10 brands. Search trend for Hijab Rabbani and Hijab Elzatta can be predicted by the ARIMA method. The FFNN method is the best model for 6 search trends for hijab and sarong brands, namely Zoya, Dian Pelangi, Gajah Duduk, Wadimor, Atlas, and Mango Gloves. ARIMAX yields the best forecast for Sapphire brand only and ERNN yields the best forecast for Shafira brand only.

Comparison to ARIMAX is done because ARIMAX is a model that consists of complete components, so that it is able to capture trends, seasonal patterns, and calendar variations in search trend for hijab and sarong compared to the simpler ARIMA. In performing an error rate analysis, we use RMSE as the evaluation criterion. If the RMSE ratio is less than 1, then the method is better than ARIMAX. Comparison of the four methods is shown in Table 10.

Table 10. Ratio of RMSE of each model to RMSE of ARIMAX

Variable	Ratio to RMSE ARIMAX		
	ARIMA	FFNN	ERNN
$Y_{1,t}$	**0.288**	0.348	1.075
$Y_{2,t}$	0.185	**0.052**	0.071
$Y_{3,t}$	0.489	**0.021**	0.318
$Y_{4,t}$	**0.280**	0.383	0.982
$Y_{5,t}$	0.800	0.926	**0.390**
$Y_{6,t}$	4.757	**0.200**	0.236
$Y_{7,t}$	0.780	**0.493**	1.132
$Y_{8,t}$	1.301	**0.603**	1.747
$Y_{9,t}$	1.301	**0.603**	1.747
$Y_{10,t}$	3.865	2.893	**2.543**

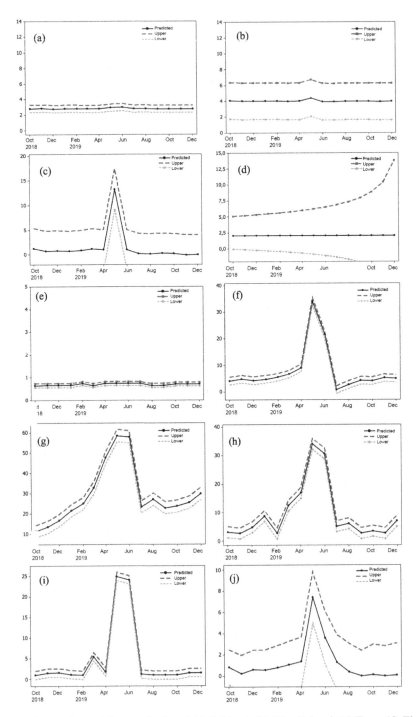

Fig. 6. The search trend forecast for brand (a) Rabbani, (b) Dian Pelangi, (c) Zoya, (d) Elzatta, (e) Shafira, (f) Gajak Duduk, (g) Wadimor, (h) Atlas, (i) Mangga, (j) Sapphire

The search trend forecast using FFNN successfully decreases the forecast error many times less than ARIMAX forecast. It only fails in forecasting the Sapphire brand ($Y_{10,t}$), where ARIMAX forecast outperforms among other models. Meanwhile, the ERNN forecast is better than ARIMAX in 5 out of 10 brands only. It implies that FFNN model, which is less complex than ERNN, can have better forecast than ERNN model.

4.4   Forecasting the Search Trend of Brands of Hijab and Sarong

After obtaining the best model from each brand, we then forecast the search trend for hijab and sarong for the period of October, November, December 2018 and the entire months in year 2019 (15-step ahead) using the best model. The forecast results are shown in Fig. 6.

5   Conclusions

Based on the results above, we can conclude that FFNN yields the best forecast in 6 out of 10 brands, namely Zoya, Dian Pelangi, Gajah Duduk, Wadimor, Atlas, and Mangga. Search trend for Hijab Rabbani and Hijab Elzatta can be well predicted by the ARIMA. While ARIMAX is the best forecasting model for brand Shappire. Lastly, the best model for forecasting brand Shafira isERNN. In the future works, we suggest to apply other forecasting methods such as hybrid method that combines statistical model and machine learning model in order to take the advantages on both methods for better forecast accuracy.

Acknowledgements. This research was supported by DRPM-DIKTI under scheme of "Penelitian Dasar Unggulan Perguruan Tinggi 2019". The authors thank to the General Director of DIKTI for funding and to anonymous referees for their useful suggestions.

References

1. USCIRF: U.S. Commission on International Religious Freedom (Annual Report). USCIRF, New York (2017)
2. Riyanto, A.D.: Pemanfaatan Google Trends Dalam Penentuan Kata Kunci Sebuah Produk untuk Meningkatkan Daya Saing Pelaku Bisnis di Dunia Internet. Seminar Nasional Informatika **1**(1), 52–59 (2014)
3. Sullivan, D.: Google now handles at least 2 trillion searches per year, 24 May 2016. https://searchengineland.com/google-now-handles-2-999-trillion-searches-per-year-250247. Accessed 10 Dec 2018
4. Choi, H., Varian, H.: Predicting the present with Google trends. Econ. Rec. **88**(S1), 2–18 (2012)
5. Lee, M.H., Hamzah, N.A.: Calendar variation model based on ARIMAX for forecasting sales data with Ramadhan effect. In: Proceedings of the Regional Conference on Statistical Sciences 2010 (RCSS 2010), Kelantan (2010)

6. Trapletti, A.: On neural networks as statistical time series models. Ph.D. thesis, Vienna University of Technology, Vienna (2000)
7. Hyndman, R.J., Athanasopoulos, G.: Forecasting: Principles and Practice. Otexts, Melbourne (2018)
8. Arsyad, L.: Peramalan Bisnis, Edisi Pertama. Universitas, Yogyakarta (2001)
9. Wei, W.W.: Time Series Analysis: Univariate and Multivariate Methods, 2nd edn. Pearson Education, Inc., Pearson (2006)
10. Cryer, J.D., Chan, K.-S.: Time Series Analysis With Applications in R, 2nd edn. Springer, New York (2008). https://doi.org/10.1007/978-0-387-75959-3
11. Lee, T.H., White, H., Granger, C.J.W.: Testing for neglected nonlinearity in time series models: a comparison of neural network methods and alternative tests. J. Econ. **56**(3), 269–290 (1993)
12. Chong, E.K., Zak, S.H.: An Introduction to Optimization. Wiley, Canada (2001)
13. Lewis, N.: Deep Learning Made Easy with R. CreateSpace Independent Publishing Platform, Australia (2016)
14. Aaker, J.L.: Dimensions of brand personality. J. Mark. Res. **34**(3), 347–356 (1997)
15. Nakama, T.: Comparisons of single- and multiple-hidden-layer neural networks. In: Liu, D., Zhang, H., Polycarpou, M., Alippi, C., He, H. (eds.) ISNN 2011. LNCS, vol. 6675, pp. 270–279. Springer, Heidelberg (2011). https://doi.org/10.1007/978-3-642-21105-8_32
16. Suhartono, S., Suhermi, N., Dedy, D.P.: Design of experiment to optimize the architecture of deep learning for nonlinear time series forecasting. Procedia Comput. Sci. **144**, 269–276 (2018)
17. Suhermi, N., Suhartono, S., Rahayu, S.P., Prastyasari, F.I., Ali, B., Fachruddin, M.I.: Feature and architecture selection on deep feedforward network for roll motion time series prediction. In: Yap, B., Mohamed, A., Berry, M. (eds.) SCDS 2018. Communications in Computer and Information Science, vol. 937, pp. 58–71. Springer, Singapore (2019)

Convolutional Neural Network Application in Smart Farming

Yudhi Adhitya[1]([✉]), Setya Widyawan Prakosa[2], Mario Köppen[1],
and Jenq-Shiou Leu[2]

[1] Kyushu Institute of Technology,
680-4, Kawazu, Iizuka, Fukuoka 820-8502, Japan
adhitya.yudhi898@mail.kyutech.jp, mkoeppen@ieee.org
[2] National Taiwan University of Science and Technology,
Taipei City 10607, Taiwan
{d10702804,jsleu}@mail.ntust.edu.tw

Abstract. The agricultural sector has a very pivotal role, furthermore very important in the global economy country in the world. The uses of machine learning become trending, and massive improvement technology has widely used in modern agricultural technology. Artificial Intelligent techniques are being used extensively in the agricultural sector as one purpose to increase the accuracy and to find solutions to the problems. As implementation of Artificial Intelligent (AI) based on Convolutional Neural Networks (CNN) application in several fields, indicates that CNN based machine learning scheme is adaptable and implemented on an agricultural area. In this contribution, we apply CNN based feature extraction on cocoa beans images. Cocoa beans images used in this study were cocoa beans (Theobroma Cacao L.) in various quality classes originating from districts in South Sulawesi, Indonesia, and we separate those images 30% for training and the remaining 70% for testing. From our assessment, the result shows that we can achieve 82.14% accuracy to classify seven classes of cocoa beans images using 5 CNN layers.

Keywords: Convolutional Neural Network · Artificial Intelligent · Classification · Feature extraction · Cocoa beans · Smart farming

1 Introduction

The rapid development of Artificial Intelligence (AI), especially conducting research related to the application of AI in several fields, becomes more popular recently. Thus, from this rapid research, there are emerging several techniques in AI domain. At the beginning of AI research, researchers used the traditional method to cope with the hardware limitation issues. Expert system and fuzzy system are pioneering these areas. Furthermore, traditional machine learning is has been used in the following era. Then, deep learning is becoming popular and booming recently.

Typically, to build a robust classifier, the selection of the feature extraction technique needs to be carefully measured to provide the representative in- formation to the classifier. Several schemes for feature extraction and feature engineering techniques are

© Springer Nature Singapore Pte Ltd. 2019
M. W. Berry et al. (Eds.): SCDS 2019, CCIS 1100, pp. 287–297, 2019.
https://doi.org/10.1007/978-981-15-0399-3_23

available. Thus, as the impact of the availability of big data, feature extraction and selection becomes essential to be addressed. In the agricultural field, there are several works to analyze images taken from the farming area. [1] proposed the utilization of Linear Discriminant Analysis (LDA) for extracting features of pig's motion. Those features obtained from the video taken by [1] to identify the aggressive interaction among pigs. [2] provided the tools named FETEX 2.0 to extract features for analyzing agricultural object- based images. [3] studied the use of K-means clustering to create an automatic strawberry grading system using image processing. The features extracted by calculating the characteristics of shape, size, and color of each strawberry.

As the introduction of deep learning for feature extraction, a robust classifier can be constructed using extracted features from neural networks. Assessment of a classifier trained using Deep Learning features has been conducted by [4]. [4] trained Support Vector Machine (SVM) using features from CNN and Hilbert-Huang Transform for recognizing the first-person action. Also, CNN- based feature extraction technique has applied in several fields. For example, using Convolutional Neural Network (CNN) features, face recognition based on CNN features is constructed and realized by [5]. Using the mitigation of CNN extraction schemes and applying it to other tasks, we call it to transfer learning also, robust and powerful machine learning scheme can be effectively and rapidly built. By applying this methodology, we can save the training time for NN models used for feature extraction.

Since several researchers have adopted the utilization of CNN to deal with classification, detection, or segmentation tasks, it relatively confirmed that we could mitigate the employment of CNN to agricultural image dataset. Thereby, the adoption of CNN to be implemented in agricultural image datasets are possible. This paper presents an assessment of the performance of features extraction based on CNN to train SVM classifier using those features. Build a cocoa beans dataset for further implementation and investigation of computer vision algorithm applied in this area. Last but not least, the assessment of several well- known CNN architectures used for feature extraction is also presented in this paper.

Subsequently, one of our paper contributions is to conduct the preliminary study for the assessment of CNN to extract features from publicly available sugarcane dataset [6]. Besides, we also train the classifier for cocoa beans images dataset. In this study, the implementation of SVM as a machine learning scheme is trained using CNN based feature extraction for cocoa beans images. To measure the advantage of transfer learning technique for daily life in an agricultural field, we evaluate the performance of our approach on classifying the quality of cocoa bean dataset taken from the photo shot of cocoa beans in the real manufactured industry. Thus, the remainder of this paper is divided into sections as follows. Section 2 discusses some conventional techniques for feature extraction, including traditional feature extraction technique as well as extraction using CNN. Section 3 presents our preliminary study for CNN based feature extraction employed to train the SVM classifier. Section 4 describes the proposed schemes for collecting cocoa bean images and creating a machine learning scheme using CNN based feature extraction. Section 5 shows and evaluates the outcomes of the proposed systems. The last part is to summarize our work and list some possible future research.

2 Related Works

2.1 Traditional Feature Extraction Technique

Employment of feature extraction was becoming essential since the utilization of machine learning schemes. [7] compared the use of Principal Component Analysis (PCA), Linear Discriminant Analysis (LDA), and kernel LDA for image classification. [8] provide the study to combine several feature extraction techniques to estimate the age of people. Combining feature extraction techniques might be useful to enhance the accuracy of the classifier. However, it also bur- dens the hardware and impacts to the long computational time. Currently, the use of a combination of feature extraction methodologies is replaced by CNN based feature extraction that presents promising results to create an accurate machine learning schemes such as classification and regression algorithm.

2.2 CNN Based Feature Extraction

Getting inspired by the success of deep learning utilization for recognizing hand-written dataset, the introduction of several techniques for developing an artificial intelligence scheme based on deep learning technology has become booming these days. Artificial intelligence based on deep learning technique is introduced by [9]. Then, the following success story has inspired researchers to investigate the usefulness of deep learning in some areas deeply. Thus, several architectures have been proposed to tackle the large dataset provided by Imagenet Challenge 2012. AlexNet [10], VGG-16 [11], and ResNets [12] are named as the favored technique and architecture to achieve state-of-the-art performance. With the addition, the architectures, as mentioned earlier, are CNN based approach deep learning.

The implementation of the feature extraction algorithm based on the well- known CNN architectures has been studied intensively by several researchers. For example, [13] proposed the idea for combining the features extracted by the pre- trained model of AlexNet structure and Hilbert-Huang transform to classify the first-person action recognition. Transfer learning technique can also be employed to detect the mispro-nunciation on Arabic phonemes [14]. For the medical field, [15] utilize CNN based feature extraction to classify MRI images for detecting Alzheimer's disease. Further-more, transfer learning is also useful for face image [16] and assessment of video quality [17].

In the agricultural sector, some approaches intensively studied to employ ma- chine learning and deep learning. Intensive survey regarding the implementation of machine learning and deep learning technique in this field has been presented by [18] and [19]. [20] proposed the use of several machine learning approaches for crop selection, while the main objective is to maximize the crop yield rate. In [21], machine learning has also been employed to estimate the mangrove biomass. From the literature studies, we can conclude that the use of deep learning and CNN based feature extraction can be one of the solutions to be implemented in the agricultural sector.

3 Preliminary Study

3.1 Transfer Learning Schemes for Cocoa Beans Dataset

As depicted in (Fig. 1), at the beginning of our scheme, training images are stored to create a database. In this study, we used an available cocoa beans image dataset [3]. Furthermore, CNN based feature extraction is employed to generate several useful features. After extracting features from images, SVM is trained to build a classification model. Eventually, our model is created and stored into storage. To test the classifier, model weights obtained from training stages are reinstating. Furthermore, performing classification after features from testing images which are different from training images are extracted. Finally, we assessed the accuracy of our classifiers.

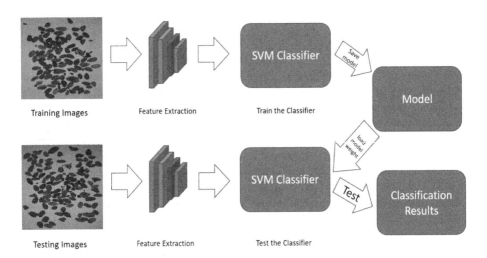

Fig. 1. Proposed scheme

3.2 Study on Cocoa Beans Dataset

Firstly, cocoa beans images obtained from [3] stored on our local storage. Furthermore, 40% of the images are separated for training stages, while the remaining images are for testing. Several features were extracted using CNN, AlexNet [4] as the well-known CNN scheme is adopted. There is 5 convolutional layers built on AlexNet. Therefore, we employ the transfer learning scheme from AlexNet trained on ImageNet. We are conducting several assessments on several convolutional layers. Results on our experiments are depicted in Fig. 2.

As depicted in (Fig. 2), Extracted features from Convolutional layer 1 are not representative enough for the classifier. Therefore, using CONV1, we achieve the lowest accuracy compared to others. From (Fig. 2), We can also assess that the increase of the accuracy can be achieved by adding more layers for extracting features. As shown in the figure, the increasing number of layers to five-layer yields better results. Using 5 CNN layers reaches 82.14% accuracy. It means that using deeper CNN structures will enhance the model of SVM classifier to classify cocoa beans images correctly. Thus, from this study, we can prove that CNN based feature extraction is also promising enough to be applied on agricultural dataset while for the preliminary study, we use publicly available cocoa beans image dataset.

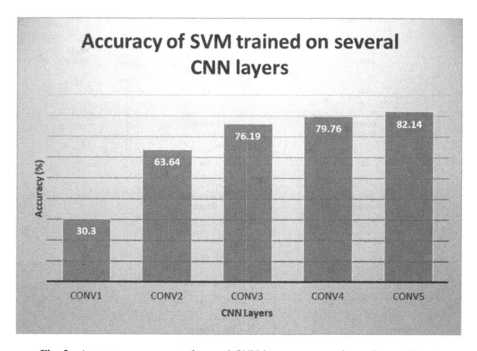

Fig. 2. Accuracy assessment of several CNN layers on cocoa beans image dataset

4 Proposed Scheme for Cocoa Dataset

As the preliminary study result shown in the aforementioned section, CNN based feature extraction can provide a promising result. Thus, the utilization of our preliminary research to our industrial-based cocoa bean images can be assessed using several CNN architecture. From accuracy assessment, there is evidence that adding deeper CNN can obtain a better result. In this study, the assessment of several CNN architecture for feature extraction are presented and discussed.

4.1 The Preparation of Cocoa Beans Images

Beans used in this study were cocoa beans (Theobroma Cacao L.) in various quality classes originating from districts in South Sulawesi, Indonesia. Sampling process of cocoa beans refers to solid sampling guidelines, Indonesian National Standard 19-04-1990 [22].

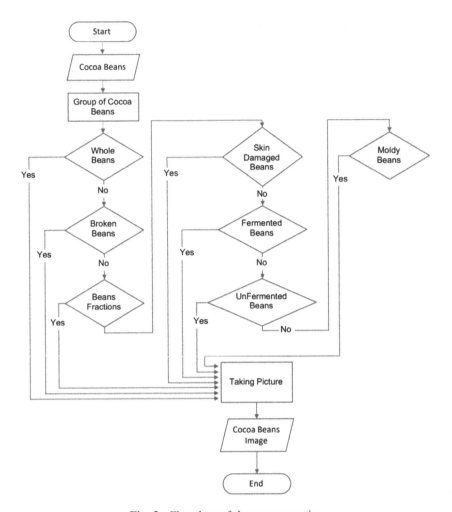

Fig. 3. Flowchart of dataset preparation

Using bright white paper as a background, we scatter samples of cocoa beans, and we controlled the lighting condition. To capture the images, we are using a compact digital camera with a specific range. Cocoa beans images then classified into 7 class [22], i.e., Whole Beans, Broken Beans, Beans Fractions, Skin Dam- aged Beans, Fermented Beans, Un-Fermented Beans, Moldy Beans. Cocoa beans images captured and save as JPG files, as shown in the flowchart of dataset preparation (Fig. 3).

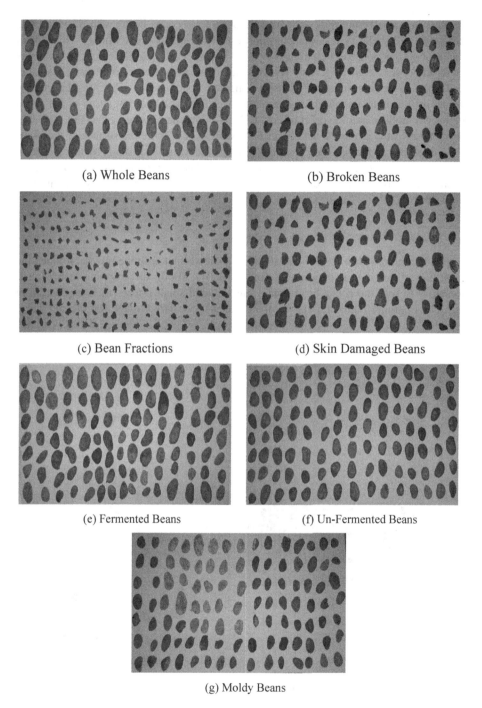

(a) Whole Beans   (b) Broken Beans

(c) Bean Fractions   (d) Skin Damaged Beans

(e) Fermented Beans   (f) Un-Fermented Beans

(g) Moldy Beans

Fig. 4. Cocoa beans sample image

Broken cocoa beans (Fig. 4(b) Broken Beans) are cocoa beans with missing parts measuring 1/2 (half) or less than the portion of the cocoa bean whole, While small fraction consider as Bean Fractions (Fig. 4(c) Beans Fractions). Defective cocoa beans are moldy cocoa beans (Fig. 4(g) Moldy Beans), no seeds, fermented (slaty), insulated seeds, flat seeds, germinating seeds. Fermented cocoa beans (Fig. 4(e) Fermented Beans) are seeds that show three-quarter or over the surface of the slices of the seeds are brown, hollow and flavorful of cocoa. Non- fermented cocoa beans (Fig. 4(f) Un-Fermented Beans) are in the act of cacao showing half or more the surface of the slices of the seed pieces grayish like slate, or grayish blue dense and robust textures, and on the noble cocoa surface dirty white. Moldy cocoa beans (Fig. 4(g) Moldy Beans) are mushroom cocoa beans on the inside and beans surface and visible by human eyes when split [23].

4.2 Data Preprocessing

To prepare the raw data for the classification task, we first crop the raw data. Since the resolution of raw data is enormous (around 1080×1060), the cropping technique is necessary to split the cocoa bean images into several small images then the feature of each class is more representative compared to just one large image (Figure 5).

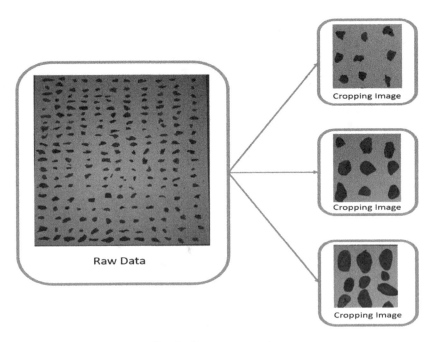

Fig. 5. Data preprocessing

4.3 Classification Algorithm Using CNN Based Feature Extraction for Cocoa Dataset

The proposed scheme of a classification algorithm for cocoa bean images is illustrated in (Fig. 6). After preparing the raw data by cropping the large image into several small images, then we split the data into training and testing data. Therefore, to extract the features from the dataset, we compare the utilization of ResNet 50, ResNet 101, Inception, and Inception-ResNet architecture. Then, features from each architecture are used to train the SVM classifier. After we get the SVM model of each feature extraction, then evaluation for each model is performed. By using the same feature extraction technique, we can assess the accuracy of the classifier trained by each feature extraction.

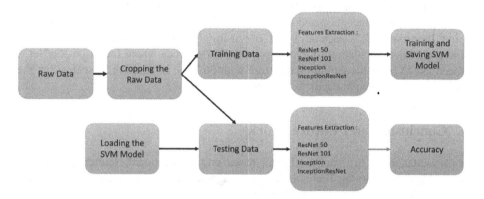

Fig. 6. Proposed schemes for classification of cocoa bean images

5 Result and Discussion

5.1 Results

Cocoa beans image dataset is used for feature selection to form a feature dataset that obtained and stored on our local storage. Raw data are preparing by cropping a large image into several small images which resulting in several small images that is more representative rather than one large image. This raw data then split into training data and testing data. Utilizing architecture from ResNet 50, ResNet 101, Inception, and Inception-ResNet to extract features from the training dataset, resulting feature that used to train the SVM classifier. Each architecture for feature extraction resulting model that apply to test dataset to evaluate each feature extraction model and give result as depicted in (Table 1). Through the feature selection layer in our model that use only one Convolutional layer, we add more Convolutional layers for extracting features that affect the accuracy.

Table 1. CNN architecture for classification tasks

CNN architecture	Accuracy	Computational time
ResNet 50	67%	4.870616 s
ResNet 101	69%	6.538103 s
Inception	72%	5.661103 s
Inception-ResNet	75%	36.376617 s

5.2 Discussion

From Table 1, it is explicitly shown that deeper CNN architecture provides more accuracy. From the results shown in Table 1, ResNet 50 is the fastest feature extraction schemes with 4.87 s in performing the extraction scheme. By using deeper architecture, more accuracy can be gained. ResNet 101 can reach 69% accuracy, and the employment of Inception and Inception-ResNet can increase the accuracy up to 3% using Inception (72% accuracy) and 6% using Inception-ResNet (75% accuracy). However, more calculation time is also needed to sacrifice where Inception-ResNet is placed as the most burdening architecture with 36.38 s. In addition, it seems that the preprocessing technique needs to have further study.

6 Conclusion and Future Works

Implementation of Artificial Intelligent based on Convolutional Neural Networks as one of Machine Learning Technologies in the agricultural sector can increase efficiency and provide a solution to the problem.

We have presented the study of CNN based features architectures for the agricultural dataset, in this paper cocoa beans images are used to conduct the experiment, adding more CNN layers yields better accuracy for a classification task and reducing computational time consumed.

References

1. Viazzi, S.: Image feature extraction for classification of aggressive interactions among pigs. Comput. Electron. Agric. **104**, 57–62 (2014)
2. Ruiz, L.A., Recio, J.A., Fernández-Sarría, A., Hermosilla, T.: A feature extraction soft- ware tool for agricultural object-based image analysis. Comput. Electron. Agric. **76**(2), 284–296 (2011)
3. Liming, X., Yanchao, Z.: Automated strawberry grading system based on image processing. Comput. Electron. Agric. **71**(Suppl. 1), S32–S39 (2010)
4. Purwanto, D., Chen, Y.T., Fang, W.H.: Temporal aggregation for first-person action recognition using Hilbert-Huang transform. In: IEEE International Conference on Multimedia and Expo (ICME) (2017)
5. Sharma, S., Shanmugasundaram, K., Ramasamy, S.K.: FAREC—CNN based efficient face recognition technique using Dlib. In: International Conference on Advanced Communication Control and Computing Technologies (ICACCCT) (2016)

6. Alencastre-Miranda, M., Davidson, J.R., Johnson, R.M., Waguespack, H., Krebs, H.I.: Robotics for sugarcane cultivation: analysis of billet quality using computer vision. IEEE Robot. Autom. Lett. **4**, 3828–3835 (2017)
7. Ye, F., Shi, Z., Shi, Z.: A comparative study of PCA, LDA and Kernel LDA for image classification. In: International Symposium on Ubiquitous Virtual Reality, ISUVR 2009. IEEE (2009)
8. Ghufran, R.S., Leu, J.-S., Prakosa, S.W.: Improving the age estimation accuracy by a hybrid optimization scheme. Multimedia Tools Appl. (MTAP) **77**(2), 2543–2559 (2018)
9. LeCun, Y., Bengio, Y., Hinton, G.: Deep learning. Nature **521**, 436–444 (2015)
10. Krizhevsky, A., Sutskever, I., Hinton, G.E.: ImageNet classification with deep convolutional neural networks. Commun. ACM (CACM) **60**, 84–90 (2017)
11. Simonyan, K., Zisserman, A.: Very deep convolutional networks for large-scale image recognition. In: International Conference on Learning Representations (ICLR) (2015)
12. He, K., Zhang, X., Ren, S., Sun, J.: Deep residual learning for image recognition. CoRR, abs/1512.03385 (2015)
13. Purwanto, D., Chen, Y.-T., Fang, W.-H.: First-person action recognition with temporal pooling and Hilbert-Huang transform. IEEE Trans. Multimedia, 1 (2019). https://doi.org/10.1109/TMM.2019.2919434
14. Nazir, F., Majeed, M.N., Ghazanfar, M.A., Maqsood, M.: Mispronunciation detection using deep convolutional neural network features and transfer learning-based model for arabic phonemes. IEEE Access **7**, 52589–52608 (2019)
15. Khagi, B., Lee, C.G., Kwon, G.R.: Alzheimer's disease classification from brain MRI based on transfer learning from CNN. In: 2018 11th Biomedical Engineering International Conference (BMEiCON) (2018)
16. Sengur, A., Akhtar, Z., Akbulut, Y., Ekici, S., Budak, U.: Deep feature extraction for face liveness detection. In: IDAP-2018, International Conference on Artificial Intelligence and Data Processing (2018)
17. Zhang, Y., Gao, X., He, L., Lu, W., He, R.: Objective video quality assessment combining transfer learning with CNN. IEEE Trans. Neural Netw. Learn. Syst. 1–15 (2019). https://doi.org/10.1109/TNNLS.2018.2890310
18. Kamilaris, A., Prenafeta-Boldú, F.X.: Deep learning in agriculture: a survey. Comput. Electron. Agric. **147**, 70–90 (2018)
19. Liakos, K.G., Busato, P., Moshou, D., Pearson, S., Bochtis, D.: Machine learning in agriculture: a review. Sensors **18**, 2674 (2018)
20. Kumar, R., Singh, M.P., Kumar, P., Singh, J.P.: Crop Selection Method to maximize crop yield rate using machine learning technique. In: 2015 International Conference on Smart Technologies and Management for Computing, Communication, Controls, Energy and Materials (ICSTM) (2015)
21. Jachowski, N.R., Quak, M.S., Friess, D.A., Duangnamon, D., Webb, E.L., Ziegler, A.D.: Mangrove biomass estimation in Southwest Thailand using machine learning. Appl. Geography **45**, 311–321 (2013)
22. Indonesian Standardization Institution: Cocoa Bean [SNI] Indonesian National Standard 2323-2008 ICS 67.140.30 (2008)
23. Attachment of the Regulation of the Minister of Agriculture Indonesian Number: 51/Permentan/OT.140/9/2012 (2012). Accessed 4 Sept 2012

Comparison of Artificial Neural Network (ANN) and Other Imputation Methods in Estimating Missing Rainfall Data at Kuantan Station

Nur Afiqah Ahmad Norazizi[1] and Sayang Mohd Deni[1,2(✉)]

[1] Centre for Statistics and Decision Science Studies,
Faculty of Computer and Mathematical Sciences,
Universiti Teknologi MARA, 40450 Shah Alam, Selangor, Malaysia
sayang@fskm.uitm.edu.my
[2] Advanced Analytic Engineering Center (AAEC),
Faculty of Computer and Mathematical Sciences,
Universiti Teknologi MARA, 40450 Shah Alam, Selangor, Malaysia

Abstract. Daily rainfall data could be considered as one of the basic inputs in hydrological (e.g. streamflow, rainfall-runoff, recharge) and environmental (e.g. crop yield, drought risk) models as well as in assessing the water quality. In Malaysia, the number of rain gauge stations with complete records for a long duration is very scarce. The occurrence of missing values in rainfall data is mainly due to malfunctioning of equipment and severe environmental conditions. Thus, the estimation of rainfall is needed, whenever the missing data happened at the principal rainfall station. In this study, daily rainfall data from eight meteorological stations located in Pahang state are considered and Kuantan is selected as the target station. The main purposes of this study is to compare the performance of the imputation methods by using Artificial Neural Network method (ANN), Bootstrapping and Expectation Maximization Algorithm method and Multivariate Imputation by Chained Equations method (MICE). Missing rainfall data has been generated randomly for Kuantan station with 5%, 10% and 15% of missingness. The three methods are compared based on Mean Absolute Error (MAE), Root Mean Square Error (RMSE) and Coefficient of Determination (R2). The findings concluded that Artificial Neural Network (ANN) is found to be the best imputation method for this study, followed by Multiple Imputation by Chained Equation (MICE) and Bootstrapping and Expectation Maximization Algorithm method.

Keywords: Daily rainfall data · Artificial Neural Network ·
Bootstrapping and Expectation Maximization Algorithm ·
Multivariate Imputation by Chained Equations · Imputation method ·
Missing data

1 Introduction

Missing data is one of the common problems arise in the process of data collection. Incomplete data occurs for a variety of reasons, such as, interruption of experiments, equipment failure, measurement limitation, attrition in longitudinal studies, censoring,

© Springer Nature Singapore Pte Ltd. 2019
M. W. Berry et al. (Eds.): SCDS 2019, CCIS 1100, pp. 298–306, 2019.
https://doi.org/10.1007/978-981-15-0399-3_24

usage of new instruments, changing methods of record keeping, lost of records, and non response to questionnaire items [1, 2, 3, 5, 8, 11, 14, 15, 18, 21–28].

There are various types of data driven models including Artificial Neural Network (ANN) which could be used for the implementation of weather forecast. The ANN provides good approximation due to the capability of the network, dynamic and works well with non-stationary data. ANN is a popular method for many hydrological data analyses as demonstrated by many researchers [6, 7, 9, 12–15, 19]. The ANN has been shown to be one of the best methods for missing data prediction at par with fuzzy logic as a fuzzy rule-based approach. The estimation of missing rainfall data using ANN were compared with the results obtained using regression and other simpler techniques such as arithmetic and inverse distance method. Based on some previous studies, the ANN was chosen to be compared with other techniques due to its adequacy and reliable in predicting missing rainfall at particular gauge stations. Meanwhile, other methods requires much longer duration of time to estimate the missing values and the process involved much more complicated if compared with ANN [3, 21].

Thus, the aim of this study is to estimate the missing rainfall data by using Artificial Neural Network (ANN), Bootstrapping and Expectation Maximization Algorithm and Multivariate Imputation by Chained Equations (MICE). These methods were chosen due to the successfulness and the capability in handling the missing value problems as mentioned by some previous researchers in the field of study [2, 4, 6, 7, 9, 12, 15, 18, 19]. In evaluating those methods, three different level of missing values such as 5%, 10% and 15% will be considered. The performance of these methods is assessed using the Mean Absolute Error (MAE), Root Mean Square Error (RMSE) and Correlation of Determination (R^2). In addition, the results of this study can provide knowledge to the researchers on several alternatives techniques for hydrological data in Malaysia and worldwide on which technique would give the best performance. Research and development in this study is conducted on continuous basis, thus there might be new findings on the issue of environmental as well.

2 Materials and Methods

2.1 Data Description

In this study, daily rainfall data were obtained from the Malaysian Meteorological Department and Drainage and Irrigation Department of Malaysia. The secondary data consists of daily rainfall amount (mm) for the period of 1975 up to 2017. Kuantan station which is located in the state of Pahang is chosen due to heavy seasonal rainfall and winds that affected most parts of Peninsular Malaysia during December 2014. The rains caused severe flooding in the East Coast region i.e. Terengganu, Pahang, and Kelantan states [17]. There were eight meteorological stations located in Pahang were selected such as Sekolah Menengah Ahmad Pekan, Felda Bukit Tajau, Felda Kampung New Zealand, Felda Sungai Pancing Selatan, Pusat Pertanian Tanaman Kampung Awah, Pusat Pertanian Bukit Goh and Mardi Sungai Baging as well as Kuantan station was chosen as the target station (Table 1).

Table 1. The geographical coordinates of the selected rainfall stations

Station	Latitude (N)	Longitude (E)
Felda Bukit Tajau	3.58°	102.73°
Felda Kampung New Zealand	3.64°	102.86°
Felda Sungai Pancing Selatan	3.80°	103.17°
Kuantan	3.76°	103.21°
Mardi Sungai Baging	4.08°	103.39°
Pusat Pertanian Tanaman Kampung Awah	3.48°	102.52°
Pusat Pertanian Bukit Goh	3.88°	103.26°
Sekolah Menengah Kebangsaan Ahmad, Pekan	3.49°	103.40°

2.2 Missing Data Imputation Analysis by Using Artificial Neural Network

In the process of imputation of missing values, the data was trained using ANN with three different learning algorithms, namely, conjugate gradient Fletcher–Reeves update (CGF), Broyden–Fletcher–Goldforb–Shanno (BFG) and Levenberg–Marquardt (LM). Three different units of analysis such as, daily, 10-day and monthly rainfall values, were used for evaluating the prediction ability of ANN. The output of the network was identified as the amount of precipitation at station X, px(t), with the inputs which was determined by the amount of the neighboring stations within the same duration in the period of time. The equation of the model assessment can be expressed as:

$$p_x(t) = f[p_1(t), p_2(t), p_3(t), p_4(t), p_5(t), p_6(t), p_7(t)] \tag{1}$$

where p1, p2, p3, p4, p5, p6 and p7 are denoted as the amount of rainfall at seven neighboring stations except Kuantan station. The ANN was trained and simulated using R Programming with the *neuralnet* package.

2.3 Missing Data Imputation Method by Using Bootstrapping and Expectation Maximization Algorithm

One of the advantages of Bootstrapping and Expectation Maximization Algorithm in Amelia package of R Programming is that, the combination of speed and the ease-of-use of algorithm with the power of multiple imputations will take into consideration.

The imputation model in Amelia package assumes that the complete data (that is, both observed and unobserved) are multivariate normal. It is denoted as the $(n \times k)$ dataset with D is defined as the observed part, D_{obs} and unobserved part, D_{mis}, where the assumption is given as the following equation:

$$D \sim N_k(\mu, \Sigma) \tag{2}$$

The state, D is defined as multivariate normal distribution with mean vector μ and covariance matrix Σ. The multivariate normal distribution is often to give a crude approximation to the true distribution of the data. Moreover, there is evidence to show that this model works as well as the other models which are more complicated and contained the mixed data. It has been reported by some previous researchers that multivariate normal model can provide valid estimate even though the assumptions is violated. This may be due to the large sample size and small percentage of missing values occurred in the dataset [3, 21]. Furthermore, transformations of many types of variables can often make this normality assumption more plausible.

2.4   Missing Data Imputation Analysis by Using Multivariate Imputation by Chained Equation

Multiple Imputation by Chained Equations (MICE) is a practical approach to generate missing rainfall values based on a set of imputation models. MICE is also known as fully conditional specification and sequential regression multivariate imputation. The three stages of MICE are described as below:

  i. Generating Multiple Imputed Data Sets
 ii. Analyzing Multiple Imputed Data Sets
iii. Combining Estimates from Multiply Imputed Data Sets

The m estimates are combined into an overall estimate and variance–covariance matrix using Rubin's rules, which are based on asymptotic theory in a Bayesian framework,

$$\hat{\theta}_j = \frac{1}{m} \sum_{j=1}^{m} \hat{\theta}_j \tag{3}$$

2.5   Performance Measure Criteria

Three performance indicators which are Mean Absolute Error (MAE), Root Mean Square Error (RMSE) and Coefficient of Determination (R2) will be used to assess the imputation methods.

$$MAE = \frac{1}{n} \sum_{i=1}^{n} |f_i - y_i| = \frac{1}{n} \sum_{i=1}^{b} |e_i| \tag{4}$$

where f_i is the prediction and y_i is the true value.

$$RMSE = \sqrt{\frac{\sum_{i=1}^{n} \left(X_{obs,i} - X_{model,i} \right)^2}{n}} \tag{5}$$

Where X_{obs} is the observed values and X_{model} is the modeled values, at time or location i.

$$R^2 = 1 - \frac{RSS}{TSS} = 1 - \frac{\sum e_i^2}{\sum y_i^2} = \frac{ESS}{TSS} = \frac{\sum \hat{y}_i^2}{\sum y_i^2} \tag{6}$$

3   Results and Discussion

Referring to Table 2, the lowest mean of rainfall amount of 5.3 mm is recorded at Pusat Pertanian Tanaman Kampung Awah. Meanwhile, the highest mean of rainfall amount of 8 mm is observed at Felda Sungai Pancing Selatan. For the period of from 1975 up to 2017, it is observed that Kuantan station is having a complete of records with no missing values. Meanwhile, 9.4% of missing values is found from Felda Bukit Tajau station which is the most missing values recorded compared to the rest of neighboring stations.

Table 2. Descriptive statistics for the target station and the neighboring stations

Station	Minimum value (mm)	Maximum value (mm)	Mean (mm)	Percentage of missing (%)
Felda B. Tajau	0.0	176.9	5.6	9.4
Felda Kg. New Zealand	0.0	184.5	5.8	3.2
Felda Sg. Pancing Selatan	0.0	280.5	8.0	1.6
Kuantan	0.0	527.5	7.9	0.0
Mardi Sg. Baging	0.0	541.5	7.7	0.7
P.P. Tanaman Kg. Awah	0.0	176.8	5.3	1.3
P.P.B. Goh	0.0	407.6	7.7	0.8
SMK Ahmad, Pekan	0.0	402.4	7.7	3.6

The visualization of the computed Artificial Neural Network is shown below. The model has 3 hidden units including the neurons and layers. The black lines showed connections with the weights of the values which will be used to impute the missing values in the target station. In addition, the blue line demonstrated the bias item which is generated by the neural network estimation. The figure showed the results for the 5% generated missing values for Kuantan station (Fig. 1).

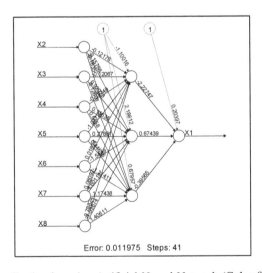

Fig. 1. Visualization by using Artificial Neural Network (Color figure online)

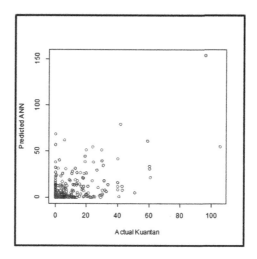

Fig. 2. Predicted values by using Artificial Neural Network versus actual values in Kuantan station (5%) (Color figure online)

Figure 2 shows the graph of the comparison of the actual data and the predicted values imputed by using Artificial Neural Network (ANN) for 5% of missing values at Kuantan station. For Fig. 2 up to Fig. 4, the blue and red dots are denoted as the actual values and the predicted values, respectively.

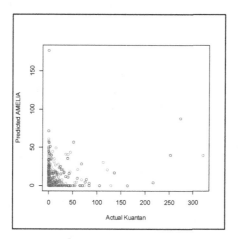

Fig. 3. Predicted values by using Bootstrapping and Expectation Maximization Algorithm versus actual values in Kuantan station (5%) (Color figure online)

Figure 3 showed the plot of comparison of the actual data and the predicted values imputed by using Expectation Maximization Algorithm estimation process for 5% of missing values at Kuantan station.

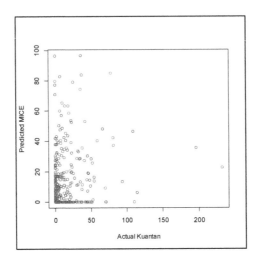

Fig. 4. Predicted values by using Multivariate Imputations by Chained Equation versus actual values in Kuantan station (5%) (Color figure online)

Figure 4 demonstrated the plot of comparison of the actual data and the predicted values imputed by using Multivariate Imputations by Chained Equation process for 5% of missing values at Kuantan station.

Table 3. Performance measures for Mean Absolute Error, Root Mean Squared Error and Coefficient of Determination

R package	MAE			RMSE			R^2		
	5%	10%	15%	5%	10%	15%	5%	10%	15%
NEURALNET	6.460	7.531	7.852	16.798	15.832	14.176	0.978	0.943	0.972
AMELIA	9.111	9.387	9.444	23.888	22.196	22.165	0.755	0.731	0.712
MICE	8.604	9.242	9.091	19.520	18.283	19.511	0.780	0.877	0.908

In order to determine the best imputation method, the lowest value of MAE and RMSE will be chosen as well as the highest value for the R square. Based on the results showed in Table 3, it could be concluded that the Artificial Neural Network (ANN) is observed to be the best imputation method followed by Multiple Imputation by Chained Equation (MICE) and Bootstrapping and Expectation Maximization Algorithm method. It is observed that the result produced by MAE and RMSE are consistently showed the lowest, and the highest value of R square for NEURALNET and followed by MICE and finally by AMELIA for the three level of missing values. Thus, it could be concluded that based on the data used in the study, the Artificial Neural Network (ANN) is found to be the best imputation method in generating missing rainfall data at Kuantan station.

4   Conclusion

The aim of this research is to compare three imputation methods in imputing the missing values at Kuantan station due to the completeness in the data set for the period of 1975 up to 2017. In evaluating the three imputations methods, missing data were created at three different levels, 5%, 10% and 15%. In addition, missing data at Kuantan station has been generated using Missing at Random (MAR) assumption. The predicted imputation results for each method were compared with the actual data at this station.

The performance for each method was evaluated using the three performance measures such as MAE, RMSE and R Squared. According to some previous studies, the most popular method for imputation of rainfall data is not necessarily to be most efficient. However, the results shown that the Artificial Neural Network (ANN) by using *neuralnet* package in R demonstrated the best method estimation for imputing the missing rainfall data compared to the two other methods. Meanwhile, Bootstrapping and Expectation Maximization Algorithm in Amelia package and Multivariate Imputation by Chained Equation (MICE) in MICE package of R Programming are found rarely been used for the estimation of missing rainfall data. However, these two methods and R programming packages which are Amelia and MICE, are both widely used in the imputation of missing data only in other extent aside from missing rainfall data.

Acknowledgement. The authors wish to thank Malaysian Meteorological Department for the data and sponsorship from Faculty of Computer and Mathematical Sciences, Universiti Teknologi MARA (UiTM). The authors are also indebted to the staff of the Drainage and Irrigation Department for providing the daily rainfall data for this study. They also acknowledge their sincere appreciation to the reviewers for their valuable suggestion and remarks in order to improve the manuscript. This research will not complete without the sponsorship from Ministry of Higher Education (600-RMI/TRGS DIS 5/3 (1/2015)).

References

1. Abuelgasim, A.A., Gopal, S., Strahler, A.H.: Forward and inverse modelling of canopy directional reflectance using a neural network. Int. J. Remote Sens. **19**(3), 453–471 (1998)
2. Amer, S.R.: Neural network imputation: a new fashion or a good tool. Unpublished Ph.D. thesis (2004)
3. Demirtas, H., Freels, S.A., Yucel, R.M.: Plausibility of multivariate normality assumption when multiply imputing non-Gaussian continuous outcomes: a simulation assessment. JSCS **78**(1), 69–84 (2008)
4. Fausett, L.: Fundamentals of Neural Networks: Architectures, Algorithms, and Applications (No. 006.3). Prentice-Hall (1994)
5. Kamaruzaman, I.F., Zin, W.Z.W., Ariff, N.M.: A comparison of method for treating missing daily rainfall data in Peninsular Malaysia. Malays. J. Fundam. Appl. Sci. (4–1), 375–380 (2017)

6. Khan, I.Y., Zope, P.H., Suralkar, S.R.: Importance of artificial neural network in medical diagnosis disease like acute nephritis disease and heart disease. Int. J. Eng. Sci. Innov. Technol. (IJESIT) **2**(2), 210–217 (2013)
7. Karunanithi, N., Grenney, W.J., Whitley, D., Bovee, K.: Neural networks for river flow prediction. J. Comput. Civ. Eng. **8**(2), 201–220 (1994)
8. Shaadan, N., Deni, S.M., Jemain, A.A.: Application of functional data analysis for the treatment of missing air quality data. Sains Malays. **44**(10), 1531–1540 (2015)
9. Kuligowski, R.J., Barros, A.P.: Using artificial neural networks to estimate missing rainfall data. J. Am. Water Resour. Assoc. **34**(6), 1437–1447 (1998)
10. Le Barbé, L., Lebel, T., Tapsoba, D.: Rain fall variability in West Africa during the years 1950–1990. J. Clim. **15**(2), 187–202 (2002)
11. Leung, H., Haykin, S.: Detection and estimation using an adaptive rational function filter. IEEE Trans. Signal Process. **42**(12), 3366–3376 (1994)
12. Livingstone, D.J., Manallack, D.T., Tetko, I.V.: Data modeling with neural networks–an answer to the maiden's prayer? J. Comp. Aid. Mol. Des. **11**, 135–142 (1996)
13. Nasr, M., Zahran, H.F.: Using of pH as a tool to predict salinity of groundwater for irrigation purpose using artificial neural network. Egypt. J. Aquat. Res. **40**(2), 111–115 (2014)
14. Rahman, N.A., Deni, S.M., Ramli, N.M.: Generalized linear model for estimation of missing daily rainfall data. AIP Conf. Proc. **1830**(1), 080019 (2017)
15. Paulhus, J.L., Kohler, M.A.: Interpolation of missing precipitation records. Mon. Weather Rev. **80**(8), 129–133 (1952)
16. Ratnayake, U., Herath, S.: Changing rainfall and its impact on landslides in Sri Lanka. J. Mt. Sci. **2**(3), 218–224 (2005)
17. ReliefWeb (2014). Malaysia: Seasonal Floods 2014 - Information Bulletin n° 1. https://reliefweb.int/report/malaysia/malaysia-seasonal-floods-2014-aaainformation-bulletin-n-1. Accessed 4 Nov 2017
18. Royston, P., White, I.R.: Multiple imputation by chained equations (MICE): implementation in Stata. J. Stat. Softw. **45**(4), 1–20 (2011)
19. Rumelhart, D.E., Hinton, G.E., Williams, R.J.: Learning representations by back-propagating errors. Nature **323**(6088), 533 (1986)
20. Rubin, D.B.: Inference and missing data. Biometrika **63**(3), 581–592 (1976)
21. Schaming, D., et al.: Easy methods for the electropolymerization of porphyrins based on the oxidation of the macrocycles. Electrochim. Acta **56**(28), 10454–10463 (2011)
22. Schafer, J.L.: Analysis of Incomplete Multivariate Data. Chapman & Hall/CRC, London (1997)
23. Burhanuddin, S.N.Z.A., Deni, S.M., Ramli, N.M.: Normal ratio in multiple imputation based on Bootstrapped sample for rainfall data with missingness. Int. J. Geomate **13**(36), 131–137 (2017)
24. Suhaila, J., Jemain, A.A., Hamdan, M.F., Zin, W.Z.W.: Comparing rainfall patterns between regions in Peninsular Malaysia via a functional data analysis technique. J. Hydrol. **411**(3), 197–206 (2011)
25. Suhaila, J., Sayang, M.D., Jemain, A.A.: Revised spatial weighting methods for estimation of missing rainfall data. Asia Pac. J. Atmos. Sci. **44**(2), 93–104 (2008)
26. Tang, W.Y., Kassim, A.H.M., Abu Bakar, S.H.: Comparative studies of various missing data treatment methods-Malaysian experience. Atmos. Res. **42**(1–4), 247–262 (1996)
27. Von Davier, M.: Imputing proficiency data under planned missingness in population models. In: Handbook of International Large-Scale Assessment. Background, Technical Issues, and Methods of Data Analysis, pp. 175–202 (2014)
28. Young, K.C.: A three-way model for interpolating for monthly precipitation values. Mon. Weather Rev. **120**(11), 2561–2569 (1992)

Social Network and Media Analytics

Anonymized User Linkage Under Differential Privacy

Chao Kong[1(✉)], Hao Li[1], Haibei Zhu[2], Yu Xiu[1], Jianye Liu[1],
and Tao Liu[1]

[1] School of Computer and Information,
Anhui Polytechnic University, Wuhu, China
kongchao315@163.com, lhthomas@163.com,
xiuyu1860@163.com, 17855368016@163.com,
liutao@ahpu.edu.cn
[2] School of Electrical Engineering, Anhui Polytechnic University, Wuhu, China
zhb877097717@126.com

Abstract. This work develops a user linkage method for anonymized social networks under differential privacy. The existing works have primarily focused on multi-dimensional features of user profile (e.g. user name, gender, ID, etc.) in general, but overlooked the special issue of privacy security. As such, the existing methods can be suboptimal for identifying users across anonymized social networks under differential privacy.

In this paper, we try to study the problem of combining differential privacy with anonymized user linkage problem simultaneously. To the best of our knowledge, none of the existing works has paid special attention to connect these two separate research problems. To tackle the challenges, we first propose a Hierarchy Differential Privacy-based Generator (HDP) to generate a social publishing graph that can well preserve differential privacy protection in the original social graph by performing anonymized user linkage purposely. We then apply the proposed AUL, short for Anonymized User Linkage, to identify users across anonymized social networks via probabilistic generative models in a semi-supervised manner. We conduct extensive experiments on several real datasets covering the tasks of social publishing graph generation and anonymized user linkage. Both quantitative results and qualitative analysis verify the effectiveness and rationality of our HDP+AUL framework.

Keywords: Anonymized user linkage · Differential privacy ·
Probabilistic generative models

1 Introduction

User linkage is a problem of identifying which users in a social network link to the same users in the other social networks. It is a well-known and paramount problem that arises in many research fields, including information retrieval, data integration, machine learning, etc. It has been widely used in many applications such as recommender systems, user behavior analysis, CTR prediction and so on. When people use these applications day by day, they generate massive and sensitive data named UGC,

© Springer Nature Singapore Pte Ltd. 2019
M. W. Berry et al. (Eds.): SCDS 2019, CCIS 1100, pp. 309–324, 2019.
https://doi.org/10.1007/978-981-15-0399-3_25

short for User Generated Content. To perform predictive analysis on anonymized social networks, it is crucial to first pre-anonymize the social network data. Traditional user linkage, link prediction or network alignment task on anonymized social networks such as UMA [1] and MNA [2] erase user profile only and are inefficient to generate true anonymized social networks with vertex, edge and structure privacy in practical applications. Unfortunately, the structure privacy is often the most likely to be over-looked in any design. For example, the explicit and implicit relations reflect someone's social community, but he or she regards it as privacy. As such, we cannot directly do the user linkage task on the raw social networks which also exist the social structure privacy issue.

To date, existing works have primarily focused on identifying users across anon-ymized social networks by erasing user profile only to protect privacy but not con-sidering potential privacy leakage of relations and social structures. In this paper, we want to identify the same users across anonymized social networks under differential privacy, which is formally defined as ULAASN (User Linkage Across Anonymized Social Networks). However, ULAASN is often a challenging task due to the following reasons: (1) the lack of enough labeled data to train the probabilistic generative models, (2) it is arduous to connect two separate research problems: privacy security and user linkage, and (3) the need of scalable algorithms for expanding social networks constantly.

To address these challenges, a novel anonymized user linkage framework named HDP+AUL is proposed in this paper. Concretely, HDP+AUL applies a three-phase solution. Firstly, calculate the vertices' degree centrality to separate the protected vertices and fringe vertices. Then, the dense areas of sub-graphs are divided by per-forming the quadtree method and injecting noise among fringe vertices. Finally, reconstruct the sub-graphs to generate a social publishing graph and apply probabilistic generative models to learn the parameters to predict the linked user pairs. Specifically, we formulate user linkage task as a semi-supervised learning problem. Our proposed approach can be performed with accuracy with few ground-truth which are usually arduous and costly to collect in Web applications. To the best our knowledge, there is so far no feasible framework to cover social publishing graph generation and anon-ymized user linkage simultaneously. We use few number of labeled user pairs and Locality Sensitive Hash (LSH) to block users and reduce the size of candidates. Moreover, we extend Fellegi-Sunter method to model the continuous distribution of social similarity, construct probabilistic generative model, learn parameters by employing EM algorithm and handle data quality such as missing value simultaneously in this general framework.

In this work, we focus on the problem of anonymized user linkage under differ-ential privacy. We propose HDP+AUL framework (short for *Hierarchy Differential Privacy and Anonymized User Linkage*), which address the aforementioned limitations and challenges of existing methods. Below we highlight our major contributions.

1. We design a novel social publishing graph generator named HDP to preserve properties of social network data as much as possible, such as the importance and distribution of vertices.

2. To account for both the few ground-truth and iterative social similarity computing, we develop a semi-supervised method named AUL based on probabilistic generative models to identify users across anonymized social networks. To avoid the explosive growth of complexity in expanding a network, we similarly resort to performing Locality Sensitive Hash (LSH) to block users and reduce the size of candidates.

3. We illustrate the performance of our algorithm against comparable baselines on several real social networks. Empirical study results manifest that the HDP+AUL outperforms baselines in user linkage across two anonymized social networks under differential privacy.

The remainder of the paper is organized as follows. We first review related work in Sect. 2. We formulate the problem in Sect. 3, before delving into details of the proposed method in Sect. 4. We perform extensive empirical studies in Sect. 5 and conclude the paper in Sect. 6.

2   Related Work

Our work is related to user linkage across anonymized social networks considering differential privacy, which can be categorized into two parts: differential privacy on social networks and anonymized user linkage.

2.1   Differential Privacy on Social Networks

Differential privacy is a strict and proven definition of privacy whose basic idea is to inject noise to the original data or statistical results, which makes it impossible for an attacker to distinguish whether a record is added to or deleted from a dataset. Differential privacy technology can resist the background knowledge of the attacker. According to the privacy parameter tuning, it can achieve a balance between privacy protection and data availability [3]. The privacy of social network data covers vertex privacy, edge privacy, and social structure privacy. In our work, we focus on social structure privacy protection. The structure privacy information of a graph is unique to the social network, such as degree of vertex, adjacency information, and cutset of graph, etc. K-anonymity [4] is applied to the processing of adjacency information, which can effectively protect the privacy information of graphs. Michael [5] described an efficient algorithm for releasing a provably private estimation of the degree distribution of a network. MB-CI [6] designed the perturbation scheme and the query function, the edge weights of the different graphs were obtained and the different Laplace noise was added to different edge weights according to the perturbation scheme.

As mentioned above, the existing works about privacy protection on social networks can be mainly divided into three categories: vertex privacy-based, edge privacy-based, and social structure privacy-based. The study of differential privacy protection on social network data has become a hot topic in recent years, and some earlier studies can go back to 2006. Researchers developed a lot of privacy protection methods on social networks, such as node clustering, node segmentation, random graph etc., most of which are based on graph structure modification. Generally speaking, graph structure modification-based methods have two main drawbacks: (1) they usually cause deviation of the original graph structure attributes, which may lead to the unavailability of social network data; (2) their performance is rather sensitive to the complexity of social network structure and background knowledge. Thus, these methods might be suboptimal for privacy protection on social network data.

2.2 Anonymized User Linkage

The study of user linkage or link prediction problem has been a long time, and some earlier study can go back to 1960s [7]. However, user linkage is also an active research problem presently and widely used in multiple applications. User linkage is most commonly employed to improve the selection of similar users in recommender systems which follow a collaborative approach to obtain better recommendations [8]. Despite effectiveness and prevalence, users on social platforms need a simple mechanism to find acquaintances among a large number of registered users. The general approach is employing user linkage or link prediction technique to identify the same users with a high degree of accuracy automatically [9].

The existing studies of user linkage can be mainly divided into two categories: supervised learning approach and unsupervised learning approach [10]. For the former, [11] first proposed a series of new centrality indices for links in the line graph. Then, three supervised learning methods are designed by using these line graph indexes and some original graph indexes to realize link weight prediction of single-layer and multi-layer networks respectively. For the latter, Carlos [12] used an unsupervised learning strategy to design and implement experiments on social networks. The experimental results illustrate that the proposed approach can find satisfactory results.

To date, some researchers have begun to apply semi-supervised learning or deep learning methods to solve the user linkage or link prediction problem [13]. It is worth pointing out that neural network-based methods which are the state-of-art vertex representation learning techniques, and they map network vertices into a low dimensional latent space by utilizing the nonlinear transformations [14]. Inspired by pioneering work, we can also perform network alignment with network embedding [15] to reconstruct the anonymized social publishing graph in this paper. The pioneering work DeepWalk [16] and Node2vec [17] extend the idea of Skip-gram [18] to model homogeneous network, whereas they may not capture the characteristics of network structure, such as the power-law distribution of vertex degrees. But it would be interesting to see the performance comparison between these network embedding methods and HDP. The semi-supervised approach should be performed with accuracy with few ground-truth and social network structure information only across anonymized social networks. These are the focus of this work.

3   Problem Formulation

The task of anonymized user linkage aims to find the same users across anonymized social networks considering differential privacy. As such, we first give the definition of ε-differential privacy[1].

Definition 1 *(ε-differential privacy). Let ε be a positive real number and A be a randomized algorithm that takes a dataset as input. The algorithm A is said to provide ε-differential privacy if, for all datasets D and D' that differ on a single element and all subsets S_A of output of A:*

$$\Pr[A(D) \in S_A] \le \exp(\varepsilon) \cdot \Pr[A(D') \in S_A],$$

where ε denotes the privacy protection level provided by algorithm A. The smaller ε is, the higher privacy protection level is.

In this work, we inject noise into original social networks with a manner of deleting edges between protected vertices and fringe vertices by tuning parameter ε to generate a social publishing graph as an anonymized social network.

Definition 2 *(Social network or social graph). Social network G is a tuple (V, E), where V denotes the set of vertices and $E = \{(v_i, v_j)|v_i, v_j \in V\}$ defines the inter-set edges, where v_i and v_j denote the i-th and j-th vertex in V respectively.*

A social network can be represented as a social graph. As such, users and relations in social networks can be represented as vertices and edges respectively. To account for anonymized user linkage under differential privacy, we may identify users considering social structure only. According to Definition 1 and 2, we use *SPG* to denote a social publishing graph which is applied to represent a social network after pre-processing by performing ε-differential privacy technique.

The ego-network plays an important role to calculate social similarity in user linkage or link prediction task. Next, we define the ego-network.

Definition 3 *(Ego-network). Given a social network G = (V, E) and vertex v ∈ V, the tuple $\alpha(v) = (N(v) \cup \{v\}, E_{N(v)})$ is represented as ego-network, where N (v) denotes the set of neighbors of v and $E_{N(v)} = \{(v_i, v_j)|v_i, v_j \in N(v) \cup \{v\}, v_i, v_j \in V\}$.*

Note that in this paper, each ego-network of v is a sub-graph which is mapped from social network G to v and N (v).

Definition 4 *(Candidate pair). Given two social networks $GA = (V^A, E^A)$ and $G^B = (V^B, E^B)$, the set $C = \left\{ \left(\alpha(v_i^A), \alpha(v_j^B) \right) |v_i^A \in V^A, v_j^B \in V^B \right\}$ is represented as the set of candidate pairs, which any tuple $(\alpha(v^A), \alpha(v^B)) \in C$ is represented as a candidate pair.*

---

[1] https://en.m.wikipedia.org/wiki/Differential_Privacy.

Actually, candidate pairs are Cartesian product of two sets of vertices V^A and V^B, that is $C = \alpha(V^A) \times \alpha(V^B)$. If two vertices are represented as the same user in a candidate pair, then this candidate pair can be represented as a linked user pair. We define the linked user pair as follows.

Definition 5 *(Linked user pair). Given two social networks $GA = (V^A, E^A)$ and $G^B = (V^B, E^B)$, the set of linked user pairs L is subset of C, that is $L \subset C$ and $L = \left\{ \left(\alpha(v_i^A), \alpha\left(v_j^B\right) \right) | v_i^A = v_i^B, \left(\alpha(v_i^A), \alpha(v_i^B) \right) \in C \right\}$. Any candidate pair $\left(\alpha(v_i^A), \alpha\left(v_j^B\right) \right) \in L$ is represented as a linked user pair.*

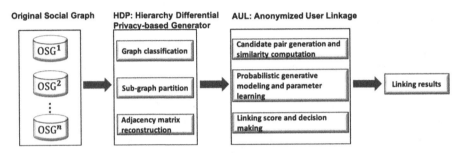

Fig. 1. HDP+AUL framework

Problem Definition. The task of ULAASN aims to identify which users in an anonymized social network link to the same users in the other one, where each user can be represented as a vertex in a social graph. In the HDP+AUL framework, both social publishing graph generation and anonymized user linkage should be developed simultaneously. Formally, the problem can be defined as:

Input: Two social publishing graphs SPG^A and SPG^B.

Output: The set of linked user pairs $L = \left\{ \left(\alpha(v_i^A), \alpha\left(v_j^B\right) \right) | v_i^A = v_j^B, \left(\alpha(v_i^A), \alpha\left(v_j^B\right) \right) \in C \right\}$.

4   HDP+AUL Framework

In a nutshell, a good anonymized user linkage algorithm should be developed on more typical and rational datasets for forecasting. To achieve this aim for a raw social graph, we first consider generating social publishing graph from two perspectives: the data availability and characteristics of network data by HDP. We then learn the parameters by AUL algorithm which consists of three components: (1) candidate pair generation and similarity computations, (2) probabilistic generative model and parameter learning, and (3) linking score and decision making. The HDP+AUL framework is shown as Fig. 1. This section presents our HDP+AUL framework along this line.

4.1   HDP: Hierarchy Differential Privacy-Based Generator

It is a common way to convert an original social graph to social publishing graph by performing reconstructing its sub-graphs with high relevance, which has been used in differential privacy method [19]. However directly performing differential privacy technique on a social graph could fail, since there is no stationary distribution of degree on sets of vertices due to overlooking their importance issue during the noise injection process, which might break the social structure seriously and lead to unavailability of data. To address this issue, we consider performing noise injection based on the importance of each vertex (i.e., protected vertices and fringe vertices).

Algorithm 1. HDP: Hierarchy Differential Privacy-based Generator

Input: Original social graph G, privacy protection level ε;
Output: Social publishing graph SPG;
1: $\varepsilon = \varepsilon_1 + \varepsilon_2 + \varepsilon_3$;
2: Calculate vertices' degree centrality: DC;
3: Rank the vertices with DC in ascending order: top-k vertices R;
4: **for** each vertex $v_i \in R$ **do**
5:     $C_i \leftarrow GraphClassification\ (G,\ \varepsilon_1)$;
6:     $A_i \leftarrow AdjacencyMatrix\ (C_i)$;
7:     $D_i \leftarrow DenseRegionPartition\ (A_i,\ \varepsilon_2)$;
8: **end for**
9: **while** $\varepsilon_{temp} < \varepsilon$ **do**
10: $\overline{A}_i \leftarrow NoiseInjection\ (D_i,\ A_i,\ \varepsilon_3)$
11: **end while**
12: **return** $SPG \leftarrow Reconstruction\ (\overline{A}_i)$;

As mentioned above, DeepWalk may not capture the characteristics of a real-world network. To reconstruct and generate a social publishing graph with high fidelity, we propose a Hierarchy Differential Privacy-based Generator (HDP), which can handle the need of privacy protection and data availability simultaneously. We highlight its core designs as follows:

1. First, we perform the graph classification on the original social graph to obtain subgraphs with high relevance. Each vertex should be dependent on its importance, which can be measured by its degree centrality. For a vertex, the greater its centrality is, the more likely a protected vertex is. In contrast to protected vertices, we choose the other fringe vertices to inject noise. As a result, the vertex importance, characteristics of graph and data availability can be preserved to some extent.

2. In contrast to pioneering work [20], we assign standard quadtree to partition the adjacency matrix of sub-graphs under differential privacy. In this work, each vertex represents the size of dense regions and the number of 1-value after noise injection.

3. Finally, we arrange the distribution of 1-value in dense regions based on quadtree to reconstruct adjacency matrix.

Generally speaking, the above generation process follows the principle of "The rich gets richer", which is physical phenomenon existing in many real-world networks, i.e., the vertex connectivity follow a scale-free power-law distribution. The workflow of our proposed HDP is summarized in Algorithm 1, where ε_1, ε_2 and ε_3 are the differential privacy protection level during three components respectively. SPG output by Algorithm 1 is the social publishing graph generated from reconstructing sub-graphs C_i. The vertex centrality can be measured by many metrics, such as PageRank, HITS and degree centrality [21], etc., and we use degree centrality in our experiments.

4.2   AUL Algorithm

We now present the full algorithm in Algorithm 2. Our proposed AUL algorithm consists of three components: candidate pair generation and similarity computation, probabilistic generative model and parameter learning and linking score and decision making. In this algorithm, $N^A(GP \cup LP)$ and $N^B(GP \cup LP)$ are represented as neighbors of linked users in SPG^A and SPG^B respectively. Each iteration of AUL algorithm (lines 1–22) consists of three steps: In step 1, we employ LSH to block the users in $N^A(GP \cup LP)$ and $N^B(GP \cup LP)$ first (lines 4–12), then compute the social similarity γ_i of each candidate pair r_i; In step 2, we construct the probabilistic generative model by employing latent variable to learn parameters with EM algorithm (lines 16–18); In step 3, we define the linking score as $\log \frac{P(r_i \in L | \gamma_i, \widehat{\theta})}{P(r_i \in U | \gamma_i, \widehat{\theta})}$ to calculate the score sc_i of each candidate pair and judge them link or not according to the linking score (lines 19–21). At the end of one iteration, the predicted linked user pairs pl which are found by the AUL will be added into the set of linked user pairs LP. The loop body is executed again until no new candidate pairs are added.

Algorithm 2. AUL: Anonymized User Linkage Algorithm
___

Input: Two social publishing graphs, SPG^A and SPG^B with partial ground-truth pairs GP, cautious parameter k;
Output: Linked user pairs LP;
$cp \leftarrow \varnothing; LP \leftarrow \varnothing$;
1: while convergence condition is not satisfied do
2:     $pl \leftarrow \varnothing$;
3:       //*Step 1: Candidate pair generation and similarity computation;*
4:     **for** each $a \in N^A(GP \cup LP)$ *or* $N^B(GP \cup LP)$ **do**
5:         shingles each vertex a or b in terms of set GP;
6:         $buckets \leftarrow a(\text{ or } b)$ //*Blocing with LSH;*
7:     **end for**
        $tempCP \leftarrow \varnothing$;
8:     **for** each bucket in LSH **do**
9:       **for** a, b exist in the bucket **do**
10:           $tempCP \leftarrow tempCP \cup \{(a,b)\}$;
11:       **end for**
12:     **end for**
        $cp \leftarrow cp \cup tempCP$;
13:     **for** each pair $r_i \in cp$ **do**
14:         computing social similarity γ_i for user pair r_i;
15:     **end for**
        //*Step 2: Probabilistic generative model and parameter learning;*
16:     **while** parameter set θ has not converged **do**
17:         $eStep (cp \cup GP \cup LP)$, $mStep(cp \cup GP \cup LP)$;
18:     **end while**
        //*Step 3: Linking score and decision making;*
19:     **for** $r_i \in cp$ **do**
20:         $sc_i \leftarrow \log \dfrac{P(r_i \in L | \gamma_i, \widehat{\theta})}{P(r_i \in U | \gamma_i, \widehat{\theta})}$;
21:     **end for**
        $pl \leftarrow k$ pairs in cp with the highest scores, $LP \leftarrow LP \cup pl$;
        remove related predicted user pairs in pl from cp;
22:     **end while**
23:     **return** LP;
___

4.3 Computational Complexity Analysis of AUL

The candidate pair generation and social similarity computation are two key processes of AUL. However, the complexity of generating candidate pairs will be increased if social networks expand. To avoid the explosive growth of complexity in expanding a network, an alternative way is to employ the two-phase blocking technique. Concretely, in the first phase, shingling is employed to users according to the ground-truth or the predicted linked user pairs, then blocking the users with Locality Sensitive Hash (LSH). Through this way, we can reduce the number of candidate pairs and obtain high-quality samples, since LSH can guarantee that dissimilar users are located in different buckets in a probabilistic way. Normally, the computational complexity of generating candidate pairs from two social networks of N users containing in b blocks is $O(\frac{N^2}{b})$ In contrast to CNL [22] and other pioneering work on n-gram-based indexing, we employ the LSH schema to speed up candidate pairs generation and social similarity computation on social feature. The universal set is the collection of users who are in the ground-truth or in the predicted linked user pairs. In the second phase, if two users of one candidate pair have at least one linked common friend, then this candidate pair is considered as a true candidate pair. The combination of the two blocking techniques slashes the number of candidate pairs, which makes a great contribution to reduce computational complexity.

In our LSH schema, there are two significant hyper-parameters: the length of a signature t and the number of signatures for each user g. If I represents the universal set of n-grams, $|I|^t$ is the number of buckets, which may be very large. If two users are mapped into same bucket, they will become a candidate pair. In the second phase, if two users of one candidate pair have at least one linked common friend, then this candidate pair is considered as a true candidate pair.

4.4 Parameters Learning and Decision Making

Likelihood. Assume that γ_i is the social similarity between two users, namely the extended Jaccard coefficient. For all unknown user pairs: $P(r_i \in L|\Theta) = p$ and $P(r_i \in U|\Theta) = 1 - p$, we employ Bayesian rules to $P(r_i \in L|\gamma_i, \Theta)$.

$$P(r_i \in L|\gamma_i, \Theta) = \frac{p \times P(\gamma_i|r_i \in L, \Theta)}{P(\gamma_i|\Theta)} \tag{1}$$

$$P(\gamma_i|\Theta) = p \times P(\gamma_i|r_i \in L, \Theta) + (1 - p) \times P(\gamma_i|r_i \in U, \Theta) \tag{2}$$

For all candidate pairs, we cannot determine in advance whether they link (L) or not (U). Therefore, we define a latent variable l_i for candidate pairs r_i. When the value of l_i equals 1, that means r_i is a linked user pair, 0 and vice versa. In addition, we define $c_i = (l_i; \gamma_i)$ as complete data vector of user set R. The probability of c_i with parameter Θ can be calculated as:

$$\begin{aligned} P(c_i|\Theta) &= [P(\gamma_i|r_i \in L|\Theta)]^{l_i}[P(\gamma_i|r_i \in U|\Theta)]^{(1-l_i)} \\ &= [p \times P(\gamma_i|r_i \in L, \Theta)]^{l_i}[(1 - p) \times P(\gamma_i|r_i \in U, \Theta)]^{(1-l_i)} \end{aligned} \tag{3}$$

The expression of the log-likelihood function is:

$$L(\Theta|X) = \sum_{i=1}^{|T|} L_i[log P(\gamma_i|r_i \in L, \Theta), log P(\gamma_i|r_i \in U, \Theta)]' + \sum_{i=1}^{|T|} L_i[log p, log(1-p)]' \tag{4}$$

One significant task in user linkage is calculating the probability of $P(r_i \in L|\gamma_i, \Theta)$ and $P(r_i \in U|\gamma_i, \Theta)$. The Eqs. 1 and 2 illustrate the premise of computing these two probability is calculating the value of $P(\gamma_i|r_i \in L, \Theta)$ and $P(\gamma_i|r_i \in U, \Theta)$.

Maximum Likelihood Estimation. In Eq. 4, L_i is a latent variable. If we cannot maximize the log-likelihood of Eq. 4, the parameters θ_1 and θ_0 of distribution $f_1(\gamma_i; \theta_1)$ and $f_0(\gamma_i; \theta_0)$ will not be estimated. Thus, we employ EM algorithm to estimate parameters p, θ_1 and θ_0.

E-step: In this step, we look for the expectation of log-likelihood, remove the latent variable l_i and prepare for maximizing log-likelihood and parameter estimation in the next step. In the k-th iteration, the conditional distribution of l_i with γ_i and $\Theta^{(k-1)}$ is $l_i|\gamma_i, \Theta^{(k-1)} \sim B(1, p_i^{(k)})$, where

$$p_i^{(k)} = P(l_i = 1|\gamma_i, \Theta^{(k-1)}) = \frac{P(r_i \in L, \gamma_i|\Theta^{(k-1)})}{P(\gamma_i|\Theta^{(k-1)})} \tag{5}$$

Thus, we can obtain the equation of expectation substituting $p_i^{(k)}$ for l_i.

M-step: In this step, the expectation of latent variable calculated in E-step which is $l_j^{(k)} = p_j^{(k)}$, $i = 1,..., m$ is employed to maximize the log-likelihood and estimate the model parameters. We take the derivative to expectation of log-likelihood as following.

$$\frac{\partial L(\Theta|X)}{\partial p} = \sum_{j=N^2+1}^{N^2+M} \left(\frac{p_j^{(k)}}{p} - \frac{1 - p_j^{(k)}}{1 - p} \right) = 0$$

$$\frac{\partial L(\Theta|X)}{\partial \theta_1} = \sum_{i=1}^{N^2+M} l_i^{(k)} \frac{\partial f_1(\gamma_i; \theta_1)}{\partial \theta_1} \frac{1}{f_1(\gamma_i; \theta_1)} = 0 \tag{6}$$

$$\frac{\partial L(\Theta|X)}{\partial \theta_0} = \sum_{i=1}^{N^2+M} (1 - l_i^{(k)}) \frac{\partial f_0(\gamma_i; \theta_0)}{\partial \theta_0} \frac{1}{f_0(\gamma_i; \theta_0)} = 0$$

Decision Making. On the one hand, we can compare the probability $P(r_i \in L|\gamma_i, \Theta)$ and $P(r_i \in U|\gamma_i, \Theta)$ to judge the candidate pair linking or not; on the other hand, the probability $P(r_i \in L|\gamma_i, \Theta)$ and $P(r_i \in U|\gamma_i, \Theta)$ can also help us making decision. For simplicity, we define the *linking score* function as: $W_i = \log \frac{P(r_i \in L|\gamma_i, \widehat{\theta})}{P(r_i \in U|\gamma_i, \theta)}$, where $P(r_i \in L|\gamma_i, \widehat{\Theta}) > P(r_i \in U|\gamma_i, \widehat{\Theta})$ when $W_i > 0$. Alternatively, we can assign r_i to the linked group if $W_i > W_0$, where $W_0 > 0$ is a threshold.

Table 1. Descriptive statistics of datasets

Social network	#Vertices	#Edges
YouTube	1157827	4945382
Foursquare	203054	5128966
Twitter	73111	1376518

5 Empirical Study

To evaluate the anonymized user linkage under differential privacy, we conduct several experiments to compare the proposed HDP+AUL with the baselines. First, in the experiment settings, we manifest the structural consistency between original and publishing graph by tuning parameter ε to provide differential privacy protection. Second, we study the performance of linking two real-world social networks. Through empirical evaluation, we aim to answer the following research questions:

RQ1. How to guarantee the structural consistency between original social graph and social publishing graph with HDP?
RQ2. How to evaluate the robustness of AUL?
RQ3. How to evaluate the scalability of AUL?

In what follows, we first introduce the experiment settings and then perform a case study, which visualizes the frequency distribution of vertices belonging to original social graph and social publishing graph respectively, to demonstrate the rationality of HDP method. Furthermore, we answer the above research questions in turn.

5.1 Experiment Settings

Datasets. (1) For social publishing graph generation task, we plot the frequency distribution of vertices in a real YouTube dataset[2], which is publicly accessible, and contains 1,157,827 vertices and 4,945,382 edges respectively. (2) For anonymized user linkage task, we use two real large-scale social networks Twitter and Foursquare. The descriptive statistics of our experimented datasets are summarized in Table 1. Note that our experimented datasets cover a wide range of anonymized user linkage under differential privacy, which can test the universality of our HDP+AUL framework.

Evaluation Protocols. (1) To evaluate the social publishing graph generation, we plot the frequency distribution of vertices. Besides, we employ the slope of standard power-law distribution and relative error of clustering coefficient to verify the structural consistency between original social graph and social publishing graph underlying ε-differential privacy protection. Moreover, following the pioneering work BiNE [15], we employ the KL-divergence as the difference measure between distributions. (2) For the anonymized user linkage task, AUL was comprehensively evaluated by Precision@K, Recall@K, F-measure@K, the number of candidate pairs and execution time on the different size of social networks.

Baseline. We compare our work with two types of baselines: (1) To evaluate the social publishing graph generation task, we try to employ a prevalent network embedding method named Random Walk Generator of DeepWalk (RWG) [17] to be the comparative baseline. (2) To benchmark the anonymized user linkage task, we extend the traditional Fellegi-Sunter (FS) [8] by performing a semi-supervised learning strategy

---

[2] http://socialnetworks.mpi-sws.org/data-imc2007.html.

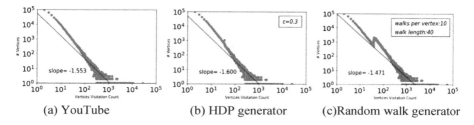

Fig. 2. The frequency distribution of nodes

for fairness consideration. Besides, we also find a semi-supervised learning algorithm named MAH (short for Manifold Alignment on Hypergraph) to be our comparative baselines.

5.2 Social Publishing Graph Generation (RQ1)

To demonstrate the effectiveness and rationality of generating social publishing graph, we compare our proposed Hierarchical Differential Privacy-based Generator (HDP, see in Algorithm 1) with RWG. We plot the frequency distribution of vertices in original social graph and generated social publishing graph in a real YouTube dataset. Moreover, we illustrate the slope of different distribution and calculate the relative error of clustering coefficient underlying the different privacy protection level ε. Besides, we calculate the difference between distributions with KL-divergence. From Fig. 2(a), we can find that the vertices display a standard power-law distribution with a slope of -1.553 in original YouTube dataset. We also use YouTube's network as the input of HDP ($\varepsilon = 0.3$) and show the distribution of vertices. As shown in Figs. 2(b) and 2(c), HDP almost generates a standard power-law distribution with a slope -1.600 which is very close to that of the original network. However, we find that the generated distribution by RWG with a slope -1.471 differs significantly from the real distribution, and it cannot be well described by a power-law distribution. Moreover, we calculate the difference between distributions of vertices in original social graph and social publishing graph by performing RWG and HDP respectively: 0.153 and 0.048, which manifest HDP retains the properties of original real network better.

5.3 Anonymized User Linkage

Robustness of AUL (RQ2). We evaluate the robustness of AUL through a self-linking task. In this task, all users come from the same social network, so we can get complete one-to-one labeled data. In order to simulate data more accurately, the users' social connections will be randomly deleted to better understanding the performance of the algorithm. For Twitter network, we create a new dataset T_N (η), which is a subnet of Twitter. In this experiment, we choose 8 users with high social relevance and their ego-networks to construct subnet. The size of the subnet is denoted as N, and η represents the probability of each edge among the fringe vertices being deleted. In order to demonstrate the robustness of the AUL algorithm, we choose 500 users as the training

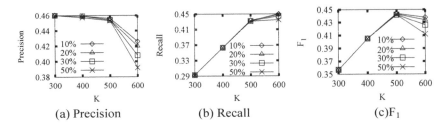

Fig. 3. The robustness of AUL

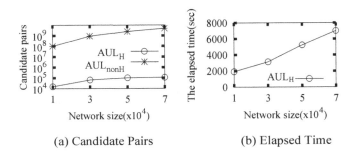

(a) Candidate Pairs   (b) Elapsed Time

Fig. 4. The scalability of AUL

set from the subnet and the remaining 564 users as the testing set. In this experiment, we observe the accuracy of AUL on $T_{564}(\eta)$ varying η from 10% to 50%. As shown in Figs. 3(a)–(c), as more and more social connections are deleted randomly, Precision@K, Recall@K, and F-measure@K drop dramatically. We observe that both of the Precision@K and Recall@K are about 41% in top-600 results. Even η is set to 50%, the precision is almost 40%.

Scalability of AUL (RQ3). We performed self-linking task on different size of social networks to evaluate the scalability of AUL. The dataset used in this experiment is TN (30%). The users are randomly sampled from the complete dataset while maintaining the social connections. Figure 4 manifest the changes in the number of candidate pairs on different size of social network and elapsed time of the algorithm in the last iteration. The user linkage on large-scale social networks with AUL employing LSH is denoted as AUL_H. While AUL_{nonH} means there is not blocking step with LSH in AUL algorithm. In Fig. 4(a), we can find that only less than 1‰ candidate pairs are remained after using LSH to block users. As shown in Fig. 4(b), AUL associated with LSH detects linked user pairs within two hours. This is rather efficient, given that we only run the algorithm in a single thread; for large-scale networks, one can easily scale up the algorithm parallelizing the candidate pair generation and social similarity computation in multiple threads.

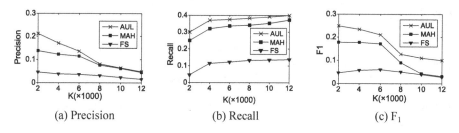

(a) Precision   (b) Recall   (c) F_1

Fig. 5. Performance of linking Twitter and Foursquare

Performance Comparison. We now turn to link two real social networks, namely complete Foursquare and Twitter. Figure 5 illustrates the performance of baselines and our proposed AUL, where we have the following key observations: (1) MAH and AUL are significantly better than FS approach. This points to the positive effect of modeling continuous distribution into probabilistic generative models. (2) AUL outperforms MAH significantly and achieves the best performance on two real social networks. This improvement demonstrates the effectiveness of our semi-supervised method.

6   Conclusion

We have presented HDP+AUL, a novel framework for anonymized user linkage under differential privacy. It combines differential privacy with anonymized user linkage simultaneously. To tackle the challenges, we first propose a Hierarchy Differential Privacy-based Generator (named HDP) to generate a social publishing graph. Then we propose Anonymized User Linkage algorithm (named AUL) to identify users across different social networks in a semi-supervised manner. To the best of our knowledge, this work is the first one to connect improper two separated research problems. Furthermore, extensive experiments on anonymized Twitter and Foursquare demonstrate the high performance of AUL, because HDP can provide more typical and rational datasets which achieve the promising balance between structural consistency and data availability under differential privacy.

In our future work, we desire to extend our work to link users on multiple networks maturely and develop a distributed algorithm to support more efficient computation and acquire better performance.

Acknowledgment. This work is supported by the Initial Scientific Research Fund of Introduced Talents in Anhui Polytechnic University (No. 2017YQQ015), Pre-research Project of National Natural Science Foundation of China (No. 2019yyzr03) and National Natural Science Foundation of China Youth Fund (No. 61300170).

References

1. Zhang, J., Yu, P.S.: Multiple anonymized social networks alignment. In: ICDM 2015, pp. 599–608 (2015)
2. Kong, X., Zhang, J., Yu, P.S.: Inferring anchor links across multiple heterogeneous social networks. In: CIKM 2013, pp. 179–188 (2013)
3. Day, W.-Y., Li, N., Lyu, M.: Publishing graph degree distribution with node differential privacy. In: SIGMOD 2016, pp. 123–138 (2016)
4. Zhou, B., Pei, J.: The k-anonymity and l-diversity approaches for privacy preservation in social networks against neighborhood attacks. Knowl. Inf. Syst. **28**(1), 47–77 (2011)
5. Hay, M., Li, C., Miklau, G., Jensen, D.D.: Accurate estimation of the degree distribution of private networks. In: ICDM 2009, pp. 169–178 (2009)
6. Li, X., Yang, J., et al.: Differential privacy for edge weights in social networks. In: Security and Communication Networks 2017, pp. 1–10 (2017)
7. Ivan, F., Alan, S.: A theory for record linkage. J. Am. Stat. Assoc. **64**(328), 1183–1210 (1969)
8. Wang, H., Hu, W., Qiu, Z., Du, B.: Nodes' evolution diversity and link prediction in social networks. IEEE Trans. Knowl. Data Eng. **29**(10), 2263–2274 (2017)
9. Zhu, L., Guo, D., et al.: Scalable temporal latent space inference for link prediction in dynamic social networks. IEEE Trans. Knowl. Data Eng. **28**(10), 2765–2777 (2016)
10. Martínez, V., Berzal, F., Cubero Talavera, J.C.: A survey of link prediction in complex networks. ACM Comput. Surv. **49**(4), 69:1–69:33 (2017)
11. Fu, C., Zhao, M., et al.: Link weight prediction using supervised learning methods and its application to yelp layered network. IEEE Trans. Knowl. Data Eng. **30**(8), 1507–1518 (2018)
12. Muniz, C.P., et al.: Combining contextual, temporal and topological information for unsupervised link prediction in social networks. Knowl.-Based Syst. **156**, 129–137 (2018)
13. Ozcan, A., et al.: Link prediction in evolving heterogeneous networks using the NARX neural networks. Knowl. Inf. Syst. **55**(2), 333–360 (2018)
14. Gao, M., Chen, L., He, X., Zhou, A.: InBiNE: bipartite network embedding. In: SIGIR 2018, pp. 715–724 (2018)
15. Wang, Z., Zhang, J., et al.: Knowledge graph embedding by translating on hyperplanes. In: AAAI 2014, pp. 1112–1119 (2014)
16. Perozzi, B., Al-Rfou, R., Skiena, S.: DeepWalk: online learning of social representations. In: KDD 2014, pp. 701–710 (2014)
17. Grover, A., Leskovec, J.: node2vec: Scalable feature learning for networks. In: KDD 2016, pp. 855–864 (2016)
18. Mikolov, T., Sutskever, I., Chen, K., Corrado, G.S., Dean, J.: Distributed representations of words and phrases and their compositionality. In: NIPS 2013, pp. 3111–3119 (2013)
19. Ayala-Rivera, V., McDonagh, P., et al.: Enhancing the utility of anonymized data by improving the quality of generalization hierarchies. Trans. Data Priv. **10**(1), 27–59 (2017)
20. Rao, A., Aljabar, P., Rueckert, D.: Hierarchical statistical shape analysis and prediction of sub-cortical brain structures. Med. Image Anal. **12**(1), 55–68 (2008)
21. Zhang, H., Fiszman, M., et al.: Degree centrality for semantic abstraction summarization of therapeutic studies. J. Biomed. Inform. **44**(5), 830–838 (2011)
22. Gao, M., Lim, E.-P., et al.: CNL: collective network linkage across heterogeneous social platforms. In: ICDM 2015, pp. 757–762 (2015)

Context Enrichment Model Based Framework for Sentiment Analysis

Nor Nadiah Yusof[(✉)], Azlinah Mohamed,
and Shuzlina Abdul-Rahman

Faculty of Computer and Mathematical Sciences, Universiti Teknologi MARA,
40150 Shah Alam, Selangor, Malaysia
nornadiah.yusof@gmail.com,
{azlinah, shuzlina}@tmsk.uitm.edu.my

Abstract. Online medium has become popular among knowledgeable society as a new emerging platform of information gathering and sharing where their thoughts and opinions are considered important in many aspects of nation building. The overwhelmed of online opinionated data has created a great challenge for researchers to mine sentiments accurately. Sentiment analysis or also known as opinion mining helps to understand and analyse those opinions due to the explosively growths of online opinionated data. This refers to the process of identifying and classifying the sentiments into several orientations, whether positive or negative opinions. Classifying sentiments based on the correct context remains a challenge, as the orientation of sentiments could change when the context changes. Existing sentiment analyser tend to classify sentiments based on keywords rather that the meaning of words. The real meaning of sentiment texts is often interpreted differently and consequently lead to poor classification results. Context plays an important role in analysing and classifying sentiment texts. This paper proposes a context enrichment model based framework for sentiment analysis in predicting the orientation of sentiments. The framework aims to provide a workflow for sentiments classification by considering the context and semantic information. There are four components in this framework: pre-processing, context enrichment model, classification and performance evaluation. This study is expected to create a new technique of classifying sentiments with more enriched semantic information.

Keywords: Context-based · Context enrichment · Sentiment classification

1 Introduction

Online opinionated data has greatly increased from day to day, as users eagerly express their thoughts and opinions about their point of interests. Many users gain benefits from the available opinions to help them in decision making, such as selecting the best products or services. However, the explosively increased volume of online opinionated data is a great challenge for users to identify the correct and related information based on the users' needs and preferences [1, 2]. Methods for handling the massive amount of these opinions have attracted the attention of researchers in sentiment analysis, also known as opinion mining. Sentiment analysis aims to analyse peoples' opinion and

© Springer Nature Singapore Pte Ltd. 2019
M. W. Berry et al. (Eds.): SCDS 2019, CCIS 1100, pp. 325–335, 2019.
https://doi.org/10.1007/978-981-15-0399-3_26

examine the score of sentiments whether positive or negative category [3]. Mining and analysing the online sentiments is a challenging task, especially to deal with a huge amount of opinionated data. Generally, this involves several main tasks, which are subjectivity detection, polarity identification and strength of polarity identification. Sentiment analysis is also classified into three main levels; document-level, sentence-level and aspect-level [4]. Document-level classifies the whole document of sentiment texts as either positive or negative review. Meanwhile, sentence-level chunks document into sentences. Each sentence is analysed and evaluated. Later, the polarity of each sentence is identified [5]. Aspect-level classifies the sentiments based on certain aspects of entities, which is used to study the existence of several different aspects of entities in a sentence. The scores of each aspect in multiple reviews are accumulated to obtain the polarity of sentiments [6].

Common sentiment analysis methods involve two main approaches: lexicon-based and machine learning [7]. The lexicon-based approach utilises dictionary of opinion words based on synonyms and antonyms of the words to identify the sentiments' polarity [8]. Meanwhile, machine learning approach identifies the text features of documents to be classified accordingly into polarities by training the labelled documents over a corpus. This requires a labelled training set to construct the supervised classifier and will utilises the available lexicons [9]. However, most lexicons are comprised as list of words rather than the meaning of words. The actual message or the meaning of sentiments is often wrongly interpreted by sentiment analysers. Context identification is essential to support the classification task in sentiment analysis, as it offers more semantic information encapsulated within documents [10]. Consequently, the accuracy of sentiment classification can be improved upon the existing context information. Thus, it is imperative to have a sophisticated approach which taken into account relevant context in document reviews. Therefore, this paper presents a framework for context enrichment model for sentiment analysis in discovering the context information to correctly classify sentiments into several orientations. The remainder of this paper is organized as follows. The next section describes related work in this area, followed by the proposed context enrichment model based framework with detail explanations for each component. Finally, this paper ends with a conclusion.

2   Related Work

Sentiment analysis offers imperative information to be discovered from the available online opinionated data provided by users. The data is significant to other users and organizations in many aspects, such as the decision-making process and marketing strategies. Grasping the challenges of analysing these sentiments, many researchers have explored and investigate the processes of sentiment analysis. Therefore, this section reviews several available frameworks in sentiment analysis. Lo and Potdar [11] studies on various frameworks of sentiment analysis that deliberates several techniques of opinion mining and sentiments classification from various literatures. The framework consists of seven main components: item extraction, feature extraction, comparison of items and features, classification of items, classification of features, classification of items and features, and finally strength of the sentiments. The evaluation of each component is based on identified parameters. Most researchers have focused on the components of feature extraction and classification of features, while few have considered comparison of items and features. Out of all frameworks reviewed, not many frameworks satisfied the parameters with context information such as the implementation of contextual clues, word n-grams and linguistic rules. Meanwhile, Wu and Tan [12] proposed a double-stage framework for cross-domain sentiment analysis. Their works focus on developing a bridge between the source domain and target domain. They focused only at ranking criteria which utilised SentiRank algorithm to classify sentiments in target domain. Thus, context information is solely based on the accurate labels of source domain data.

Hassan et al. [13] suggested a bootstrap ensemble framework for Twitter sentiment analysis. The nature of the Twitter environment, being diverse and dynamic, motivated them to tackle the sparse and imbalance data in Twitter analysis. The proposed framework is capable to construct time series which assists to reflect events in extracting high influence opinions from users. The semantic features of the proposed framework rely on lexical resources that provide scores for various word senses. Meanwhile, He et al. [14] proposed a novel framework of social media competitive analytics for the need of business marketing strategies. The main concern is to provide sentiment benchmark for organizations to analyse customers' sentiments and utilised it for companies marketing intelligence. Besides, they also develop an innovative social media tools to collect, monitor, and compare social media data for several brands and retail companies to provide recommendations and actions for marketing proposes. They utilised available tools to extract key concepts and generate categories to gain fast insights of textual data without concerning on the contextual information.

Although there have been some researches in developing the framework of sentiment analysis, there are not many researches focusing in the development of context-based framework in this area. Agarwal et al. [15] proposed a novel sentiment analysis model based on common sense knowledge and context information. They utilised three important factors, which are domain specific ontology, importance of features and contextual information, in determining the overall sentiment texts. Katz et al. [16] introduced a model to generate features of context terms based on key terms for supervised classification. The context terms are used to model the relation between identified key terms in order to identify the most suitable context for each key term. Muhammad et al. [17] proposed an approach to hybridise general purpose lexicons. Their strategies to capture contextual polarity from local and global contexts significantly improve the sentiment classification. Sharma et al. [10] proposed a contextualized sentiment scoring by utilising interrelationships between words in a review document to evaluate their semantic relevance to a particular sentiment. The scoring approach is based on neighbouring words and their influences on one another. Meanwhile, Yusof et al. [18] presented the state of art of context in sentiment analysis according to the level, type, and representation of context. The identification of contextual information in resolving ambiguities can help to identify the sentiment orientations. Therefore, considering the importance of recognising and utilising contextual knowledge in sentiment analysis, we proposed a context enrichment model based framework in sentiment analysis. The proposed approach acts as a bridge to fill the gap of classifying the sentiments with more enrich semantic information.

3   Context Enrichment Model

The proposed approach aims to develop a context enrichment model which provides contextual information in sentiment texts for document-level sentiment analysis and is illustrates as in Fig. 1. There are four main components in this framework; pre-processing, context enrichment model, classification and performance evaluation. A description of each component is provided in the following section.

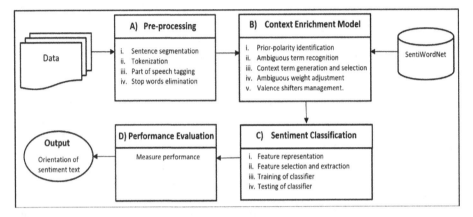

Fig. 1. Context enrichment model based framework for sentiment analysis.

3.1   Pre-processing

The pre-processing component is an important task in sentiment analysis. It is the process of cleaning and transforming the raw and unstructured data into understandable format [19]. This phase produces a clean dataset which is ready for further processing such as classification [20]. There are several benchmark datasets used in sentiment analysis from various domains such as movie reviews, travel destination reviews and product reviews [17, 21]. From the collections of benchmark datasets, the first step is to undergo sentence segmentation task, in which a document or a review is chunked into sentences. At this stage, the punctuation mark of full stop acts as the segmentation mechanism to break up the sentences in a document. Each sentence is then being tokened into the word level. Subsequently, part of speech tagging classifies words in sentences into classes such as noun, verb, adjective and adverb. This step helps to identify the possessions of words and become one of the most important indicators for further action in context enrichment model. There are 16-tag set of POS according to Penn treebank word classification. However, this research reclassifies the 16-tag into 4-tag word classes. Later, the stop words elimination process takes place. This removes common or unnecessary words and significant for dimension reduction purposes.

3.2   Context Enrichment Model

The context enrichment model acts as a bridge between the pre-processing and classification phases. The purpose of this model is to infuse semantic information for context identification and evaluation of context in the sentiment texts. At this stage, the construction of comprehensive context component is a key factor in this study. Basically, there are six main elements involves in the development of context enrichment model; prior-polarity identification, ambiguous term recognition, context term generation, context term selection, ambiguous weight adjustment, and valence shifters management. The identification of prior-polarity is by utilizing a lexicon called SentiWordNet. SentiWordNet is a well establish lexicon resource and widely used in sentiment analysis. It is a lexical resource that is derived from WordNet, which shows the positivity, negativity and neutrality of terms in a synset. Each synset consists of three numerical scores; Pos(s), Neg(s) and Obj(s) that has a value between 0 and 1 as shown in Eq. 1. The scores indicate how positive, negative and objective (neutral) the terms in the synset [23].

$$Pos(s) + Neg(s) + Obj(s) = 1 \qquad (1)$$

The structure of terms in SentiWordNet is represented in the following form: lemma#PoS#sense-number, where lemma refers to the term itself, PoS refers to part of speech tagging, and sense-number refers to the senses of term. A term in SentiWordNet may have several senses, which different senses of the same term may have different opinion-related properties and thus, it can have different polarities [24]. The smallest sense-number indicates the most common or the most frequent sense for the term. There are several approaches in selecting the senses of term in SentiWordNet such as random selection, first sense, mean and Lesk algorithm [25–27]. The second element in context enrichment model is ambiguous term recognition. An ambiguous term (AT) is a term which has the same positive and negative polarity values [15]. This situation shows that the term is neither incline to positive nor negative value. Thus, it creates an ambiguous situation of the term's orientation as the term's belonging of positivity and negativity value is equal. A term is considered as ambiguous if it fulfils both of these criteria; it has positive and negative scores which are more than zero and have equal values of positive and negative score.

For each AT identified in the previous step, it is crucial to generate the context terms (CT) in order to identify the most suitable context for AT and indirectly helps in ambiguity resolution [28]. The generation of CT can be done by considering the surrounding words of the AT in a sentence. Words located nearby in a sentence may have similar semantic concept. One method to generate CT is considering the adjacent terms for each ambiguous term in a sentence. In paradigmatic model of distributional hypothesis, the direction in which the context region is extended to the preceding or succeeding neighbours [29]. To generate the CT, the distributional information is collected by measuring the adjacent terms with threshold value of n = 3, or also known as 3 + 3 context window. The threshold value shows the searching space for CT, which means both the terms preceding and proceeding of each ambiguous term. When the threshold value is set to 3, there are three terms preceding the AT and three terms following the AT selected as candidates for context terms as shown in Fig. 2.

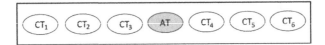

Fig. 2. The position of potential context terms for the identified ambiguous term.

Then, each candidate of CT with both positive and negative value will be assessed. Therefore, 12 candidates are selected as set of context terms in order to identify the most similar context for each ambiguous term. Generally, no research has identified a suitable threshold value or relevant context window in gaining the distributional information. However, several studies have indicated that a narrow context window, for example 2 + 2, is preferable to use for acquiring context information [29, 30]. The next step is the selection of the context term. At this stage, the distance between ambiguous term and context terms is measured by the term similarity. Similarity measure describes how much the ambiguous terms are close to positive or negative value in the sentiment texts. Term with highest similarity value is selected for further step as it gives the most similar semantic meaning to the ambiguous term. To measure the similarity value, cosine similarity is carried out, as shown in Eq. 2.

$$similarity_{cosine}(x, y) = \frac{\sum_{i=1}^{n} x_i y_i}{\sqrt{\sum_{i=1}^{n} x_i^2} \times \sqrt{\sum_{i=1}^{n} y_i^2}} \tag{2}$$

where x_i = polarity of ambiguous term and y_i = polarity of neighbouring term according to the context window. All senses for each CT for both positive and negative values are taken into account and assess as shown in Fig. 3. The term with highest similarity value is selected for weight adjustment of ambiguous term. After CT is selected for each AT, the weight (polarity) of AT is adjusted accordingly. Weight adjustment for AT is to tune the AT with positive or negative values based on the CT identified. The proposed adjusted weight for AT is by summation of current polarity of AT with average polarity value between AT and CT.

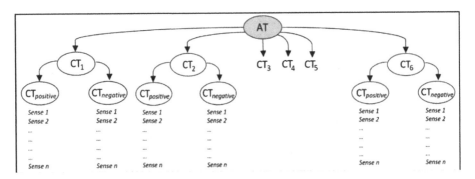

Fig. 3. The alignment which measure the similarity between AT and CT by taken into account all senses of each CT for both positive and negative value.

Finally, the last element in context enrichment model is valence shifters management. Valence shifters are words which modify the degree of positivity or negativity of opinion words [31]. They are typically used to increase, decrease, or negate sentiment polarity strength of terms, sentences or documents. There are two main tasks at this stage; modifiers and negation handling. Modifiers enhance or reduce the polarity strength of sentiment words in a sentence [32]. It changes the semantic intensity in two ways; intensifier and diminisher. Let M_i be a list of intensifier and M_d be a list of diminisher. M_i and M_d are subsets to modifiers Mod_s represented as $M_i = \{list\ of\ intensifier\} \in Mod_s$ and $M_d = \{list\ of\ diminisher\} \in Mod_s$. If a word is found in Mod_s, then the modifier's adjacent word polarity is computed as shown in Eq. 3.

$$Polarity_{adj\_word} = polarity\ (w_m) + (polarity\ (w_m) \times modifier_{strength}\ (w_s))$$
$$if\ (r \in R \wedge (w_s \in Mod_s)) \tag{3}$$

where $polarity\ (w_m)$ is the polarity of modifier's adjacent word, $modifier_{strength}\ (w_s)$ is the strength of Mod_s and r is a review from set of reviews R.

Meanwhile, negation handling is one of sentiment analysis challenges that highly affect the accuracy of sentiment analyser [3]. Negation words affect document's polarity as it inverts the meaning of both terms and sentences in a document [33]. With the existence of negation words in a sentence, it's important to identify the scope of negation. The scope of negation denotes the sequence of terms in a sentence that are affected by negation words [34, 35]. The scope of negation in this study is based on the nearby sequence of terms in a sentence when negation terms occur. Let Neg be a list of negation terms, represent as $Neg = \{list\ of\ negation\ words\}$. If a word is found in Neg, then the negation's adjacent word polarity is computed as shown in Eq. 4.

$$Polarity_{adj\_word} = (polarity\ (w_n) \times (-1)),$$
$$if\ ((r \in R \wedge ((w_n - 1) \in Neg)) \vee (r \in R \wedge ((w_n - 1) \in Mod_s) \wedge ((w_n - 2) \in Neg)))$$
$$\tag{4}$$

where $polarity\ (w_n)$ is the polarity of negation's adjacent word, Mod_s is a list of modifiers terms, Neg is a list of negation terms and r is a review from set of reviews R.

3.3   Classification

The output from context enrichment model is infused into supervised machine learning algorithms for classification task. Therefore, a task called feature representation is required before classification process takes place. In feature representation task, sentiment texts are vectorised into term-document matrix format, in which the terms for each document are transformed in the following format:

$$Label_{instance\_n}\quad feature_1 : polarity_1\quad feature_2 : polarity_2 \ldots\ldots\ldots feature_n : polarity_n$$

where $Label_{instance\_n}$ refers to the document polarity, whether positive or negative document review, $feature_n$ is the terms in the document review, and $polarity_n$ is the weight for the particular term based on the polarity value extracted from the context enrichment model as explained in the previous step. Then, feature selection is performed in order to extract the most important features with high discriminating power for term-document matrix representation. Feature extraction is then performed to reduce the number of terms in a document by transforming the large number of attributes into a reduced set of features. The number of features or the size of term-document matrix for each document is experimentally evaluated, such as 100, 500, 1000 and 3000 terms for each document [36, 37].

Consequently, the classification task, which consists of training and testing phase, is carried out. In the training phase, the goal is to construct a classification model that fits the parameters in the classifier and to find the optimal weight parameters [38]. The labelled dataset will be trained into the defined class labels. In the testing phase, the performance of the trained classifier will be tested using another set of data in order to evaluate the classification model. Generally, there are several types of supervised machine learning approach to undergo the classification task. The algorithms of supervised machine learning approach consist of Naïve Bayes, Bayesian Network, Artificial Neural Network, Support Vector Machine (SVM), J48 and Random Forests [7, 39]. Moreover, many researchers had investigated the capabilities and performance of the learning algorithms. SVM delivers the most stable and highest accuracy as compared to other machine learning algorithm in sentiment analysis area [40, 41]. The output from the classifier is the polarity of sentiment texts. Additionally, the strength of sentiment polarities will be identified.

3.4 Performance Evaluation

The performance evaluation is an important aspect in measuring the ability of classifier that correctly classifies sentiments. The performance of classifier can be measured based on parameters in contingency table as shown in Table 1. TP denotes for true positive, TN is true negative, FP is false positive and FN is false negative. TP and TN measure the values of positive and negative values that are correctly identified according to actual class. Contrarily, FP and FN indicate the condition of positive and negative values of predicted class that are incorrectly classified according to the actual class. The most common performance evaluation is based on accuracy and is calculated based on the formula $P = (TP + TN)/(TP + TN + FP + FN)$ [42]. Accuracy shows the probability of true value of the class label.

Table 1. Contingency table.

	Predicted class	
Actual class	*Positive*	*Negative*
Positive	TP	FP
Negative	FN	TN

4 Conclusion

Classifying sentiments into orientations is a challenging task due to the exponential growth of online opinionated data. Many studies have explored various approaches of analysing and classifying the opinions in sentiment analysis. However, most approaches tend to ignore the importance of having context information in classifying sentiments. The proposed framework aims to incorporate semantic information by considering the context of sentiment texts to assist the classification algorithm. All four components in the framework, including pre-processing, context enrichment model, classification, and performance evaluation are briefly discussed and elaborated. The significant of this study is to provide an enhanced approach to classifying sentiments with correct context information. It is expected that the performance of sentiment analyser may further be improved.

Acknowledgement. The authors would like to thank Faculty of Computer and Mathematical Sciences and Research Management Centre of Universiti Teknologi MARA (LESTARI 0019/2016).

References

1. Mohamed, A., Najafabadi, M.K., Wah, Y.B., Zaman, E.A.K., Maskat, R.: The state of the art and taxonomy of big data analytics: view from new big data framework. Artif. Intell. Rev. 1–49 (2019)
2. Najafabadi, M.K., Mohamed, A., Onn, C.W.: An impact of time and item influencer in collaborative filtering recommendations using graph-based model. Inf. Process. Manag. **56**(3), 526–540 (2019)
3. Hussein, D.M.E.D.M.: A survey on sentiment analysis challenges. J. King Saud Univ.-Eng. Sci. **30**(4), 330–338 (2018)
4. Kaur, A., Gupta, V.: A survey on sentiment analysis and opinion mining techniques. J. Emerg. Technol. Web Intell. **5**(4), 367–371 (2013)
5. Appel, O., Chiclana, F., Carter, J., Fujita, H.: A hybrid approach to the sentiment analysis problem at the sentence level. Knowl.-Based Syst. **108**, 110–124 (2016)
6. Schouten, K., Frasincar, F.: Survey on aspect-level sentiment analysis. IEEE Trans. Knowl. Data Eng. **28**(3), 813–830 (2015)
7. Yusof, N.N., Mohamed, A., Abdul-Rahman, S.: Reviewing classification approaches in sentiment analysis. In: Berry, M., Mohamed, A., Wah, Y.B. (eds.) SCDS 2015. CCIS, vol. 545, pp. 43–53. Springer, Singapore (2015). https://doi.org/10.1007/978-981-287-936-3_5
8. Feldman, R.: Techniques and applications for sentiment analysis. Commun. ACM **56**(4), 82–89 (2013)
9. Medhat, W., Hassan, A., Korashy, H.: Sentiment analysis algorithms and applications: a survey. Ain Shams Eng. J. **5**(4), 1093–1113 (2014)
10. Sharma, S., Chakraverty, S., Sharma, A., Kaur, J.: A context-based algorithm for sentiment analysis. IJCVR **7**(5), 558–573 (2017)
11. Lo, Y.W., Potdar, V.: A review of opinion mining and sentiment classification framework in social networks. In: 2009 3rd IEEE International Conference on Digital Ecosystems and Technologies, pp. 396–401. IEEE (2009)

12. Wu, Q., Tan, S.: A two-stage framework for cross-domain sentiment classification. Expert Syst. Appl. **38**(11), 14269–14275 (2011)
13. Hassan, A., Abbasi, A., Zeng, D.: Twitter sentiment analysis: a bootstrap ensemble framework. In: 2013 International Conference on Social Computing, pp. 357–364. IEEE (2013)
14. He, W., Wu, H., Yan, G., Akula, V., Shen, J.: A novel social media competitive analytics framework with sentiment benchmarks. Inf. Manag. **52**(7), 801–812 (2015)
15. Agarwal, B., Mittal, N., Bansal, P., Garg, S.: Sentiment analysis using common-sense and context information. Comput. Intell. Neurosci. **2015**, 30 (2015)
16. Katz, G., Ofek, N., Shapira, B.: ConSent: context-based sentiment analysis. Knowl.-Based Syst. **84**, 162–178 (2015)
17. Muhammad, A., Wiratunga, N., Lothian, R.: Contextual sentiment analysis for social media genres. Knowl.-Based Syst. **108**, 92–101 (2016)
18. Yusof, N.N., Mohamed, A., Abdul-Rahman, S.: A review of contextual information for context-based approach in sentiment analysis. Int. J. Mach. Learn. Comput. **8**, 399–403 (2018)
19. Rusli, M.F., Aziz, M.A., Aris, S.R.S., Jasri, N.A., Maskat, R.: Understanding Malaysian English (Manglish) jargon in social media. J. Fundam. Appl. Sci. **10**(2S), 116–125 (2018)
20. Kansal, H., Toshniwal, D.: Aspect based summarization of context dependent opinion words. Procedia Comput. Sci. **35**, 166–175 (2014)
21. Weichselbraun, A., Gindl, S., Scharl, A.: Extracting and grounding contextualized sentiment lexicons. IEEE Intell. Syst. **28**(2), 39–46 (2013)
22. Ye, Q., Zhang, Z., Law, R.: Sentiment classification of online reviews to travel destinations by supervised machine learning approaches. Expert Syst. Appl. **36**(3), 6527–6535 (2009)
23. Baccianella, S., Esuli, A., Sebastiani, F.: Sentiwordnet 3.0: an enhanced lexical resource for sentiment analysis and opinion mining. Lrec **10**, 2200–2204 (2010)
24. Guerini, M., Gatti, L., Turchi, M.: Sentiment analysis: how to derive prior polarities from SentiWordNet. In: Proceedings of the 2013 Conference on Empirical Methods in Natural Language Processing, pp. 1259–1269 (2013)
25. Li, X., Dong, Y., Wang, G.G., Hou, M.: Prior polarity dictionary derived from SentiWordNet based on random forest algorithm. In: 2017 2nd International Conference on Automation, Mechanical Control and Computational Engineering (AMCCE 2017). Atlantis Press (2017)
26. Lesk, M.: Automatic sense disambiguation using machine readable dictionaries: how to tell a pine cone from an ice cream cone. In: Proceedings of the 5th Annual International Conference on Systems Documentation, pp. 24–26. ACM (1986)
27. Pandit, R., Sengupta, S., Naskar, S.K., Sardar, M.M.: Improving Lesk by incorporating priority for word sense disambiguation. In: 2018 Fifth International Conference on Emerging Applications of Information Technology (EAIT), pp. 1–4. IEEE (2018)
28. Abd-Rashid, A., Abdul-Rahman, S., Yusof, N.N., Mohamed, A.: Word sense disambiguation using fuzzy semantic-based string similarity model. Malays. J. Comput. **3**(2), 154–161 (2018)
29. Sahlgren, M.: The distributional hypothesis. Ital. J. Disabil. Stud. **20**, 33–53 (2008)
30. Akhtar, M.S., Gupta, D., Ekbal, A., Bhattacharyya, P.: Feature selection and ensemble construction: a two-step method for aspect based sentiment analysis. Knowl.-Based Syst. **125**, 116–135 (2017)
31. Cheng, R., Loh, J.M.: Learning-based method with valence shifters for sentiment analysis. In: 2017 IEEE International Conference on Data Mining Workshops (ICDMW), pp. 357–364. IEEE (2017)

32. Farooq, U., Mansoor, H., Nongaillard, A., Ouzrout, Y., Abdul Qadir, M.: Negation handling in sentiment analysis sentiment classification using enhanced contextual valence shifters at sentence level. J. Comput. **12**, 470–478 (2017)
33. Phu, V.N., Tuoi, P.T.: Sentiment classification using enhanced contextual valence shifters. In: 2014 International Conference on Asian Language Processing (IALP), pp. 224–229. IEEE (2014)
34. Sharif, W., Samsudin, N.A., Deris, M.M., Naseem, R.: Effect of negation in sentiment analysis. In: 6th International Conference on Innovative Computing Technology, INTECH 2016, pp. 718–723 (2017)
35. Diamantini, C., Mircoli, A., Potena, D.: A negation handling technique for sentiment analysis. In: 2016 International Conference on Collaboration Technologies and Systems (CTS), pp. 188–195. IEEE (2016)
36. Moraes, R., Valiati, J.F., Neto, W.P.G.: Document-level sentiment classification: an empirical comparison between SVM and ANN. Expert Syst. Appl. **40**(2), 621–633 (2013)
37. Kim, K.: An improved semi-supervised dimensionality reduction using feature weighting: application to sentiment analysis. Expert Syst. Appl. **109**, 49–65 (2018)
38. Tripathi, G., Naganna, S.: Feature selection and classification approach for sentiment analysis. Mach. Learn. Appl. Int. J. **2**(2), 1–16 (2015)
39. Razak, Z.I., Abdul-Rahman, S., Mutalib, S., Abdul Hamid, N.H.: Web mining in classifying youth emotions. Malays. J. Comput. **3**(1), 1–11 (2018)
40. Ghiassi, M., Lee, S.: A domain transferable lexicon set for Twitter sentiment analysis using a supervised machine learning approach. Expert Syst. Appl. **106**, 197–216 (2018)
41. Zou, H., Tang, X., Xie, B., Liu, B.: Sentiment classification using machine learning techniques with syntax features. In: 2015 International Conference on Computational Science and Computational Intelligence (CSCI), pp. 175–179. IEEE (2015)
42. Tripathy, A., Agrawal, A., Rath, S.K.: Classification of sentiment reviews using n-gram machine learning approach. Expert Syst. Appl. **57**, 117–126 (2016)

2019 Thai General Election: A Twitter Analysis

Chamemee Prasertdum[1] and Duangdao Wichadakul[2(✉)]

[1] Department of Computer Engineer, Faculty of Engineering,
Chulalongkorn University, Bangkok, Thailand
6070439021@student.chula.ac.th
[2] Chulalongkorn Big Data Analytics and IoT Center (CUBIC),
Department of Computer Engineer, Faculty of Engineering,
Chulalongkorn University, Bangkok, Thailand
duangdao.w@chula.ac.th

Abstract. Elections are the most important part in the democracy. The general election in Thailand was held on 24 March 2019. It was the first vote in the last five years since the seizure of power by the military coup leaded by General Prayut Chan-o-cha. A social media has become a part of daily life with a real time message distribution including sharing an opinion in politics. Twitter becomes the most widely used tool for creating trends by political parties and politicians. In this paper, we propose an analysis of election result according to data analysis from Twitter and Election Commission of Thailand (ECT). We used Twitter Advanced Search to collect data within three months from January 1 to March 31, 2019. We found the top keywords and hashtags trended on Twitter as the Future Forward Party (FFP). Meanwhile, the winner of the popular vote was the Palang Pracharat Party. We found no correlation between the number of retweets and votes as only one party got the outstanding number of retweets. From the analysis of retweeters' accounts, the characteristics of retweeters of the Future Forward Party differed from the retweeters of other parties. We then refined the analysis based on the sentiment of messages mentioning the political parties tweeted by the retweeters. The correlation coefficient between the number of positive mentions and votes became 0.615 with the removal of the FFP from the analysis. In addition, we found that the number of messages with a specific political party mentioned might reflect the election results. Finally, the TF-IDF analysis of words from the tweet messages during the campaign showed that words delivered by each party had different contextual meanings such as paying attention to people, party's policy, complaint, and blame.

Keywords: Twitter · Social media mining · Election

1 Introduction

Online social networks are easily and quickly disclosed on the internet. Twitter is like another online tool, which is used to communicate with people. Users can comment with the 280 characters in real time [1]. Most user groups are among the 16–24-year-olds [2]. Based on the forecast number of Twitter consumer in Thailand, the number of active Twitter consumer is expected to grow by 4.1 million in 2019, up from 2.7 million in 2014.

© Springer Nature Singapore Pte Ltd. 2019
M. W. Berry et al. (Eds.): SCDS 2019, CCIS 1100, pp. 336–350, 2019.
https://doi.org/10.1007/978-981-15-0399-3_27

Twitter data analysis has received a lot of attention over the past few years. It has played an important role in the elections such as predicting the outcomes of political events. For example, in the 2016 US presidential election, plenty of people supported Donald Trump without speaking it out loud [3]. Therefore, the prediction of US election depended on social media analysis and generated new results with more accurate than traditional approach. In Thailand, almost 52 million people were eligible voters in the election on March 24, 2019. This was the first Thai election in 8 years with about 8 million new voters. The first-time voters were more enthusiastic about the election than ever. The survey resulted from the Siam Internet Poll Research Institute, Siam Technology College (STC) found about 61% of the samples admitted that they were excited for their first election voting [4]. Such samples also posted many comments and tweets on Twitter. For example, in early January 2019, after the deputy prime minister announced the deferral of the election from 24th February, there were numerous responses on Twitter. The hashtag #เลื่อนแม่มึงสิ ("Delayed again, you motherfucker") was trended on Twitter as the first rank in Thailand and 1 of the top 10 global trends on Twitter.

In our research, we analyzed Twitter data from the retweets, tweets, hashtags, and keywords. Twitter is an online media that focuses on interaction. User can comment to show their political power. The paper is organized as follow: Sect. 2 presents the literature studied. Section 3 explains the approach of data collection and technique. Section 4 presents our analysis results with the discussions. We conclude the paper in Sect. 5.

2   Related Work

The occurrence of social media has highlighted the significance of interpersonal communicators and relations in the diffusion of information and behavior in public areas. Through the use of tweeting and mentioning functions, Twitter allows shared chat context by exchanging messages [5]. In Twitter, there are many specific topics such as political events, disasters, influenza or important events in the time period.

Pudtal, et al. [6] gathered Twitter data associated with His Majesty King Bhumibol Adulyadej passing away for the period from 1 October to 31 December 2016. The authors analyzed and compared this specific event with other events that were mentioned, for example, earthquakes off the west coast of Sumatra in Indonesia. The authors proposed a method which could extract events from tweets, by considering the changing ratio of the number of retweets and distinct tweets of each word.

Culotta [7] analyzed more than 500-million Twitter messages in the eight-month period from September 2009 to May 2010. This work proposed methods to track influenza's epidemic rates from a large number of Twitter messages, with the best method achieving 95% correlation with national statistics data.

Yoshida, et al. [8] analyzed the retweeters of each political party in the 48th parliamentary election of Japan to identify their characteristics. The correlations of the retweet rates among the parties were calculated. A high correlation between two parties indicated the similar characteristics of their retweeters. They that found the retweeters' personality of @jimin koho (The Liberal Democratic Party) and @CDP2017 (The Constitutional Democratic Party) were similar. Meanwhile, the retweeters' characteristics of @jcp cc (The Japanese Communist Party) and @osaka ishin (The Japan Innovation Party) were similar.

Buccoliero, et al. [9] analyzed the influence of social media in the US presidential election in 2016. The authors found that both candidates generated tweets differently in both quality and quantity. These symbolized their personality and attitudes. The authors tried to emphasize the main individuality between the Twitter tactics of Clinton and Trump. The quantitative analysis result of tweets reported the candidates' average daily tweets in each month. Clinton tweeted 21.56 times a day on average while Trump tweeted 13.25 times. From word frequency analysis: Trump wrote '@realdonaldtrump' 787 times while Clinton wrote 'Hillary' 1464 times.

Soler, et al. [10] presented a tool to analyze conversations on Twitter, called TaraTweet. The TaraTweet is a web application for predicting the results of elections. They used Spanish elections during the year 2011 and 2012 to investigate the use of Twitter to find out if the stored conversation can predict the outcome of the election. They analyzed 500,000 tweets with three experiments. They identified the correlation between the Twitter's mentions and actual votes. This paper indicated that Twitter may be the right tool to predict political election results.

3   Methodology

Twitter has more than 300 million new subscribers per month [11]. It is a popular social network, especially for short messages called tweets. In this research, the 2019 Thai general election data were retrieved from Selenium & Twitter Advanced Search - https://twitter.com/search-advanced). We collected election data from January 1, 2019 to March 31, 2019. During the period of study, the total number of downloaded tweets were 668,799 with 64 million retweets. PyThaiNLP library [12] was used to perform Thai word tokenization. We also created a dictionary of stop words used in Twitter. After that, we used TF-IDF to find the most important words among all tweets from the political parties. Also, we performed the sentiment analysis using PyThaiNLP on the tweets mentioning the political parties. These tweets, which contained either the @party name or @leader name, were collected from the retweeters for the same time period and separated into groups according to the mentioned party.

4   Analysis of Twitter Data

4.1   Change of the Keywords in Each Political Party

There were approximately 22,737 keywords from the tweet messages related to Thailand's election in 2019. Keywords included hashtags and words used to represent phrases in the election. The hashtag #ฟ้ารักพ่อ (Fah Loves Daddy) represented the adoration for the Future Forward Party, which has Thanathorn Juangroongruangkit as the leader. We collected all keywords related to 20 political parties including Future Forward Party, Phalang Pracharath Party, Phue Thai Party, New Economics Party, Democrat Party, People Power Party, Chartthaipattana Party, Seri Ruam Thai Party, Prachachart Party, Action Coalition for Thailand Party, Bhumjaitha Party, Thai Local Power Party, Thai Civilized Party, Prachaniyom Party, Thai Nation Power Party, and Puea Chat Party as shown in Table 1. The number of retweets of keywords was directly related to each specific party.

Table 1. The number of retweets of keywords related to each political party

Top keywords	Political parties	#retweets
พรรคอนาคตใหม่, อนาคตใหม่, ฟ้า, ธนาธร, ปิยบุตร, อนาคต, พรรณิการ์, หมอเก่ง, ฟ้ารักพ่อ, ทิม, วาโย, ใหม่, บั้งไฟ, พรรคส้ม, คุณช่อ, ธ, ทุกท่านกับ, อนค, ติ่งส้ม, พรรคอนาคต, จุ้ย, ธนากร, FWPthailand	อนาคตใหม่ (Future Forward)	10,403,073
พลังประชารัฐ, พรรคพลังประชารัฐ, ประยุทธ์, ลุงตู่, พปชร, พลังประชา, สนธิรัตน์, บิ๊กตู่, มาดามเดียร์, ประยุทธ, ประชารัฐ, ตุ่งตู่	พลังประชารัฐ (Phalang Pracharath)	2,529,782
เพื่อไทย, ชัชชาติ, พรรคเพื่อไทย, คุณหญิง, สุดารัตน์, แม่หน่อย, พท	เพื่อไทย (Pheu Thai)	1,907,475
เศรษฐกิจใหม่, มิ่งขวัญ, ลุงมิ่ง, เศรษฐกิจ, พีมิ่ง, พรรคเศรษฐกิจใหม่, ลุงมิ่งใจดี, มิ่ง	เศรษฐกิจใหม่ (New Economics)	1,408,315
อภิสิทธิ์, ประชาธิปัตย์, หมอเอ้ก, ไอติม, พรรคประชาธิปัตย์, พริษฐ์, ประชาธิปัตย์, ปชป, คุณชวน, แมว, คุณ, พรรคประชาธิปัตย์, เอ้ก, ป	ประชาธิปัตย์ (Democrat)	584,284
พลังปวงชนไทย, นิคม, ปวงชนไทย, พรรคพลังปวงชนไทย	พลังปวงชนไทย (People Power)	466,198
เพื่อชาติ, พรรคเพื่อชาติ, สงคราม, จตุพร	เพื่อชาติ (Puea Chat)	399941
เสรีพิศุทธ์, เสรีรวมไทย, เสรี, พรรคเสรีรวมไทย, วีรบุรุษนาแก, ป๋าเสรี	เสรีรวมไทย (Seri Ruam Thai)	386136
ประชาชาติ, วันมูหะมัดนอร์, พรรคประชาชาติ, วันนอร์, ทวี	ประชาชาติ (Prachachart)	360,266
ภูมิใจไทย, พรรคภูมิใจไทย, ภูมิใจ, พรรคภูมิใจ, อนุทิน, กัญชา, ภท	ภูมิใจไทย (Bhumjaithai)	173,861
รวมพลังประชาชาติไทย, สุเทพ, พรรครวมพลังประชาชาติไทย, เอนก, ลุงกำนัน	รวมพลังประชาชาติไทย (Action Coalition for Thailand)	122,202
พลังท้องถิ่นไท, พรรคพลังท้องถิ่นไท, ฟิล่ม	พลังท้องถิ่นไท (Thai Local Power)	26,441
ไทยศรีวิไลย์, ไทย, พรรคไทยศรีวิไลย์, มงคลกิตติ์	ไทยศรีวิไลย์ (Thai Civilized)	26,282
ชาติพัฒนา, พรรคชาติพัฒนา, สุวัจน์, ชพน	ชาติพัฒนา (Chartpattana)	13786
ประชานิยม	ประชานิยม (Prachaniyom)	3,202
พลังไทยรักไทย, พรรคพลังไทยรักไทย	พลังไทยรักไทย (Phalang Thai Rak Thai)	3,158
พรรครักษ์, รักษ์ผืนป่าประเทศไทย, รักษ์, ดำรงค์	รักษ์ผืนป่าประเทศไทย (Thai Forest Conservation)	2,477
พลังชาติไทย, พรรคพลังชาติไทย, ทรงกลด	พลังชาติไทย (Thai Nation Power)	1,568
ประชาภิวัฒน์, พรรคประชาภิวัฒน์, วรวงศ์	ประชาภิวัฒน์ (People Progressive)	1,442
ชาติไทยพัฒนา, พรรคชาติไทยพัฒนา, หนูนา	ชาติไทยพัฒนา (Chartthaipattana)	1,251

4.2 Analysis of Retweets, Tweets and Votes

We considered 5 parties with the highest votes including the Future Forward Party (FFP), Phalang Pracharath Party (PPRP), Phue Thai Party (PTP), Democrat Party (DP), and Bhumjaithai Party (BP) and computed the number of retweets according to the context of each political party. In this aspect, we intended to test if the total number of retweets could be used to predict the outcome of the election. The election and specific information of the election system might lead to too many or too few votes. The number of retweets from each political party and votes for each party earned in the election were collected as shown in Table 2.

Table 2. The number of retweets and votes

Political parties	# of retweets	# of votes
FFP	8,733,877	6,265,950
PPRP	2,074,872	8,433,137
PTP	941,865	7,920,630
DP	503,288	3,947,726
BP	23,319	3,732,883
Total	12,277,221	30,300,326

The values in Table 2 was then converted into percentages as shown in Table 3. We noticed that the party of the Future Forward Party received a much higher percentage of retweets than votes while the percentages of the two values of the other four parties were on the opposite (Table 3). There are many ways to measure a forecast accuracy or find the difference between two variables. Here, we used the Mean Absolute Error [13], also known as MAE, to evaluate the overall relationship between the percentage of retweets and votes. If the MAE value is low, the two values can be estimated as close to each other. In our prediction, the mean absolute error (MAE) between the two values was 20.18%. The mean absolute error (MAE) in our case study was higher than reported by Marko, et al. [14] and Tumasjan, et al. [15] with an average error of 5.23 and 1.65 respectively.

Table 3. Percentage of retweets and votes for each political party

Political parties	% retweets	% votes	% error
FFP	71.14	20.68	50.46
PPRP	16.90	27.83	−10.93
PTP	7.67	26.14	−18.47
DP	4.10	13.03	−8.93
BP	0.19	12.32	−12.13
Total			20.18

Figure 1 Illustrates the top 5 political parties with R^2 equaled to 0.0761. We summarized that two variables are were not related.

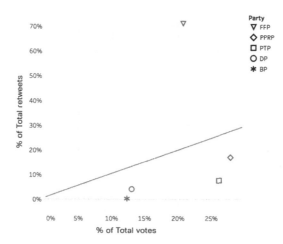

Fig. 1. The correlation coefficient of retweets and votes of the top 5 political parties

We further performed the sentiment analysis on tweet messages that mentioned each political party, i.e., including either @party name or @leader name, to investigate the retweeters' opinions on the parties. We also examined the relationship between the number of positive and negative mentions and votes. The total number of collected mentions was 1729 with the 1182 positives and 547 negatives. Examples of the positive mentioning tweets were the giving encouragement to Mr. Thanathorn, the leader of the Future Forward Party, after receiving a summons from the prosecutor. The #savethanathorn hashtag was the top trend within the mentioning tweets. Examples of the negative tweets were about political harassments such as the state officials violating civil liberties. The number of the positive and negative mentions and votes for each party were collected as shown in Tables 4 and 5.

Table 4. The number of positive mentions and votes

Political parties	# mention positive	% mention positive	% votes	% error positive
FFP	899	47.77	20.68	27.09
PPRP	403	21.41	27.83	−6.42
PTP	242	12.86	26.14	−13.28
DP	232	12.33	13.03	−0.7
BP	106	5.63	12.32	−6.69
Total				10.84

Table 5. The number of negative mentions and votes

Political parties	# mention negative	% mention negative	% votes	% error negative
FFP	257	46.98	20.68	26.30
PPRP	122	22.30	27.83	−5.53
PTP	53	9.69	26.14	−16.45
DP	98	17.92	13.03	4.89
BP	17	3.11	12.32	−9.21
Total				12.47

Figure 2 shows the linear association between the number of positive and negative mentions and the number of votes in the election, with R^2 equaled to 0.0996 and 0.0588 respectively. The positive and negative mentions of the Democrat Party and Phalang Pracharath Part were closely related with their received votes. When we removed the Future Forward Party from the analysis, the correlation coefficient of the positive mentions and votes became higher with R^2 equaled to 0.615 while R^2 of the negative mentions and votes was 0.196.

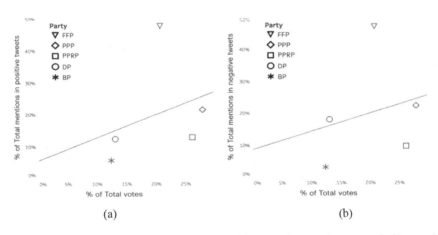

(a)   (b)

Fig. 2. The correlation coefficient of the (a) positive mentions and votes and (b) negative mentions and votes of the top 5 political parties

4.3   Analysis of Political Tweets

We observed the number of tweets and retweets during the campaign and evaluated how they evolved over time. The number of political tweets and retweets increased from the date of the official election announcement and maximized before the election day. The maximum of tweets and retweets were not in election day because Thailand has a law not allowing any political-related campaign and discussion in public starting from 6 p.m. of the day before the election day. This prohibition ended at 8 p.m. of the election day. The rate of tweets and retweets increased again when the ballot counts had been announced. From the poll result of 1,266 students from Chulalongkorn University, the Future Forward Party leader, Thanathorn Juangroongruangkit, was recognized as a politician with clear visions. Meanwhile, they also liked his ideas the most [16].

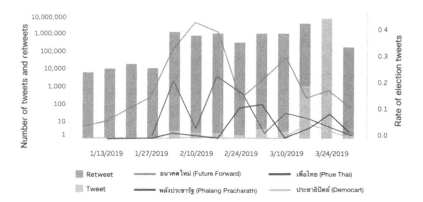

Fig. 3. The number of election-related tweets and retweets during the election period (from January 1 to March 31, 2019)

In addition, we found that the first-time voter group, which was the most social media familiarized group, made the new political party like the Future Forward Party very popular and being mentioned. Therefore, the rate of the Future Forward Party retweets was higher than other parties. This, however, might cause the Swing Vote phenomenon [17] which will affect the change of the votes. In Fig. 3, the Future Forward party generated the highest tweets followed by the Phalang Pracharath Party, the Phue Thai Party, and the Democrat Party. On March 24, 2019, the election day, the tweet rate of Phue Thai Party was higher than the Phalang Pracharath Party's, starting after the election closed. These two parties got a close number of votes.

4.4   Grouping User in Each Party

The distribution of information on Twitter can be done via the retweet function. We grouped the rate of retweets of each party's tweets to analyze the similarities of the tweeting users of each party. In Thailand, there are only four parties that the party leader has his or her own account. We collected the tweet and retweet data from the four accounts of the political party leaders as shown in Table 6.

Table 6. The number of tweets, retweets and followers of the 4 political party leaders: Thanathorn (Future Forward Party), Sudarat (Phue Thai Party), Prayuth (Phalang Pracharat Party) and Abhisit (Democrat Party)

Screen name	# of retweets	# of tweets	# of followers
@Thanathorn_FWP	2,091,209	88	376,927
@sudaratofficial	501,832	55	93,059
@prayutofficial	33,859	39	38,215
@Abhisit_DP	29,177	226	392,110

The political party leaders were not equally mentioned throughout the campaign. General Prayut, who is the current prime minister tweeted a small number of messages from account @prayutofficial. However, he was mentioned by 62.30% on March, 24. The percentage of mentions computed from the number of daily tweets mentioning one of the four specific leaders divided by the total number of tweets mentioning any of them. The account of Thanathorn (@Thanathorn_FWP) had the largest number of retweets. He was mentioned about 38.25% on March, 24. Figure 4 shows details about the evolved number of tweet messages mentioning each political leader.

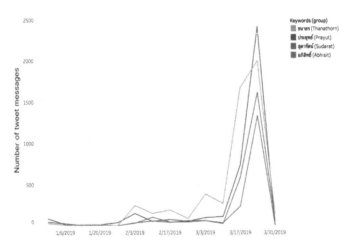

Fig. 4. The number of tweets mentioning the four political parties from January 1 - March 31, 2019

To find the similarity of accounts that retweeted from a specific party, we first deployed K-means in WEKA [18] to cluster the retweeters from all four parties by their number of tweets and retweets. The suitable k was determined by the Elbow method [19], which was 5 in our experiments. The rate of the number of retweets divided by the number of tweets representing each cluster was calculated (see Table 7). The proportion of retweeters belonged to each of the 5 groups of each political party is visualized in Fig. 5.

Table 7. Users (retweeters) clustered by their political retweets and tweets

Cluster	#Retweets/#Tweets
Cluster 1	0.0073
Cluster 2	0.1751
Cluster 3	0.4189
Cluster 4	0.5697
Cluster 5	0.9620

From Fig. 5, the proportion of retweeters within each group of the political party leaders @sudaratofficial @prayutofficial and @Abishit_DP were similar. The proportions of retweeters in cluster 2 and cluster 5 of @Thanathorn_FWP, however, was much lower and much higher compared with the results of other party leaders. The cluster 5 represented the retweeters who mainly retweeted from the party with very a low number of their own tweets.

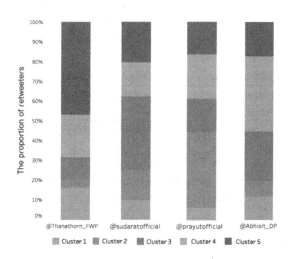

Fig. 5. The proportion of retweeters in 5 clusters within each political party

4.5   Analysis Election-Related Events

During the three months of data collection, there were many events occurring each day. Therefore, we also tried to extract important events that were hidden in the tweet messages. To do this, we considered the originality ratio of the number of retweets and the number of tweets in which the word appeared as shown in Eq. (1) and described in [6]. The #Retweet is the number of retweets in which the word appeared in each day. The #distinctTweet is the number of tweets that the word appears on each day. We ranked the significant events based on the calculated ratio. The discovered significant events are summarized in Table 8.

$$Originality\ Ratio = \frac{\#Retweet}{\#distinctTweet} \qquad (1)$$

Table 8. Top events derived from word's originality ratio

Date	Event
1 Jan 2019	เปิดตัว ชัชชาติ (Debut, Chatchart)
5 Jan 2019	เลื่อนเลือกตั้ง (Postpone election)
6 Jan 2019	ชุมนุม นัดหมาย ราชประสงค์ (Rally, Making Appointments, Ratchaprasong)
8 Jan 2019	ชุมนุม ราชประสงค์ (Rally, Ratchaprasong)
22 Jan 2019	การเลือกตั้ง ประกาศ พระราชกฤษฎีกา พ.ร.ฎ ราชกิจจานุเบกษา (Election, Announce, Royal decree, Government Gazette)
23 Jan 2019	ประกาศ เลือกตั้ง (Announce, Election)
28 Jan 2019	เลือกตั้งล่วงหน้า ลงทะเบียน เขต สิทธิ์ (Early Voting, Register, Constituency, Eligible voter)
6 Feb 2019	ไทยรักษาชาติ เซอร์ไพรส์ ข่าวลือ (Thai Raksa Chart, Surprise, Rumor)
7 Feb 2019	แคนดิเดต ทูลกระหม่อม พระสเลนเดอร์ (Candidate, Princess UbolRatana Rajakanya, Slender)
8 Feb 2019	ราชโองการ ประวัติศาสตร์ พลิกล็อค แคนเซิล (History, King's command, Upset, Cancle)
9 Feb 2019	พ่อ ฟ้า รัก (Sky Loves Daddy)
14 Feb 2019	ศาลรัฐธรรมนูญ กกต. ยุบ (Constitutional court, Election Commission of Thailand, Dissolve)
17 Feb 2019	หนักแผ่นดิน (Scum of the earth)
18 Feb 2019	เลือกตั้งล่วงหน้า วันสุดท้าย (Early Voting, The last day)
7 Mar 2019	ยุบพรรค พรรคไทยรักษาชาติ (Dissolve political party, Thai Raksa Chart Party)
11 Mar 2019	ประชาธิปัตย์ เวที คนใต้ (Democrat, Stage, Southern people)
17 Mar 2019	เลือกตั้งล่วงหน้า (An election in advance)
19 Mar 2019	คลิป โป๊ะแตก เนชั่น (Video clip, Caught lying, Nation)
22 Mar 2019	ประเทศไทย ปราศรัย (Thailand, Speech)
24 Mar 2019	เลือกตั้ง คะแนน อภิสิทธิ์ ลาออก (Election, Votes, Abhisit, Resign)
25 Mar 2019	กกต. โป๊ะแตก โกง เครื่องคิดเลข นิวซีแลนด์ (Election Commission of Thailand, Caught lying, Cheat, Calculator, New Zealand)
27 Mar 2019	ตั้งรัฐบาล (Set up the government)
30 Mar 2019	อนาคตใหม่ เดินสาย ขอบคุณ (Future Forward, Thanks for Every Vote)

4.6   Analysis of Word with TF-IDF

We examined words most frequently used by the party leader. These words were then compared with the words most often used by others when talking about the leaders of these political parties. To find the most frequently used words, the tweet messages from the accounts of the political party leaders, @Thanathorn_FWP from the Future Forward Party, @sudaratofficial from the Phue Thai Party, @prayutofficial from the Phalang Pracharat Party, and @Abhisit_DP from the Democrat Party were analyzed with TF-IDF [20]. From word clouds in Figs. 6 and 7, the bigger word size, the more im- portant. The trending words standing out from the tweets of the Future Forward Party focused on หาเสียง (campaign) หลักการ (principle) and ประสาน (coordination). There were also words that were in the negative context such as สาดโคลน (distort), บิดเบือน (besmirch), and ฝ่ายค้าน (opposition). From these words, Thanathorn was accused of besmirching other parties. Trending words of the Phalang Pracharat Party emphasized the story of daily life and their concerns for the people like สมปรารถนา (fulfill), สตรี (women), and ซิงเกิล (single music). In comparison, standing out words of the Democrat Party were in line with their policy such as สวัสดิการ (welfare), เรียนจบ (graduating) and ค่าแรง (wages). From the Pheu Thai Party, tweets showed a strong and decisive way such as มีโอกาส (strategy), ให้กำลังใจ (encourage), and ข้อบังคับ (regulations).

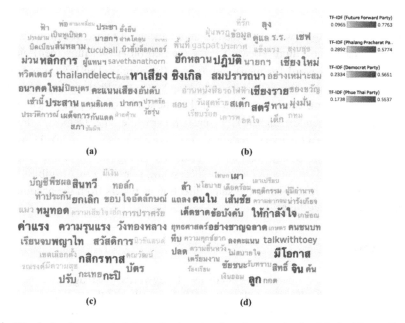

Fig. 6. Word clouds generated from the tweets of each party leader (a) the Future Forward Party, (b) the Phalang Pracharath Party, (c) the Democrat Party, (d) the Phue Thai Party

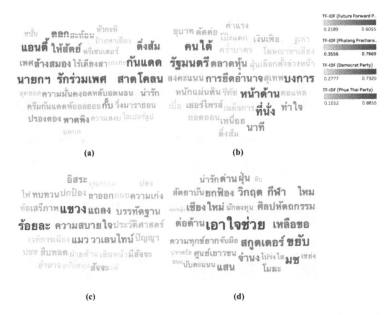

Fig. 7. Word clouds generated from tweets that mentioned each party leader (a) the Future Forward Party, (b) the Phalang Pracharath Party, (c) the Democrat Party, (d) the Phue Thai Party

Word clouds in Fig. 7 were analyzed as above using all tweets mentioning a party leader. Mentions of the Future Forward Party focused on กันแดด (Sunscreen), กับ or ครับ (kráp is a polite Thai word for men) and สาดโคลน (besmirch). The kráp was from the parody of Thanathorn's speech "Everybody kráp". For the Phalang Pracharat Party, tweets showed grousing and blame mentioned to the leader of the party such as หน้าด้าน (shameless), บงการ (dictate), and ที่นั่ง (a seat). Meanwhile, the Democrat Party got mentioned words as แขวง (district), ร้อยละ (percent), and วาเลนไทน์ (Valentine). These words were not related to the policy like words observed in Fig. 6. The Phue Thai Party mentioned words focused on เอาใจช่วย (wish success to), เหลือขอ (be incorrigible), and ขยับ (move). These words remained in the same direction as tweets from the account of the party leader, @sudaratofficial.

5   Conclusions

This paper represents a quantitative analysis of keywords, hashtags, tweets, and re-tweets of Twitter on the Thailand election 2019. We collected these data from January 1 to March 31, 2019. The data were collected using the Twitter advanced search with Selenium. Overall, we found that Twitter was an open social media platform for political consideration. While the number of retweets may indicate the voting trend, we found no relation between the two variables. This needs to further investigate. From the sentiment analysis of the messages mentioning each political party, tweeted by the re-tweeters, the positive mentions showed some correlation with the number of votes after we removed the Fast Forward Party (FFP) from the analysis. Finally, with the analysis of words and contexts from the tweets of each party leader's account, each party had their own strategies for the political campaign.

References

1. Praciano, B.J.G., et al.: Spatio-temporal trend analysis of the Brazilian elections based on Twitter data. In: IEEE International Conference on Data Mining Workshops (ICDMW), pp. 1355–1360. IEEE, Singapore (2018)
2. matichonweekly. อัพเดตข้อมูลผู้ใช้ "โซเชียลมีเดีย" ล่าสุด แพลตฟอร์มไหน "แรง" แพลตฟอร์มไหน "แผ่ว"? และพัฒนาการขั้นต่อไปของ "สื่อสังคมออนไลน์" 2018. https://www.matichonweekly.com/col umn/article_88648
3. Wang, L., Gan, J.Q.: Prediction of the 2017 French election based on Twitter data analysis. In: 9th Computer Science and Electronic Engineering (CEEC), pp. 89–93. IEEE, Colchester (2017)
4. News, S.: Poll shows more than half of teenagers excited to be elected for the first time in life. https://www.sanook.com/news/7674526/. Accessed 10 Feb 2019
5. Koylu, C.: Modeling and visualizing semantic and spatio-temporal evolution of topics in interpersonal communication on Twitter. Int. J. Geogr. Inf. Sci. **33**(4), 805–832 (2019)
6. Pudtal, A., Sinthupinyo, S.: An analysis of Twitter in the passing of his majesty king Bhumibol Adulyadej. **7**, 187–193 (2017)
7. Culotta, A.: Detecting influenza outbreaks by analyzing Twitter messages, pp. 1–11 (2010). arXiv:1007.4748
8. Yoshida, M., Toriumi, F.: Analysis of political party Twitter accounts' retweeters during Japan's 2017 Election, pp. 736–739 (2018)
9. Buccoliero, L., et al.: Twitter and politics: evidence from the US presidential elections 2016. J. Mark. Commun. 1–27 (2018)
10. Soler, J.M., Cuartero, F., Roblizo, M.: Twitter as a tool for predicting elections results. In: IEEE/ACM International Conference on Advances in Social Networks Analysis and Mining, pp. 1194–1200. IEEE, Istanbul (2012)
11. Aslam, S.: Twitter by the numbers: stats, demographics & fun facts (2019). https://www. omnicoreagency.com/twitter-statistics/
12. PyThaiNLP: pythainlp 2.0.4 (2019). https://pypi.org/project/pythainlp/
13. Wang, W., Lu, Y.: Analysis of the mean absolute error (MAE) and the root mean square error (RMSE) in assessing rounding model, vol. 324, pp. 1–10 (2018)
14. Skoric, M., Poor, N., Achananuparp, P., Lim, E.-P., Jiang, J.: Tweets and votes: a study of the 2011 Singapore general election. In: 45th Hawaii International Conference on System Sciences, pp. 2583–2591. IEEE, Maui (2012)
15. Tumasjan, A., Sprenger, T.O, Sandner, P.G., Welpe, I.M.: Predicting elections with Twitter: what 140 characters reveal about political sentiment. In: Proceedings of the Fourth International AAAI Conference on Weblogs and Social Media, pp. 178–185 (2010)
16. Matichon Online: m.: โพลนิสิตจุฬาฯ 1,266 ตัวอย่าง "อนาคตใหม่" ครองใจนิสิตจุฬาฯ ธนาธร-ชัชชาติ นายกฯ. https://www.matichon.co.th/politics/news_1406479. Accessed 15 Mar 2019
17. Hill, S.: Changing votes or changing voters? How candidates and election context swing voters and mobilize the base **48**, 131–148 (2017)

18. Patil, R., Deshmukh, S., Rajeswari, K.: Analysis of simple k-means with multiple dimensions using WEKA. Int. J. Comput. Appl. **110**, 14–17 (2015)
19. Bholowalia, P., Kumar, A.: EBK-means: a clustering technique based on elbow method and k-means in WSN. Int. J. Comput. Appl. **105**, 17–24 (2014)
20. Góralewicz, B.: The TF∗IDF algorithm explained. https://www.onely.com/blog/what-is-tf-idf/. 06 Mar 2018

Science Lab Repository Requirements Elicitation Based on Text Analytics

Norhaslinda Kamaruddin[1(✉)], Abdul Wahab[2], Mohammad Bakri[3], and Muhammad Hamiz[3]

[1] Advanced Analytics and Engineering Centre,
Faculty of Computer and Mathematical Sciences,
Universiti Teknologi MARA, 40450 Shah Alam, Selangor, Malaysia
norhaslinda@fskm.uitm.edu.my
[2] Kulliyah of Information and Communication Technology,
International Islamic University Malaysia,
P.O. Box 10, 50728 Kuala Lumpur, Malaysia
[3] Faculty of Computer and Mathematical Sciences,
Universiti Teknologi Mara Cawangan Melaka Kampus Jasin,
77300 Merlimau, Melaka, Malaysia

Abstract. Requirements elicitation is an important task before any development of system repository can be conducted. Typically, traditional methods such as interview, questionnaire and observation are made to gauge the users' needs. However, the users may not be able to spell out specifically of their need especially if there is no available system to compare resulting to outrageous demands and unrealistic expectations to the repository developer. An alternative approach to gauge the user needs from users' reviews of the on-the-shelf software may be a good starting point. In this paper we attempt to extract requirements from the users' independent reviews gathered from the internet using text analytics approach. The keywords are visualized based on its relevance and importance to the user. Then, it is used as a benchmark for the user to alter to their specific repository needs. From the experimental results, it is observed that there are functions that are very much needed by the user and yet there are also functions that are not used at all. Hence, this proposed approach may give insight to the user and developer about the actual needs of the respective system. It is envisaged that such approach can be a guide to the novice user and the developer in order to shorten the time to agree on the development of the repository system.

Keywords: Requirements elicitation · User review · Business rules · Text analytics · Word cloud

1 Introduction

Repository is defined as a shared database of information about engineered artifacts produced or used by an enterprise [1]. It is necessary to bring together the discrete data items that are resided in different databases as the local aggregation of data items into one centralized location for a specific purpose that can operate these data as one. The

© Springer Nature Singapore Pte Ltd. 2019
M. W. Berry et al. (Eds.): SCDS 2019, CCIS 1100, pp. 351–360, 2019.
https://doi.org/10.1007/978-981-15-0399-3_28

development of a good repository system to keep track of several databases that collect, manage and store data sets is essential for data analysis, sharing and reporting. Informed decision making can be derived based on available data. It allows better problem tracing because the data in the repository are compartmentalized for better data perseverance and archive. Hence, proper repository design and planning is important to ensure the need of the users can be fulfilled.

Typically, traditional methods such as interview, questionnaire, document analysis and observation are conducted to extract information from the system users and stakeholder of the activities they are going to perform and the nature of the data they are going to work on. However, Mulla and Girase [2] stated that 40% of defects in software projects are due to incorrect recorded requirements. Such finding is supported by Atagoren and Chouseinoglou [3] who observed that 7,499 out of 36,424 defect data are originated from requirement phase. This is not surprising because users may have outrageous demands and unrealistic expectations especially if they are not familiar with the system and if there is no system to compare with. The inability to explicitly express the need is giving difficulty for the developer to anticipate the actual needs for specific function development. To complicate matters, the usage of on the shelf software may not be ideal for a specific business. Some of the functions are never used resulting to unoptimized space for storage and unnecessary learning enforced to the users.

In this work, we propose an alternative approach to gather the users' needs from the independent users' reviews gathered from the on the shelf software that are available online. The evaluation of current software by the independent user can be used as a guidance to get the requirements from the different users' experience with almost similar repository system characteristics. Keywords frequency are measured using text analytics technique coupled with word cloud data visualization technique give better insight on the feedbacks. The unique and word pair frequencies are also compared and analyzed.

This paper is organized in the following manner. Section 2 describes the literature review of relevant requirement elicitation methods that are commonly used. Concept of text analytics is also discussed. The detailed methodology is presented in Sect. 3. In Sect. 4, the experimental result and discussion are reported. This paper is concluded with summary, conclusion and future work.

2   Literature Review

System Development Life Cycle (SDLC) is a series of phases used to develop a system according to expectation and requirements given by users. There are six main phases, which are; Planning, Analysis, Design, Implementation, Testing & Integration and Maintenance. Each phase contributes to different tasks and focuses.

In the second phase of SDLC, the users' requirements should be determined and documented so that the expectations after delivering the system will be met [4]. Since requirements can be gathered not only from the system's users, requirements elicitation (RE) process is mostly done prior to the execution of analysis and specification steps. Requirement elicitation can be defined as practices done to extract information from any group of users in the system [5], namely; user, customer and other involved stakeholders. It is a process to understand the domain problem and to identify several solutions for the stated problem. This is a hard task as user normally will never know what they actually want and what can the system exactly provides.

Technically, there are lots of techniques that can be used for RE, however, different kind of RE process requires different techniques. Table 1 summarizes the common techniques, its description and methods inside the techniques as suggested by [6].

Table 1. Comparison of requirement elicitation techniques [6].

Techniques	Description	Methods
Traditional	Most commonly used technique	Interview, questionnaire/ survey, and document analysis
Contextual	Requirements are collected at the users' workplace	Observation, ethnography and protocol analysis
Collaborative/group	A group of stakeholders from different expertise has equal power to decide on specific issue in getting the requirements	Prototyping, joint application development, brainstorming, and group work
Cognitive	Information is gathered and analyze up to human thinking level which could understand the problems in depth	Laddering, card sorting, repertory grids, and class responsibility collaboration

Yousof et al. group the different RE techniques into four groups as illustrated in Table 1. The most popular RE technique is the traditional methods due to its simplicity and practicality. However, even though the traditional methods are commonly used, these methods require a specific and detailed questions so that it could be understood by user. Besides, users may give answers that they think are socially desirable even if the answer are not fully accurate [7]. The analysts are required to be at the users' workplace to yield the requirements for contextual methods. It is time consuming because multiple sessions with different users may be needed because it is not easy to extract complete requirements in one go [8]. For collaborative technique, experts from different fields are required. The identification of an expert is hard and time consuming. Moreover, to get a consensus in a session is also a challenge as everyone may have their own opinions and preferences. Subsequently, cognitive technique offers a solution by stacking the questions about the criteria of user preferences in a hierarchical manner. Analyst can extract information in a structured way that offer directive answers from the users. However, no technique can claim superiority above the other because all of these methods are adjustable according to what kind of environment is preferred by the analyst.

Nowadays, customers review become a vital part in representing good product image to the public especially in small businesses [9]. A research by [10] agreed that customer reviews can be an important source in understanding the needed requirements from users. In the world of Internet, online customer reviews can be defined as any positive, neutral, or negative evaluation of product, services or even a person posted by former customers on websites that host consumer reviews [11]. A good review could lead to an increase of the services subscription. As online reviews are popular among users, researchers tried to have a full understanding on the impact of the reviews. One of the most popular technique is using text analytics [12–14].

Text analytics is a technique to get useful information through text. Useful information could be in the form of pattern gathered from the topic of interest. Once the pattern is learned, appropriate actions can be applied to extract the topic of interest. Most of the time, online user reviews are considered big data and in the form of unstructured text, hence, text analytics could be the way to learn the meaning behind the unstructured text. For example, according to [12], content such as user's opinion on subject matter can be crawled from the webpage. Once the data is crawled, it can be represented in any kind of visualization technique. One of the popular text analytics visualization technique is word cloud [15, 16]. Word cloud or tag clouds will provide first impression on what are the most used text in documents [17]. Hence, by having a word clouds visualization of online user reviews on subject matter, some requirements could be elicited which could help the analyst to learn better about the topic.

3 Methodology

The research method is divided into two parts, namely; review extraction and keyword analysis using text analytics. The reviews from users are extracted from the website. Relevant keywords and word pairs are extracted. Then, the word cloud diagrams are developed based on the features of each software to get better insight of the features that contributed to the score in the reviews.

3.1 User Reviews Analytics

The user comments from the website are extracted using web scraping method and analyzed using text analytics. The comments are gathered from https://www.capterra.com/medical-lab-software/ website that compare 138 medical systems online with total number of review of 42. The review consists of information such as product rating, number of users, deployment type and focused features. A program was written using Python programming to do the processes as presented in Fig. 1.

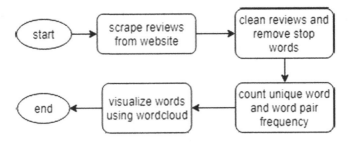

Fig. 1. General flow of user review extraction processes.

The reviews contain of three major parts of pro, cons and comments. User are expected to fill in information as accurately as possible. However, it is not compulsory for the user to fill in every part of the review. Example of the review is illustrated in Fig. 2. For the purpose of analysis, each part of the review is stored separately for individual analysis.

Pros: Very easy to adapt to your needs, you can have separate databases belonging to single Groups, very flexible System. Good support.

Cons: It can get very slow if you add too many records to 1 single module, we have 1 with 6000 records and it is a bit too slow. The Import of tubes through a box Reader should be improved (we have the scanner but do not use it), it is complicated to Import big data from another database.

Overall: We started using this Software 3 years ago for all the Groups in our Department, we use it for cells in LN, reagents in freezers, slides, equipment (pipettes, instruments, etc...), plasmids, and General documents (SOPs).

Fig. 2. An example of user review

Then, the reviews extracted need to be preprocessed to ensure only relevant keywords are extracted. Any HTML and symbols from the reviews are removed as it can hinder the analysis process. Stop words are also removed so that only significance words are analyzed. To get an overview of the words that are frequently used in each part of the reviews, frequency of unique words and word pair is calculated. Finally, the word cloud diagrams are generated for each part of the review to make it easier to compare the frequency of each word with one another.

3.2 Features Analytics

Medical lab features for each software will be analyzed. The list of features is taken from the website. The software features listed in the website are: Data Analysis Auditing, Data Security, EMR Interface, Fax Management, Lab Instrument Interface, Multi-Location Printing, Online Instrumentation, Physician Test Panels, Procedure-Based Billing, and Sample Tracking. The correlation between the features and the score will be determined by using heatmap and boxplot.

4   Experimental Results and Discussion

The outputs for user review analytics and features analytics are analyzed to gauge the criteria that make a good system or software.

4.1 User Reviews Analytics

The unique word frequency, the word pair frequency and the word cloud diagrams for pros, cons and comments section are derived. The pros section is describing the good features that are recommended by the user. On the contrary, the cons section is more related to the bad features that are the least favorable by the user. In addition, the comment section is an unstructured data that allow the user to give their feedback freely. The output for pros and cons section is provided in Tables 2 and 3 respectively.

Table 2. The output for pros section.

Unique Word Frequency	Word Pair Frequency
('easy', 70), ('use', 52), ('user', 28), ('time', 23), ('like', 22), ('friendly', 21), ('reports', 19), ('good', 18), ('support', 18), ('also', 17)	('easy', 'use', 23), ('user', 'friendly', 16), ('ease', 'use', 9), ('friendly', 'easy', 7), ('customer', 'service', 5), ('everything', 'need', 5), ('easy', 'learn', 4), ('like', 'easy', 4), ('would', 'recommend', 4), ('makes', 'easy', 3)
Wordcloud	

Table 3. The output for cons section.

Unique Word Frequency	Word Pair Frequency
('use', 21), ('would', 17), ('support', 16), ('get', 15), ('like', 15), ('features', 14), ('cons', 14), ('sometimes', 13), ('patient', 13), ('able', 13)	('customer', 'service', 7), ('tech', 'support', 3), ('user', 'friendly', 3), ('would', 'like', 3), ('per', 'workstation', 3)
Wordcloud	

Based on Table 2, it is observed that most of the good reviews of the software recommending the features of easy to use, user friendly, good customer support and good reporting module. The results are validated and confirmed by manual comparison of the reviews that are using the words. Such result shows that the success of the software is highly correlated with its usability. Reporting functionality also plays a big part in the software success. On contrary, most of the bad reviews on the software use the word 'use', 'support', 'features' and 'patient'. By manually searching the reviews that use these words, most of the user are making a complaint about difficulty of using the software, not user friendly or having too much features that are not used. The users are also complained about the bad customer support, lack of patient management module, and bad licensing (per workstation).

The last part of the review is comment section. Most of the review that use the comment section did not fill the other part of the review (pros and cons). The comment section also mostly contains positive review to the software as shown in Table 4. Among words that are frequently used in this section are 'good', 'support', 'easy', and 'use' which represents good support and easy to use or user-friendly system. The software also has a good sample management module, medical tests module and easy learning curve. The comment part is essential to be included in the analysis because users can write their review without restriction and not bounded by the pre-selected features description.

Table 4. The output for comment section.

Unique Word Frequency	Word Pair Frequency
('easy', 28), ('use', 28), ('good', 25), ('support', 20), ('management', 16), ('service', 15), ('data', 15), ('work', 14), ('also', 12), ('great', 11)	('easy', 'use', 11), ('customer', 'service', 6), ('sample', 'management', 6), ('also', 'good', 4), ('user', 'friendly', 4), ('medical', 'tests', 4), ('suitemed', 'ims', 3), ('learning', 'curve', 3), ('helps', 'us', 3), ('single', 'screen', 3)
Wordcloud	

4.2 Features Analytics

Each software will be tagged manually according to the features listed in the website. The data is then used to generate heatmap to see the correlation between the features and average score of the software. The output of the heatmap is shown in Fig. 3.

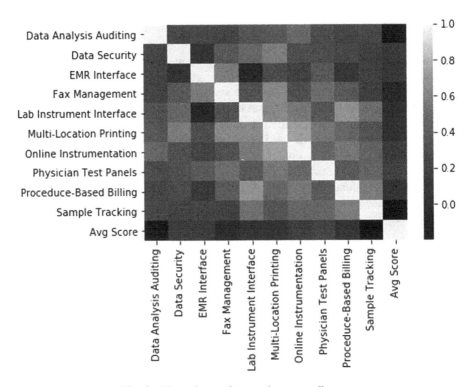

Fig. 3. The software features heat map diagram

Based on the heatmap in Fig. 3, there are no feature that is highly correlated with the average score. Features that has no correlation at all with average score are 'Data Analysis Auditing' and 'Sample Tracking'. On the other hand, features that has slight correlation with average score is 'Physician Test Panels', 'Online Instrumentation', 'Data Security' and 'EMR Interface'. In addition, a boxplot diagram is used to see the average score range for each feature. The boxplot for the data is shown Fig. 4. Based on the boxplot, software with EMR Interface and Physician Test Panels are mostly having high review scores. The other features show no clear correlation with the score. When developing the software, the developer should focus on the EMR Interface and Physician Test Panels. The Sample Tracking and Data Analysis Auditing can be marked as low priority.

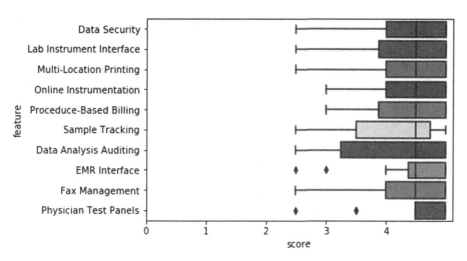

Fig. 4. The software features boxplot diagram

5 Conclusion

In this paper, we proposed the use of user review on the on-the-shelf system that is currently available in the market. This is to give an insight of what features that are highly appreciated by the user as well as the least favorable features. The user reviews are gathered from independent website that is freely available online. User review and features analytics are implemented, and it is found that there are few keywords that can help user and developer to agree on the selection of features. Typically, users do not really know what they want until they have used the system. Such approach is tedious and may create a lot of difficulty for the software developer to create the mock system. Hence, the proposed approach can give an indicator of the features that should be retained, added or excluded especially in the development of science lab repository [18]. The experimental results show potential of such approach to be implemented. It is envisaged that such approach can be a guide to the novice user and the developer in order to shorten the time to agree on the development of the repository system.

Acknowledgements. The authors would like to thank Universiti Teknologi MARA (UiTM), International Islamic University Malaysia (IIUM) and Ministry of Higher Education Malaysia (MOHE) for providing financial support through the MITRA grant (600-IRMI/PERDANA 5/3/MITRA (007/2018)-3) to conduct the work published in this paper.

References

1. Bernstein, P.A., Dayal, U.: An overview of repository technology. In: VLDB, vol. 94, pp. 705–713 (1994)
2. Mulla, N., Girase, S.: A new approach to requirement elicitation based on stakeholder recommendation and collaborative filtering. Int. J. Softw. Eng. Appl. **3**(3), 51 (2012)

3. Atagoren, C., Chouseinoglou, O.: A case study in defect measurement and root cause analysis in a turkish software organization. In: Lee, R. (ed.) Software Engineering Research, Management and Applications, vol. 496, pp. 55–72. Springer, Heidelberg (2014). https://doi.org/10.1007/978-3-319-00948-3_4

4. Network ID. QuickStudy: system development life cycle. Computerworld, 14 May 2002

5. Dick, J., Hull, E., Jackson, K.: Requirements Engineering, p. 115. Springer, Heidelberg (2017). https://doi.org/10.1007/978-3-319-61073-3

6. Yousuf, M., Asger, M.: Comparison of various requirements elicitation techniques. Int. J. Comput. Appl. **116**(4), 8–15 (2015)

7. Patten, M.L.: Questionnaire Research: A Practical Guide, p. 10. Routledge, Abingdon (2016)

8. Keller, T.: Contextual requirements elicitation. In: Seminar in Requirements Engineering, Spring 2011, Department of Informatics (2011)

9. Capoccia, C.: Online Reviews are the Best Thing That Ever Happened to Small Businesses [Internet]. Forbes. Forbes Magazine (2018). https://www.forbes.com/sites/forbestechcouncil/2018/04/11/online-reviews-are-the-best-thing-that-ever-happened-to-small-businesses/#658d186d740a. Accessed 28 May 2019

10. Filieri, R., Hofacker, C.F., Alguezaui, S.: What makes information in online consumer reviews diagnostic over time? The role of review relevancy, factuality, currency, source credibility and ranking score. Comput. Hum. Behav. **80**, 122–131 (2018)

11. Huang, Y., Li, C., Wu, J., Lin, Z.: Online customer reviews and consumer evaluation: the role of review font. Inf. Manag. **55**(4), 430–440 (2018)

12. Liu, B.: Sentiment analysis and subjectivity. In: Handbook of Natural Language Processing, vol. 2, pp. 627–666 (2010)

13. Kamaruddin, N., Wahab, A., Lawi, R.A.M.: Jobseeker-industry matching system using automated keyword selection and visualization approach. Indones. J. Electr. Eng. Comput. Sci. **13**(3), 1124–1129 (2019)

14. Zamani, N.A.M., Kamaruddin, N., Wahab, A., Saat, N.S.: Visualization of job availability based on text analytics localization approach. Indones. J. Electr. Eng. Comput. Sci. **16**(2), 744–751 (2019)

15. Salloum, S.A., Al-Emran, M., Monem, A.A., Shaalan, K.: Using text mining techniques for extracting information from research articles. In: Shaalan, K., Hassanien, A.E., Tolba, F. (eds.) Intelligent Natural Language Processing: Trends and Applications. SCI, vol. 740, pp. 373–397. Springer, Cham (2018). https://doi.org/10.1007/978-3-319-67056-0_18

16. Hearst, M., Pedersen, E., Patil, L.P., Lee, E., Laskowski, P., Franconeri, S.: An evaluation of semantically grouped word cloud designs. IEEE Trans. Vis. Comput. Graph. 1–14 (2019). https://doi.org/10.1109/TVCG.2019.2904683

17. Bakri, M., et al.: Insight extraction on cross-cultural interaction through astronomy online labs using data analytics. Indones. J. Electr. Eng. Comput. Sci. **16**(1), 508–515 (2019)

18. Kamaruddin, N., Wahab, A.: Interlaboratory data fusion repository system (InDFuRS) for tocotrienols-based treatment. Indones. J. Electr. Eng. Comput. Sci. **13**(3), 1130–1135 (2019)

A Gillespie Algorithm and Upper Bound of Infection Mean on Finite Network

Sapto Wahyu Indratno$^{(\boxtimes)}$ and Yeftanus Antonio

Statistics Research Division, Institut Teknologi Bandung,
Center for Advanced Sciences Building, Ganesha 10, Bandung 40132, Indonesia
sapto@math.itb.ac.id, yeftanus@students.itb.ac.id

Abstract. Cyber-attacks are expected to increase in the next few years. This condition requires an estimate of the amount of risk that will occur. Cyber risk can be reflected in the number of infected computers obtained by models that can explain the process of spreading viruses on computer networks. Mathematical models on epidemiology can be used to understand the process of spreading viruses on computer networks inspired by the process of spreading diseases in biological populations. Stochastic susceptible-infectious-susceptible (SIS) model is a simple epidemic model will be used to estimate the risk (number of infected computers) on several computer networks. Based on a fixed population and homogeneous mixing assumptions, we get the upper bound of infection mean from the model. Mean of the sample path in dynamic processes is generated by the Gillespie Algorithm or Simulation Stochastic Algorithm (SSA) to compare with the upper bound of the infection mean. The computational result confirms the mean of sample paths always less than the upper bound of infection mean.

Keywords: Gillespie algorithm · Stochastic SIS model · Upper bound · Infection mean

1 Introduction

Biology systems has inspired many computational algorithms such as neural networks in machine learning, evolutionary theories on genetic algorithms, ant colonies on optimization, and many others [16]. Likewise, the process of spreading malicious software (viruses, worms, and trojans) is identical to the process of transmitting disease in the biology population. Various deterministic and stochastic epidemic models are used to understand this process [6, 10, 11, 13, 15, 18–21]. A simple classic epidemic model to explain the disease transmission process is known as the Susceptible-Infectious-Susceptible (SIS) model. This model divides the population into two states, namely vulnerable or susceptible and infected [5]. In the perspective of the network, each computer unit contacts with others on a network structure that remains with the connection between computer units expressing the relationship between computer units. Homogeneous mixing occurs in this compartment model, where each computer has the same possibility of making contact with an infected computer [4, 12].

© Springer Nature Singapore Pte Ltd. 2019
M. W. Berry et al. (Eds.): SCDS 2019, CCIS 1100, pp. 361–374, 2019.
https://doi.org/10.1007/978-981-15-0399-3_29

Suppose the numbers of vulnerable and invective computers at t are denoted by $S(t)$ and $I(t)$ with the fixed number of computer unit population $N = S(t) + I(t)$. Let β be the probability of successful transmission viruses from an infected computer to vulnerable computer so that $\beta \in [0, 1]$. The rate of contact $\bar{\beta}$ is a function of population size $\bar{\beta} = \beta N$ under assumption perfect mixing. Infected computers contact another N computer and transmit viruses to βN computers per time unit. The probability for one computer unit to recover from the infection and return its status to a vulnerable computer is given by γ with $\gamma \in [0, 1]$ so that the number of recovered computers per unit time or the recovery rate is γI. Basic reproductive number is the numbers of second infections produced by an infected computer unit in the population. Analytical solution of the deterministic model is given by

$$I(t) = \frac{i_0(\bar{\beta} - \gamma)\exp\left[(\bar{\beta} - \gamma)t\right]}{(\bar{\beta} - \gamma) + \bar{\beta}i_0/N\left(\exp\left[(\bar{\beta} - \gamma)t\right] - 1\right)} \tag{1}$$

where i_0 is the number of infected computers at $t = 0$.

This study uses the assumption of homogeneous mixing by assuming that each computer unit can make contact as much as the average degree of the network structure. We built a stochastic model to capture population variety and carried out theoretical analysis and simulations to obtain information related to the influence of contacts on different graph topologies, where the previous results used deterministic models [4, 17]. The specific graph structure that has been studied influences the dynamics of the infection process [8, 22]. The specificity of the network structure is very interesting to develop the discussion of epidemics in the network [12, 17] which is as mathematical objects in the field of graph theory. Thus, the characteristics of a graph that represents effective contact of the graph topology can be defined and used to estimate the upper bound of infection mean.

We compare thousand random paths from the stochastic model using Gillespie algorithm [9] for evidence for the upper bound of infection means. Gillespie algorithm is a numerical simulation for a continuous-time Markov chain by decomposing the dynamic process into independent processes [7]. This algorithmic program is created firstly for the simulate biochemical system. Then the algorithm until now is developed for many branches of science, including epidemiology system model. Five classes of graphs namely complete graph (fully connected network), cyclic graph (ring network), star graph (star network), wheel graph (star and ring network), and path graph (line network) are considered as a network topology to estimate for.

The next section of this article is written with the following organization. The process of building the SIS stochastic model, its properties, and solutions of its dynamic equations is given in Sect. 2. Section 3 is the core that explains the upper bound of infection mean, parameter sensitivity, and its relationship with the probability of outbreaks. In Sect. 4, numerical evidence of the upper bound is shown using the Gillespie Algorithm on a predefined graph topology. In the last section, we give conclusions about the results that have been obtained.

2 Stochastic SIS Model

In this section, we will discuss the SIS stochastic model. Modifications of contact parameters will be done to discuss the SIS stochastic model on the network. Some structural properties of the network like average degrees can be included as a parameter [22]. The deterministic model discussed earlier does not capture the variation of the population [6]. Variation of computer units in each compartment is formed by built the dynamics of the stochastic process. Let $I(t)$ denotes the number of infected computers at time t, with $I(t) \in \{0, 1, 2, \ldots, N\}$, in other word state space $I(t)$ is discrete random variable. The time index t is defined by $t \in [0, \infty)$ which is a continuous-time Markov chain. The infection and recovery processes follow an independent Poisson process with a rate of infection and recovery. The transition process is given in Table 1 based on the assumption that infection and recovery processes follow the Poisson process. Property of the Poisson process is in a short interval of time h only one event can occur. Let $p_x(t)$ be the probability at time t defined by $p_x(t) = P(I(t) = x)$. Probability differential equation or known as Kolmogorov equation from the transition probability in Table 1 is given by

$$\frac{dp_x(t)}{dt} = \frac{\bar{\beta}(x-1)(N-(x-1))}{N}p_{x-1}(t) + \gamma(x+1)p_{x+1}(t) - \left[\frac{\bar{\beta}(N-x)}{N} + \gamma x\right]p_x(t).$$

(2)

Table 1. The transition processes of the stochastic SIS model.

Initial condition	Probability	Transition process
(S, I)	$\bar{\beta}SI/N$	$(S-1, I+1)$
(S, I)	γI	$(S+1, I-1)$

The solution of differential equation in (2) can be obtained using the generator matrix form [3]. We can write that equation in matrix form, $dp(t)/dt = Qp(t)$ where $p(t)$ is probability vector defined as $p(t) = (p_0(t), p_1(t), \ldots, p_N(t))'$ and Q is the generator matrix with a coefficient of Kolmogorov equation as entries and characterized by the sum of all entries in each column is zero. The solution of Kolmogorov equation in (2) is

$$p(t) = e^{Qt}p(0)$$

(3)

where $p(0) = (p_0(0), p_1(0), \ldots, p_N(0))'$ is the initial probability vector. Dynamic of first moment equation can be obtained by definition of expectation $dE[I(t)]/dt = \sum_{x=0}^{N} x(dp_x(t)/dt)$ and depend on dynamic of infection probability in (2). After some algebraic operation, we get (see [2])

$$\frac{dE[I(t)]}{dt} = (\bar{\beta} - \gamma)E[I(t)] - \frac{\bar{\beta}}{N}E[I^2(t)]. \tag{4}$$

The differential equation for the first moment depends on the second moment and the second moment depends on the higher moment. Then the Eq. (4) cannot be solved directly. SIS epidemic model discussed previously is assumed having a perfect contact mixing, which means that each computer communicates with other computers. One infected computer contacts with N another computer and successfully transmit the virus with $\bar{\beta} = \beta N$ computer units per time unit. The number of infected computers per infective per time unit is $\bar{\beta}(S/N)$ so that the number of new infections per unit time is the rate of infection given by $\beta N(S/N)I = \beta SI$. This form defined as the rate of occurrence of mass action [3]. The number of populations N can be approached with contact parameter η, which is the rate of any one computer is connected to other [14]. The probability of successful transmission is given by β so that for contact on a specific network structure the rate of infection is given by $\beta \eta(S/N)I$. Each infective generates a new infection with a $\beta \eta$ per unit of time with an average of $1/\gamma$ in time units. The next averages of infection R_0 is given by $\bar{R}_0 = \beta \eta / \gamma$. Suppose contact rate of undirected graph η is given by average degree $\eta = \langle k \rangle$. Every computer unit contact with $\langle k \rangle = (1/N)\sum_{i=1}^{N} k_i$ another in average per time units where k_i is degree of i-th computer unit in computer network structure. In this case, we have

$$\frac{dE[I(t)]}{dt} = (\beta \langle k \rangle - \gamma)E[I(t)] - \frac{\beta \langle k \rangle}{N}E[I^2(t)] \tag{5}$$

The other modification can be done using the fact that in undirected graph, density of graph is $2|E| = \sum_{i=1}^{N} k_i$, so that $D = \sum_{i=1}^{N} k_i/[N(N-1)]$ where $|E|$ is the number of edges in graph and D is density of graph. Density of graph is ratio between the number of edges in a graph and the number of edges in complete graph. Density of graph express the quantity to identify characteristic of graph structure. For large N we can approximate $\langle k \rangle/(N-1) \approx \langle k \rangle/N$ and $D = \langle k \rangle/N$ then $\eta = \langle k \rangle$ can be approximated by $\eta = DN$.

3   Upper Bound of Infection Mean

In the previous section, the dynamics of infection mean is obtained by the dynamic of infection probabilities via SIS stochastic model in Eq. (5) with modification contact parameter. The equation also depends on the second moment $E[I^2(t)]$ so that the dynamics of the second moment are also needed. Dynamic of the second moment also depends on the third moment and so on. In this section, we estimate the first moment using the upper bound of infection mean. Consider the number of vulnerable computers and the number of infection computers is negatively correlated under fixed population assumption. The results of these facts lead us to the following proposition

Proposition 1. *Let $I(t)$ be the random variable that states the number of infected computer units at t. Under the fixed population assumption, the following relationship is fulfilled*

$$E\left[I^2(t)\right] \geq E[I(t)]^2 \tag{6}$$

where $E[I^2(t)]$ is the second moment and $E[I(t)]$ is the first moment.

Proof. Due to the relationship $I(t)$ and $S(t)$ are negatively correlated and under fixed population assumption $S(t) = N - I(t)$ then $Cov(I(t), N - I(t)) \leq 0$. Using definition of covariance between two random variables we get $Cov(I(t), N - I(t)) = E[I(t)]^2 - E[I^2(t)]$ and finally, the relationship is proven.

Consider dynamic equation of mean with modified constant parameter in (5), inequality from Proposition 1 gives us the inequality of this differential equation

$$\frac{dE[I(t)]}{dt} \leq (\beta\langle k\rangle - \gamma)E[I(t)] - \frac{\beta\langle k\rangle}{N}(E[I(t)])^2. \tag{7}$$

Inequality of differential Eq. (7) has a dynamic upper bound that can be solved to get upper bound of the mean of the number of infections. Take the upper bound of inequality (7) then we get

$$\frac{dE[I(t)]}{dt} = (\beta\langle k\rangle - \gamma)E[I(t)] - \frac{\beta\langle k\rangle}{N}(E[I(t)])^2 \tag{8}$$

which is Ordinary Differential Equation (ODE) that can be solved analytically. The solution of that ODE gives us relationship between upper bound and deterministic solution in (1). The following theorem provides an explanation regarding the relationship

Theorem 1. *Let $m(t) = E[I(t)]$ be the expectation of random variable $I(t)$ which explains the number of infected computers at time t, and $\tilde{I}(t)$ be a solution of a deterministic model. Let $P(I(0) = i_0) = 1$ be the infection probability at $t = 0$ and $m(0) = i_0$ then*

$$m(t) \leq \tilde{I}(t) \tag{9}$$

so $\tilde{I}(t)$ is the upper bound of the mean of the number of infections.

Proof. Consider the upper bound of mean dynamic in Eq. (8) and using the separation variable method then the solution can be determined as

$$m(t) = \frac{i_0(\beta\langle k\rangle - \gamma)\exp[(\beta\langle k\rangle - \gamma)t]}{(\beta\langle k\rangle - \gamma) + \frac{\beta\langle k\rangle}{N}i_0(\exp[(\beta\langle k\rangle - \gamma)t] - 1)} \tag{10}$$

where i_0 is assumed at $t = 0$, $p_{i_0}(0) = P[I(0) = i_0] = 1$ thus $P[I(0) \neq i_0] = 0$ and has implications $m(0) = i_0$. By the modified contact parameter, $\bar{\beta}$ can defined as $\eta = \beta\langle k\rangle$

where $\eta = \langle k \rangle$ is contact parameter or contact rate. Substitution term $\bar{\beta}$ to Eq. (10) generates the equation in (1) for $\tilde{I}(t)$ that is the solution of deterministic model. Because the solution in Eq. (10) is obtained by solving the dynamic equation of the upper bound of the infection mean, outbreak then it will always be less than a deterministic solution, i.e. $m(t) \leq \tilde{I}(t)$.

The Gillespie algorithm described by Algorithm 1 will be used to confirm Theorem 1 by performing a stochastic simulation which the discussion of results is given in Sect. 4.

Algorithm 1: Gillespie Algorithm for Stochastic Simulation

Input: Network contact rate $\langle k \rangle$; T=365; Probability β and γ.
for *i=1 to 1000* do
   while *Number of infection>0 and* $t(j) \leq T$ do
      Generate uniform random number u_1 and u_2
      Define total transition probability P and probability of infection P_i
      Determine time step to next transition $T_x = -\frac{ln(u_1)}{P}$
      if $u_2 \leq P_i$ then
         $I \leftarrow I + 1$
         $S \leftarrow S - 1$
      else
         $I \leftarrow I - 1$
         $S \leftarrow S + 1$
      end
      j=j+1
   end
end
Calculate the mean of $I(t)$ for every inteval t.
Output: Mean of $I(t)$ for every time step t

This algorithm will be run for the network topology previously given. The following subsection explains the steps to analyze the upper bound in each graph topology. The first step is to do a parameter sensitivity analysis were to determine the parameters that make the upper bound defined for the maximum N given. Furthermore, the effect of the number of initial infections on computer networks was carried out to analyze the upper bound. This result will be compared with the probability analysis of an outbreak.

3.1 Parameter Sensitivity

The upper bound of the mean depends on two parameters, namely probability of successful transmission β and the probability of recovery γ. We will look for the value of parameter β and γ so that $m(t)$ is defined. Figure 2(a)-(d) shows the value of $m(t)$ in different value of $\beta \in [0, 1]$ and $\gamma \in [0, 1]$ as parameters of $m(t)$ in the four different topologies discussed earlier. Parameter sensitivity applies for maximum $N = 100$ and fixed $i_0 = 2$. Different colors in contour graph give the different values of $m(t)$ explained by color bars in the left of contour graph. White color indicates the values of parameters β and γ which make $m(t)$ undefined.

Sensitivity of the parameters is done for maximum time $t = 365$ where upper bound defined for each value of time $t = 0$ to $t = 365$. For complete graph topology in Fig. 1(b) with $\eta = N - 1$, $m(t)$ is defined if $0 < \beta < 0.01$ and $0 < \gamma < 1$ also for $0.01 < \beta < 0.02$ and $0.05 < \gamma < 1$. Figure 2(b) gives the results of parameter sensitivity for cycle graph with $\eta = 2$.

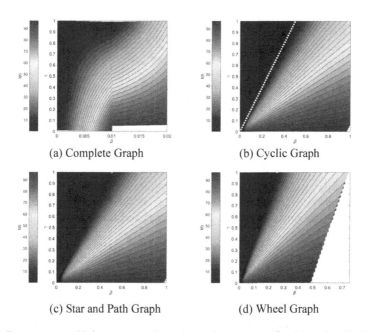

(a) Complete Graph   (b) Cyclic Graph

(c) Star and Path Graph   (d) Wheel Graph

Fig. 1. Parameter sensitivity on several graph topologies using fixed $i_0 = 2$ and $N = 100$.

In this graph topology, $m(t)$ is defined if $2\beta \neq \gamma$ and the value of β not close to one when γ approaches zero. Requirements for star graph topology and path graph topology given by Fig. 2(c). Approximation of contact rates on these two graphs to cycle graph makes $m(t)$ defined similarly to the cycle graph. Wheel graph topology has average contact $\eta = 4(N - 1)/N$ so $m(t)$ is defined if $0 < \beta < 0.5$ for $0 < \gamma < 1$ and $0.5 < \beta < 0.7$ for specific γ values. This result will be used as a consideration for determining parameters for stochastic simulations using Gillespie Algorithm in the following discussion.

3.2 Upper Bound for Several Graph Structure

Suppose we have a fixed and specific graph structure, the upper bound for each structure is given by

1. *Complete Graph (Fully Connected Network)*. A complete graph with N nodes denoted by K_N have $N - 1$ degrees at each node and $\langle k \rangle$ equal to $N - 1$. The upper bound of mean on fully connected network follows

$$m(t) = \frac{i_0(\beta(N-1) - \gamma)\exp[(\beta(N-1) - \gamma)t]}{(\beta(N-1) - \gamma) + \frac{\beta(N-1)}{N}i_0(\exp[(\beta(N-1) - \gamma)t] - 1)}. \tag{11}$$

2. *Cyclic Graph (Ring Network)*. A cyclic graph with N nodes denoted by C_N having 2 degrees at each node and $\langle k \rangle$ equal to 2. The upper bound of mean on ring network follows

$$m(t) = \frac{i_0(2\beta - \gamma)\exp[(2\beta - \gamma)t]}{(2\beta - \gamma) + \frac{2\beta}{N}i_0(\exp[(2\beta - \gamma)t] - 1)}. \tag{12}$$

3. *Star Graph (Star Network) and Path Graph (Line Network)*. A star and path graph with N nodes denoted by S_N and P_N have $\langle k \rangle$ equal to $2(N-1)/N$. The upper bound of mean on star network and line network follows

$$m(t) = \frac{i_0(2\beta - 2\beta/N - \gamma)\exp[(2\beta - 2\beta/N - \gamma)t]}{(2\beta - 2\beta/N - \gamma) + \frac{2\beta(N-1)}{N^2}i_0(\exp[(2\beta - 2\beta/N - \gamma)t] - 1)}. \tag{13}$$

4. *Wheel Graph*. A wheel graph with N nodes denoted by W_N have $\langle k \rangle$ equal to $4(N-1)/N$. The upper bound of mean on star and line network follows

$$m(t) = \frac{i_0(4\beta - 4\beta/N - \gamma)\exp[(4\beta - 4\beta/N - \gamma)t]}{(4\beta - 4\beta/N - \gamma) + \frac{4\beta i_0(N-1)}{N^2}(\exp[(4\beta - 4\beta/N - \gamma)t] - 1)}. \tag{14}$$

Figure 1 shows the effect of the number of initial infections i_0 on the upper bound. The parameter β is selected with the same value, $\beta = 0.01$, which means that in one day there is one successful infection computer from 100 computers. Parameter $1/\gamma$ states the time needed to recover one computer unit. In the complete graph topology selected $\gamma = 0.25$, which means that the time needed to recover one computer is four days, thus with significant contacts, this case is representing high infection or with $R_0 > 1$. In cyclic and star graph topologies considered $\gamma = 0.025$ which means the time needed to recover one computer is 40 days, thus representing a case with a low spread with $R_0 \leq 1$ but with considerable damage because it requires 40 days of recovery time. Whereas in the wheel graph the same thing happened in the complete graph, for the case of $R_0 > 1$ but the case of moderate infection.

Based on the results in Fig. 1(a) can be shown that the upper bound of infection mean on the complete graph is not influenced by i_0. The upper bound quickly to equilibrium point of $m(t)$ around $m(t) = 70$ in same time in the case fixed population $N = 100$ and different size of i_0. Different size of i_0 differentiate at the starting point but do not have a significant effect when time is change. The epidemic threshold used is $R_0 = 4$ that is the complete graph with $\beta = 0.01$ and $\gamma = 0.25$ there are four new infections generated by an infected computer.

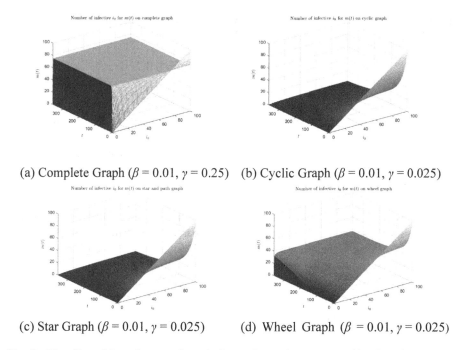

(a) Complete Graph ($\beta = 0.01$, $\gamma = 0.25$)   (b) Cyclic Graph ($\beta = 0.01$, $\gamma = 0.025$)

(c) Star Graph ($\beta = 0.01$, $\gamma = 0.025$)   (d) Wheel Graph ($\beta = 0.01$, $\gamma = 0.025$)

Fig. 2. The effect of i_0 on the upper bound of mean in graph structures with selected parameters.

Figure 1(b) gives the result that the initial value of i_0 affects the speed of time reaching the value of upper bound equilibrium $m(t)$ where there is no infection. Greater value of i_0 makes $m(t)$ slower to reaches the equilibrium value at around $m(t) = 0$. Study of effect i_0 in cyclic graph is done for $\beta = 0.01$ and $\gamma = 0.025$. The upper bound $m(t)$ on the star and path topology have the same behavior and almost identical form with cyclic graph if using the same value of β and γ (see Fig. 2(b)-(c)). This is because the contact between two graphs only around $2(N - 1)/N \approx 2$, so for large N and same value of parameter, both graph topologies will give the same result. It has been found that the wheel graph has twice the degree average than the star graph and path graph. Figure 1(d) shows the implication of increasing the size of i_0 that make $m(t)$ bigger than the case in star and cyclic graph.

3.3   Probability of Epidemic

On a network with N which is relatively large, it is possible for an outbreak. In the history of the spreading of computer viruses, example in the emergence of Meet Chernobyl Virus in 1998 which infected more than 60 million computers and resulted in commercial losses of around 1 billion US dollars. Furthermore, a computer virus outbreak occurred in 2004 by the My Doom virus via e-mail which quickly infected 2 million computers. Therefore, the probability of an outbreak (epidemic) is significant to know. Probability of an outbreak is defined as (see [2])

$$1 - p_0(t) \approx \begin{cases} 0; & if R_0 \leq 0 \\ 1 - \left(\frac{1}{R_0}\right)^{i_0}; & if R_0 > 0 \end{cases} \tag{15}$$

where $R_0 = \beta\eta/\gamma$ and η as contact parameter. In fully connected network topology $R_0 = \beta(N-1)/\gamma$ and other topologies by substitution $\eta = \langle k \rangle$, where $\langle k \rangle$ is a degree average of network. Probability of no infection $P(I(t) = 0)$ can be obtained by the solution of the stochastic model using matrix exponential approximation in Eq. (5) by Pade approximation (see [1]).

Numerical results for fully connected, cyclic, ring, path, and wheel network topology described in Fig. 3. Probability of an outbreak is calculated based on the selected parameters according to the results of the parameter sensitivity analysis in the previous section. The same parameters will be used to compare the probability of an outbreak with the results at the upper bound. The probability of an outbreak is close to 1 for complete graph topology as in Fig. 3(a) if $i_0 = 3$ that indicate high probability spreading of the virus. Figure 3(b)-(c) indicates the chance of a low spread of the virus where the probability goes to zero for every i_0 given. The case on the wheel graph represents the moderate process of virus spreading. The higher value of i_0 makes the probability of infection getting higher and slower towards virus-free conditions.

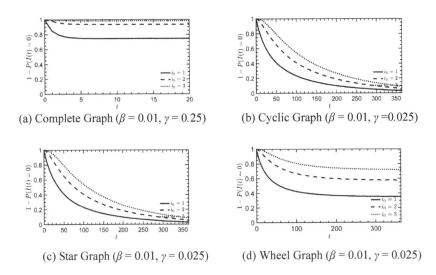

(a) Complete Graph ($\beta = 0.01, \gamma = 0.25$)   (b) Cyclic Graph ($\beta = 0.01, \gamma = 0.025$)

(c) Star Graph ($\beta = 0.01, \gamma = 0.025$)   (d) Wheel Graph ($\beta = 0.01, \gamma = 0.025$)

Fig. 3. Probability of an outbreak in each graph topology with fixed $N = 100$ and given parameter β and γ where $i_0 = 1, 2, 3$.

4   Gillespie Algorithm Result

Let a random variable $I(t)$ be the number of infected computers at time t and assume $I(t) = x$. A continuous random variable that expresses the time for the next event is given a process in the condition x that is the inter-arrival time T_x. Suppose that $G_x(t)$ is the probability for the process to remain in state x for t with $G_x(t) = P(T_x > t)$. At a short interval time h, the probability of the process remains in the state x at time $t + h$ is given by $G_x(t + h) = G_x(t)p_{xx}(t)$.

Consider $p_{xx}(t) = 1 - [\bar{\beta}x(N - x)/N + \gamma x]h + o(h)$ so that it is obtained

$$\frac{dG_x(t)}{dt} = -\left[\frac{\bar{\beta}x(N - x)}{N} + \gamma x\right]G_x(t). \tag{16}$$

The solution of Eq. (16) will be obtained by separating variable with the initial value $G_x(0) = P(T_x > 0) = 1$ as $G_x(t) = \exp\left[-\left(\bar{\beta}x(N - x)/N + \gamma x\right)t\right]$, where $G_x(t)$ is the survival function. Distribution function of T_x is $F_x(t) = P(T_x \leq t)$ and $F_x(t)$ is equal to $1 - \exp\left[-\left(\bar{\beta}x(N - x)/N + \gamma x\right)t\right]$, which is an exponential distribution function with rate $\theta = [\bar{\beta}x(N - x)/N] + \gamma x$. Numerical simulation using Gillespie Algorithm is done by the fact that the distribution function in interval $[0, 1]$ so that for $U \sim Uniform(0, 1)$ by specifying

$$P(U \leq F_x(t)) = P\left(F_x(F_x^{-1}(U)) \leq F_x(t)\right) = P\left(F_x^{-1}(U) \leq t\right). \tag{17}$$

This equation gives Inter-arrival time for the process (see [2])

$$T_x = F_x^{-1}(U) = -\frac{\ln(1 - U)}{\left[\frac{\bar{\beta}x(N-x)}{N} + \gamma x\right]} = -\frac{\ln(U)}{\left[\frac{\bar{\beta}x(N-x)}{N} + \gamma x\right]} \tag{18}$$

where U is Uniform random variable over $(0, 1)$ and by the property of uniform distribution, the distribution of $1 - U$ is equal to U.

Stochastic simulation is done by generating a thousand random paths through the Gillespie [9] algorithm then comparing with deterministic solutions as the upper bound of infection mean. This algorithm is also known as the Stochastic Simulation Algorithm (SSA) for continuous-time Markov chain simulation. Two random numbers $u_1, u_2 \in U(0, 1)$ are needed for inter-arrival time T_x that has been previously defined and for the incidence of infection (see Algorithm 1) in Sect. 3. In Sect. 3, the same value β has been used to analyze the upper bound. Simulations in this section will be carried out by considering different cases. Figure 4 shows a thousand sample path from the Gillespie Algorithm for fixed $N = 100$ with same recovery probability $\gamma = 0.25$ and a different probability of β. Graph topology in Fig. 4(a)-(d) have illustrated different values of $\bar{\beta} = 2$, $\bar{\beta} \approx 1$, $\bar{\beta} = 0.5$ and $\bar{\beta} = 0.4$. The greater value of $\bar{\beta}$ makes the upper bound and mean of SSA get closer. If the contact parameter and probability of infection β produces the same value of $\bar{\beta}$ then, we get the same result. This means that the closeness between the upper bound and the simulation mean is influenced by the

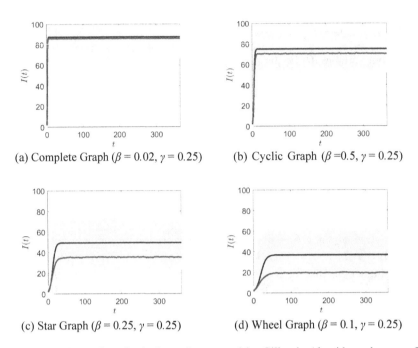

(a) Complete Graph ($\beta = 0.02$, $\gamma = 0.25$)   (b) Cyclic Graph ($\beta = 0.5$, $\gamma = 0.25$)

(c) Star Graph ($\beta = 0.25$, $\gamma = 0.25$)   (d) Wheel Graph ($\beta = 0.1$, $\gamma = 0.25$)

Fig. 4. Thousand sample paths (red areas) generated by Gillespie Algorithm where $\gamma = 0.25$, $N = 100$, $T = 365$ and $i_0 = 2$, compared with deterministic solutions (blue curve) and the mean of a thousand random steps (red curve). (Color figure online)

magnitude of the probability of infection and the number of contacts on computer networks.

Cases in numerical simulations are shown in Table 2 for $t = 365$ days. Distance is defined as the difference between the upper bound and the mean of the stochastic simulation. Based on the table, the difference between the upper bound and the simulation gets closer when $\bar{\beta}$ gets bigger. Thus, the spread of the virus is influenced by the number of contacts on the network. Networks with fewer contacts will have a lower spread rate than networks with higher contacts. The upper bound gives the maximum value of the risk that can be generated from a topology even though with the same parameters, the realization can be less than that value. The smallest topology will produce a low spread. Ring topology has the smallest contact of the other four topologies so that the same parameters will result in stronger resistance. The number of computers N on a network also affects the distance between the upper bound and means a simulation that represents random events. Table 2 shows for $N = 50$ and the same parameters give smaller distances than $N = 100$. Thus, if we have a network with N, which is relatively small, the upper bound can be used as an estimate of the risk, namely the number of units of the infected computer.

Table 2. Comparison of Gillespie Algorithm (GA) Mean and Upper Bound (UB) for $t = 365$ and parameter in Fig. 4.

N	100			50		
Topology	GA mean	UB	Distance	GA mean	UB	Distance
Full	86.15	87	1.23	34.41	37	2.84
Cyclic	69.289	75	5.71	34.955	38	2.55
Star	36.97	49	12.53	16.94	24	7.55
Wheel	20.00	37	16.87	5.34	18	12.78

5 Conclusion

We have treated the average degree as a contact parameter in the SIS stochastic model that was previously built on a deterministic model [17]. The fact that the number of vulnerable and infected computers are negatively correlated produces the upper bound of infection mean in Theorem 1. As evidence for this theorem, a thousand random steps are simulated. The mean of a thousand random steps is compared to a deterministic solution as the upper bound of infection mean through the Gillespie algorithm [9]. That algorithm has confirmed the mean of the number of infections in stochastic models is always less than or equal to the deterministic solution. Analysis and simulation are also carried out at the upper bound and the probability of endemic. Based on this analysis with the same parameters, more significant contact parameters on a computer network increase the number of infections and the quickness to reach the equilibrium of the number of infections.

This study limited for the fixed population assumptions and homogeneous mixing assumptions. Heterogeneous contact cases or complex networks such as scale-free networks or heterogeneous Poisson processes that use different rates of infection can be considered for further discussion. This research is also still limited to the compartment level. Development to the individual level and also for two or more joint topologies such as the discussion of the metapopulation is also quite interesting.

Acknowledgement. We would like to thank the Institute for Research and Community Services (LPPM-ITB) for funding this research through the P3MI program given to the Statistics Research Division.

References

1. Al-Mohy, A.H., Higham, N.J.: A new scaling and squaring algorithm for the matrix exponential. SIAM J. Matrix Anal. Appl. **31**(3), 970–989 (2010). https://doi.org/10.1137/09074721x
2. Allen, L.J.S.: An introduction to stochastic epidemic models. In: Brauer, F., van den Driessche, P., Wu, J. (eds.) Mathematical Epidemiology, pp. 81–130. Springer, Heidelberg (2008). https://doi.org/10.1007/978-3-540-78911-6_3
3. Allen, L.J.S.: An Introduction to Stochastics Processes with Applications to Biology. CRC Press Tylor and Francis Group, Boca Raton (2010)

4. Barabási, A.L.: Network Science. Cambridge University Press (2016). http://networksciencebook.com/
5. Daley, D.J., Gani, J.: Epidemic modelling (1999). https://doi.org/10.1017/cbo9780511608834
6. Dangerfield, C.E., Ross, J.V., Keeling, M.J.: Integrating stochasticity and network structure into an epidemic model. J. R. Soc. Interface **6**, 761–774 (2009). https://doi.org/10.1098/rsif.2008.0410
7. Deng, X., Wang, X.: The application of Gillespie algorithm in spreading. In: Proceedings of the 3rd International Conference on Mechatronics Engineering and Information Technology (ICMEIT 2019). Atlantis Press (2019). https://doi.org/10.2991/icmeit-19.2019.110
8. Ganesh, A., Massoulie, L., Towsley, D.: The effect of network topology on the spread of epidemics. In: Proceedings of the IEEE 24th Annual Joint Conference of the IEEE Computer and Communications Societies, vol. 2, pp. 1455–1466, March 2005. https://doi.org/10.1109/INFCOM.2005.1498374
9. Gillespie, D.T.: Exact stochastic simulation of coupled chemical reactions. J. Phys. Chem. **81**(25), 2340–2361 (1977). https://doi.org/10.1021/j100540a008
10. Grishunina, Y., Manita, L.: Stochastic models of virus propagation in computer networks: Algorithms of protection and optimization. Lobachevskii J. Math. **38**(5), 906–909 (2017). https://doi.org/10.1134/s1995080217050122
11. Kaluarachchi, P.K.: Cybersecurity: stochastic analysis and modelling of vulnerabilities to determine the network security and attackers behavior. Doctoral dissertations, University of South Florida (2017)
12. Kiss, I.Z., Miller, J.C., Simon, P.L.: Mathematics of Epidemics on Network. Springer, Heidelberg (2017). https://doi.org/10.1007/978-3-31950806-1
13. Liu, J., Bianca, C., Guerrini, L.: Dynamical analysis of a computer virus model with delays. Discret. Dyn. Nat. Soc. **2016**, 1–21 (2016). https://doi.org/10.1155/2016/5649584
14. Lloyd, A.L., Valeika, S.: Network models in epidemiology: an overview. In: Complex Population Dynamics: Nonlinear Modeling in Ecology, Epidemiology and Genetics, pp. 189–214. World Scientific (2007)
15. Mishra, B.K., Ansari, G.M.: Differential epidemic model of virus and worms in computer network. Int. J. Netw. Secur. **14**(3), 149–155 (2012)
16. Navlakha, S., Bar-Joseph, Z.: Algorithms in nature: the convergence of systems biology and computational thinking. Mol. Syst. Biol. **7**(1), 546 (2011)
17. Newman, M.E.J.: The spread of epidemic disease on networks. https://doi.org/10.1103/PhysRevE.66.016128
18. Nguyen, B.: Modelling cyber vulnerability using epidemic models (2017). https://doi.org/10.5220/0006401902320239
19. Omair, S.M., Kumar, S.: e-Epidemic on the computer viruses in the network. Eur. J. Adv. Eng. Technol. **2**(9), 78–82 (2015)
20. Pokhrel, N.R., Tsokos, C.P.: Cybersecurity: a stochastic predictive model to determine overall network security risk using Markovian process. J. Inf. Secur. **08**, 91–105 (2017). https://doi.org/10.4236/jis.2017.82007
21. Qin, P.: Analysis of a model for computer virus transmission. Math. Probl. Eng. **2015**, 1–10 (2015). https://doi.org/10.1155/2015/720696
22. Szabó-Solticzky, A., Simon, P.L.: The effect of graph structure on epidemic spread in a class of modifies cycle graph. Math. Model. Nat. Phenom. **9**(2), 89–107 (2014). https://doi.org/10.1015/mmnp/20149206

Sentiment Analysis in Social Media Based on English Language Multilingual Processing Using Three Different Analysis Techniques

Nor Saradatul Akmar Zulkifli[(⊠)] ⓘ and Allen Wei Kiat Lee

Soft Computing & Intelligent Systems (SPINT), Faculty of Computing,
Universiti Malaysia Pahang, 26300 Gambang, Kuantan, Pahang, Malaysia
saradatulakmar@ump.edu.my

Abstract. Numerous numbers of companies have utilized the web to offer their services and products. Web customers dependably look through the reviews of other customers towards a product or service before they chose to buy the things or viewed the films. The company needs to analyse their customers' sentiment and feeling based on their comments. The outcome of the sentiment analysis makes the companies easily discover either the expression of their users is more to positive or negative. There are numerous numbers of sentiment analysis techniques available in the market today. However, only three (3) techniques will be used in this research which are the Python NLTK Text Classification, Miopia and MeaningCloud. These techniques used to analyse the sentiment analysis of the reviews and comments from English language in social media. 2400 datasets from Amazon, Kaggle, IMdB, and Yelp were used to analyse the accuracy of these techniques. From this analyses, average accuracy for sentiment analysis using Python NLTK Text Classification is 74.5%, meanwhile only 73% accuracy achieved using Miopia technique. The accuracy achieved when using MeaningCloud technique is 82.1% which is the highest compared to other techniques. This shows that hybrid technique offers a greatest accuracy for sentiment analysis on social reviews.

Keywords: Multilingual sentiment analysis · English language · Text classification

1 Introduction

Nowadays, the web has been broadly utilized by the general population universally. Individuals utilize the web to discover films, recreations and books. In the meantime, they additionally express their emotions towards those perspectives with the goal that other people can without much of a stretch to recognize what happens precisely and whether there is any item worth to purchase or attempt through the perception of the surveys. The company can carry out sentiment analysis to analyse their customers' sentiment and feeling toward their products and services. There are different sentiment analysis techniques that exist in the market today to help the companies to conduct the sentiment analysis for their customers' reviews and comments. This help to improve

© Springer Nature Singapore Pte Ltd. 2019
M. W. Berry et al. (Eds.): SCDS 2019, CCIS 1100, pp. 375–385, 2019.
https://doi.org/10.1007/978-981-15-0399-3_30

the company's products and services. It is very troublesome and tedious if enlist a man to dissect the information to know whether the greater part of clients give a positive or negative audit from a huge number of remarks. In this manner, machine learning dependent on sentiment analysis has been presented. It makes the data mining quicker.

However, previous studies reported that the sentiment analysis techniques construct their model only consist of one language [1–3]. The multilingual people can comprehend numerous languages. They can give the customer reviews in social media using different languages. If the model of sentiment analysis technique only consists of one language, it is quite hard for the sentiment analysis technique to extract the data. For example, Twitter produces enormous amounts of opinion tweets that consist of different languages. It is quite hard for single language model sentiment analysis technique to extract the text and make sentiment analysis. The company cannot interpret their customer reviews from different languages [2, 4, 5]. Next, using of words in casual languages and emoji when commenting in social media. For example, some of the sentiment analysis technique cannot detect the "can't, wouldn't" in casual language. While some of the sentiment analysis technique can detect it. This will cause the positive and negative classification in sentiment analysis will be different. Besides that, some of the users tend to use the emoticons to express their expression in social media (Table 1).

Table 1. The summarization of multilingual processing in sentiment analysis

Multilingual processing in sentiment analysis	Explanation
Corpus-based with interpretation and machine learning [6, 7]	• Sentiment analysis with Python NLTK Text Classification • Solve the issue of knowledge procurement issue of rule-based machine interpretation
Lexicon with rule-based machine translation (RBMT) approach [7, 8]	• Denotes machine translation systems based on linguistic information about target and source languages • Miopia
Hybrid machine interpretation	• Mixing the two other frameworks • Using MeaningCloud

Using the emoji to comment is a trend for the media social users. This causes some sentiment analysis technique hard to detect the emojis. This will impact on the evaluation of sentiment analysis technique. This will affect the output of the sentiment analysis. This will cause the positive and negative grouping in sentiment analysis will be different [2, 5]. Lastly, the accuracy of the yield is the issue of the sentiment analysis. There isn't any sentiment analysis technique that can accomplish the 100% precision. The different sentiment analysis technique will be used in different situations to obtain the better accuracy of sentiment analysis. The general population will ponder which approach should they utilize, and in which circumstance they are required to utilize and bringing about the time-consuming in choosing for the correct approach [1, 9] (Table 2).

Table 2. The comparison of sentiment analysis techniques

Sentiment analysis technique	Explanation
Naïve Bayes [10]	• Apply Bayes theorem to arrange the information with assumption that the traits are restrictively autonomous
SVM [11]	• Divide information into paired classes • It uses the hyperplane • It determines linear separators in the search space
Neural network [12]	• Hardware or programming designed after the assignment of neurons in the human mind in data innovation • Deep learning technology
Python NLTK text classification	• Open source natural language processing (NLP) platform • Capable of textual tokenisation, parsing, classification, stemming, tagging and semantic reasoning
General Bayesian Network [13, 14]	• It is the probabilistic graphical model
MIOPIA	• Evaluate the dependencies among sentences • Do the calculation on their sum weighting
SentiStrength	• Estimate the sentiment analysis polarity in short messages
Semantic orientation calculator	• Calculate entire sentiment analysis polarity and strength of messages
MeaningCloud	• Facebook/Twitter (hybrid machine interpretation) • Classification model • Use hybrid algorithm

Natural Language Processing (NLP) makes human to speak with the machine viably. There are various applications made in in NLP late decades. A large portion of these is to a great degree profitable in regular daily existence. For instance, a machine that takes rules by voice. NLP is a piece of man-made brainpower. It manages breaking down, understanding and producing the languages that individuals utilize normally with the end goal to bond with Personal Computers (PCs) in both written and talked settings utilizing human languages [15]. Normally the word limits are mixed, and the sentences appreciated are unique. At the accompanying level, the sentence structure of the dialect causes us to pick how the words joined to make greater implications. Working up a program that appreciates regular dialect is a troublesome issue. Quantities of common dialects are enormous. They contain vastly various sentences. Moreover, there is much equivocalness in normal language. Numerous words have a couple of suggestions and sentences have implications diverse in different settings. This makes creation of projects that fathoms a characteristic dialect, a testing errand.

Normally the word limits are blended, and the sentences comprehended are unique. At the accompanying level, the sentence structure of the language causes us to pick how the words joined to make greater implications. Working up a program that appreciates regular language is a troublesome issue. Quantities of common languages are enormous. They contain vastly various sentences. Moreover, there is much equivocalness in normal languages. Numerous words have a couple of suggestions and sentences have implications diverse in different settings. This makes creation of projects that fathoms a characteristic language, a testing errand. There are 3 types of multilingual processing approaches which are Corpus-based with Interpretation and Machine Learning, Lexicon with Rule based Machine Interpretation and Hybrid Machine Interpretation [16–19].

The aim of this research is to study the multilingual processing used in the NLP analysis especially in sentiment analysis. The comparison between different patterns of sentiment analysis in social media will be considered. NLP makes human to speak with the machine viably. There are various applications made in in NLP late decades. A large portion of these is to a great degree profitable in regular daily existence. For instance, a machine that takes rules by voice [20]. There are several objectives to achieve which is to explore and evaluate Social Review in English language using 3 different sentiment analysis techniques which is the Python NLTK Text Classification, Miopia and MeaningCloud tools in term of their polarity text classification (positive or negative classification) and to make comparison of the accuracy performance between these three techniques.

2 Methodology

The scope of this research is separated into five (5) categories; (i) The accuracy of the outcome using three sentiment analysis techniques (Python NLTK Text Classification, Miopia and MeaningCloud) in English language will be used to make comparison. (ii) The classification is based positive, neutral and negative classification. (iii) The datasets utilized are from the sentiment polarity dataset version 2 which consists of 32938 positive reviews and 31783 negative reviews. The datasets in Bo Pang and Lillian Lee reviews is utilized as well. (iv) The datasets used are also from Amazon, Yelp and IMDb datasets. (v) There are 3 sentiment analysis techniques (Python NLTK Text Classification, Miopia and MeaningCloud) used in this research.

Figure 1 shows the methodology flow in completing this research. The first step is that the English data are stored in the database for structuring the model. This is because all the part-of-speech tag (POS-tagging) and the parser together with the control of information are formed. The datasets used in this research are Kaggle, Amazon, Yelp and IMDb datasets. There are 3 sentiment analysis techniques used in this research which is the Python NLTK Text Classification, Miopia and MeaningCloud tools.

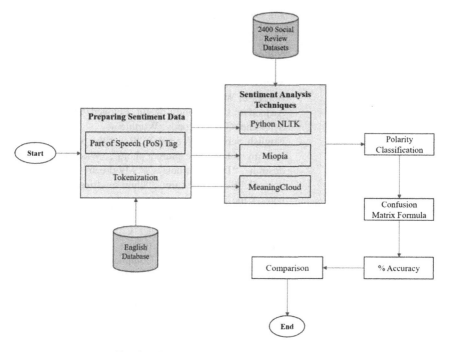

Fig. 1. Flow chart for the implemented study

Table 3 depict the list of sentiment data which had been extract from each Sentiment Analysis. These sentiments data will be used to test the techniques in the pre-preliminary analysis. The second step is inputting the English language reviews for testing. The sentiment analysis is carried out. After the sentiment analysis process, the classification of the reviews (positive, negative and neutral) are collected. The sentiment analysis accuracy is calculated using the confusion matrix formula.

The output of sentiment analysis accuracy by using three sentiment analysis techniques in English language are gathered. The last step is the assessment method, which is comparing the accuracy of different sentiment analysis techniques in English language.

3 Result and Discussion

Based on the Table 4, it shows that the accuracy of the Python NLTK Text Classification (Corpus-based approach) is higher than Miopia tool (Lexicon-based approach) with 74.50% and 73.00% separately. This is because the corpus-based approach consists of knowledge automatically acquired from text using machine learning. It needs to have consistent of sample corpus. It can used for testing for wide coverage of datasets. But for the Lexicon-based approach, it suffers the knowledge acquisition bottleneck.

Table 3. List of sentiment data extract from each sentiment analysis techniques

Techniques	Sentiment data
Python NLTK text classification & MeaningCloud	**Noun:** Perfection, Genius, Mastery, Advantage, Decay, Love, Damage **Verbs:** Culminate, Boost, Ruin, Excel, Betray, Assist, Decrease, Waste **Adjective:** Calm, Finest, Right, Beloved, Slim, Fierce, Scientific, Clean, Bloody, Worthless **English Sentences (corpus):** "Boost the performance", "I love my father", "I capture a photo", "Waste food is in the dustbin", "The flight delays", "My dream is become doctor", "He is quite calm", "I hurt my friend", "My friends betrays me"
MIOPIA	**Noun:** Masterpiece, Perfection, Excellence, Angel, Zenith, Attempt, Complexity, Conflict, Congestion **Verbs:** Culminate, Exalt, Conquer, Assist, Capture, Comfort, Battle, Decrease, Complain, Mangle, Curse **Booster Words:** The least, Nowhere near, Not that, Noticeably, Totally, Remarkably, Entirely **Adverbs:** Pleasingly, Enjoyable, Courteously, Professionally, Intimately, Rarely, Plainly, Indifferently, Clumsily **Adjectives:** Beloved, Accessible, Elegant, Worthy, Infectious, Intentional, Foolish, Freaky, Rusty, Unsympathetic, Malicious **Emoticons:** :E, :\, XO, :_(, : [, x3?, =], >:D, ^_^

Even though the Lexicon-based approach is quite accurate in analyzing the sentiment, but it faces problems when they are wide range of datasets are tested. The knowledge acquisition bottleneck means that the transform of datasets in that approach are quite time consuming, incomplete and insufficient. Therefore, in the testing work, it shows that the accuracy of Python NLTK Text Classification tool (Corpus-based approach) is higher than Miopia tool (Lexicon-based approach) with Kaggle datasets, Amazon datasets, Yelp datasets and IMDb datasets.

Table 4. Accuracy percentage for three (3) sentiment analysis techniques

Sentiment analysis technique	Accuracy (testing work)
Python NLTK text classification	74.50%
Miopia	73.00%
MeaningCloud	82.13%

Even though the Lexicon-based approach is quite accurate in analysing the sentiment, but it faces problems when they are wide range of datasets are tested. The knowledge acquisition bottleneck means that the transform of datasets in that approach are quite time consuming, incomplete and insufficient.

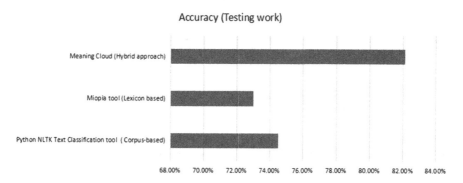

Fig. 2. Accuracy percentage (in bar graph) difference between three different techniques

Therefore, in the testing work, it shows that the accuracy of Python NLTK Text Classification tool (Corpus-based approach) is higher than Miopia tool (Lexicon-based approach) with Kaggle datasets, Amazon datasets, Yelp datasets and IMDb datasets.

Table 5. Comparison between reference and the testing work using Miopia tools

Miopia tools	Accuracy
Testing	73%
Reference	74%

The MeaningCloud is the hybrid approach for sentiment analysis technique. The accuracy percentage of MeaningCloud tool is 82.13%. It is the highest among the 3 sentiment analysis techniques. This is because its hybrids the characteristics of corpus based and lexicon- based approach. It can explore advantages of other approaches. It can achieve the more accurate result in analysing the sentiment in wide coverage (datasets). It can achieve the maximum classification accuracy. Therefore, the accuracy percentage of MeaningCloud tool is highest (82.13%) as compared to the Python NLTK Text Classification (Corpus-based approach) (74.50%) and the Miopia tools (Lexicon-based approach) (73.00%). The result for this technique is shown in Fig. 2.

As shown in Table 5 and Fig. 3, the accuracy from the reference work is higher than testing work with 74.00% and 73.00% separately. This may because the number of datasets used in reference work are more than the numbers of datasets tested in testing work. The number of datasets used in the reference work including 2 252 adjectives, 1 142 nouns, 903 verbs, 745 adverbs and 177 intensifiers as listed in Table 3.

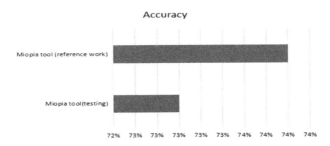

Fig. 3. Comparison between reference work and testing work for Miopia tools

When the number of datasets to be tested with sentiment analysis techniques increases, the more exact the result would obtain. Overall the accuracy of the Miopia sentiment analysis technique is proven in the testing work with the accuracy percentage around 73% is likely similar with the accuracy of Miopia sentiment analysis technique in the reference work with the accuracy percentage around 74%.

Table 6. Comparison between reference and testing works using Python NLTK text classification and Turkey tools

	Sentiment analysis technique	Accuracy
Testing	Python NLTK text classification tool	74.5%
Reference	Turkey author tool	74%

As depicted in Table 6 and Fig. 4, the accuracy of the testing work is higher than reference work with 74.50% and 74.00% separately. The is because the datasets used in the reference work only focus on the products reviews. But for the testing work, it focusses on wide range of datasets (Kaggle, Amazon, Yelp and IMDb datasets). Another reason is that the Turkey author used the method of computing the word PMI. But for the testing work it used the Python NLTK Text Classification tool. It is the reliably decipher extensive substance that will help to achieve more better in the accuracy of sentiment analysis technique. These two factors cause the difference in the accuracy of the corpus-based approach sentiment analysis techniques. Overall the accuracy of the Python NLTK Text Classification tool is proven in the testing work with the accuracy percentage around 74.5% is likely similar with the accuracy of Turkey author reference work with the accuracy percentage around 74%.

The tests dependent on the proposed information (datasets) was conducted successfully. The sentiment grouping and calculation of the accuracy of the sentiment analysis techniques was conducted successfully in the testing. The outcomes getting for this testing is robust as there no any major issue happened during conducting the experiment. The accuracy of sentiment analysis obtained by this three sentiment analysis techniques (Python Nltk Text Classification, MIOPIA and MeaningCloud tool) are calculated and recorded.

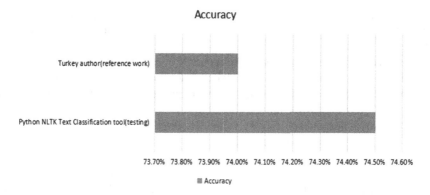

Accuracy

Fig. 4. Comparison between reference work and testing work for Python NLTK Text classification

The accuracy for Python Nltk Text Classification, MIOPIA and MeaningCloud tool are 74.5%, 73% and 82.125% respectively. The MeaningCloud tool shows the highest accuracy among these three sentiment analysis techniques. This is due to the hybrid approach that combine the characteristics of corpus based and lexicon-based approach in order to achieve optimum classification accuracy.

4 Conclusion

This research was the sentiment classification on the Kaggle, Amazon, Yelp and IMDb datasets. The first objective is to evaluate the polarity classification for Social Media Review in English language using three (3) sentiment analysis techniques which is the Python NLTK Text Classification, Miopia and MeaningCloud tools. And the second objective is to compare the accuracy performance between these techniques. The Python Nltk Text Classification is the Corpus-based with interpretation and machine learning. The Miopia tool is the Lexicon with Rule-Based Machine Translation (RBMT) Approach meanwhile, the MeaningCloud tool is the Hybrid Machine Interpretation. The datasets used in this research were the Kaggle, Amazon, Yelp and IMDb datasets. There were 800 datasets which is randomly chosen to be analysed using these techniques. The text classification analyses for these datasets were recorded using the Microsoft Excel.

The output obtained by Python Nltk Text Classification, Miopia and MeaningCloud tools were discussed and compared. The MeaningCloud tool shows the highest accuracy among these techniques. This is due to the hybrid approach that combine the characteristics of the corpus based and lexicon-based approach in order to achieve optimum classification accuracy. Programmed calculator, using different datasets in sentiment analysis techniques and offline sentiment analysis techniques advanced version are a few improvements that can be conducted in this research for future work.

There are a few constraints faced in this research. The number of sentiment datasets used is a critical element, where the total numbers of datasets used can affect the evaluation of the result. The total number of sentiment datasets required for sentiment analysis techniques should be as many as possible to obtain a more accurate evaluation of the result. However, for this research only 2400 datasets were analysed. Secondly, manual calculation is done to calculate the accuracy of each technique. After getting the total numbers of true positive and true negative that have been analysed, the users must calculate the accuracy for each technique manually using the confusion matrix formula. The Python Nltk Text Classification is only able to calculate the positive polarity value, negative polarity value and subjectivity value. Besides that, the Miopia tool is only able to calculate the sum of sentiment score. Next, the MeaningCloud tool is only capable to get the global sentiment, confidence level, polarities and subjective/objective. Therefore, the users need to calculate the total numbers of true positive and true negative and the sentiment analysis techniques' accuracy manually and record it using the Microsoft excel.

There are a few improvements that can be conducted in this research for future work. First, a programmed calculator for the entire number of true positive, true negative, false positive, false negative of the sentiment datasets will be a gigantic help for calculating the sentiment analysis techniques' accuracy. Second, datasets other than Kaggle, Amazon, Yelp and IMDb datasets can be used in this research. This can help the users to know the robustness of the sentiment analysis techniques for analysing the sentiment datasets. This also can help the sentiment analysis techniques to obtain higher accuracy in analysing the sentiment datasets.

Acknowledgement. A high appreciation to Faculty of Computer Systems and Software Engineering, Universiti Malaysia Pahang, under grant RDU1703287.

References

1. Bhuta, S., Doshi, A., Doshi, U., Narvekar, M.: A review of techniques for sentiment analysis of twitter data. In: 2014 International Conference on Issues and Challenge in Intelligent Computing Techniques (ICICT) (2014)
2. Sarlan, A., Nadam, C., Basri, S.: Twitter sentiment analysis. In: 2014 International Conference on Information Technology and Multimedia (ICIMU), Putrajaya, Malaysia (2014)
3. Cardona-Grau, D., Sorokin, I., Leinwand, G., Welliver, C.: Introducing the Twitter impact factor: an objective measure of urology's academic impact on Twitter. Eur. Urol. Focus. **2**(4), 412–417 (2016)
4. Dashtipour, K., et al.: Erratum to: multilingual sentiment analysis: state of the art and independent comparison of techniques. Cogn. Comput. **8**(4), 772–775 (2016)
5. Neethu, M.S., Rajasree, R.: Sentiment analysis in Twitter using machine learning techniques. In: 4th ICCCNT, Tiruchengode, India (2013)
6. Bhavitha, B.K., Anisha, P.R., Chilunkar, N.N.: Comparative study of machine learning techniques in sentimental analysis. In: International Conference on Inventive Communication and Computational Technologies (2017)
7. Hailong, Z., Wengan, G., Jiang, B.: Machine learning and lexicon based methods for sentiment classification: a survey. In: 11th Web Information System and Application Conference (2014)

8. Ehsan Basiri, M., Kabiri, A.: Translation is not enough: comparing lexicon-based methods for sentiment analysis in Persian. In: International Symposium on Computer Science and Software Engineering Conference (CSSE) (2017)

9. Okpor, M.: Machine translation approaches: issues and challenges. Int. J. Comput. Sci. Issues **11**(5), 159–165 (2014)

10. Al-Aidaroos, K., Abu Bakar, A., Othman, Z.: Naive Bayes variants in classification learning. In: International Conference on Information Retrieval & Knowledge Management (CAMP) (2010)

11. Jadav, B.M., Vaghela, V.B.: Sentiment analysis using support vector machine based on feature selection and semantic analysis. Int. J. Comput. Appl. **146**(13) (2016)

12. Al-Saffar, M., Sabri, B., Tao, H., Awang, S., Abdul Majid, M., Al-Saiagh, W.: Sentiment analysis in arabic social media using association rule mining. J. Eng. Appl. Sci. **11**(2), 3239–3247 (2016)

13. Ang, S.L., Ong, H.C., Low, H.C.: Classification using the general bayesian network. Pertanika J. Sci. & Technol. **24**(1), 205–211 (2016)

14. Minn, S., Shunkai, F., Desmarais, M.C.: Efficient learning of general Bayesian Network Classifier by local and adaptive search. In: International Conference on Data Science and Advanced Analyrics (DSAA) (2014)

15. Noorhuzaimi, M.N., Ab Aziz, M.J., Mohd Noah, S.A., Hamzah, M.P.: Anaphora resolution of malay text: issues and proposed solution model. In: International Conference on Asian Language Processing (2010)

16. Ballabh, A., Jaiswal, U.C.: A study of machine translation methods and their challenges. Int. J. Adv. Res. Sci. Eng. **4**(1), 423–429 (2015)

17. Al-Saffar, A., Awang, S., Tao, H., Omar, N., Al-Saiagh, W., Al-Bared, M.: Malay sentiment analysis based on combined classification approaches and Senti-lexicon algorithm. PLoS ONE **13**(4), e0194852 (2018)

18. Yusof, N.N., Mohamed, A., Abdul-Rahman, S.: Reviewing classification approaches in sentiment analysis. In: Berry, M., Mohamed, A., Yap, B. (eds.) SCDS 2015. CCIS, vol. 545, pp. 43–53. Springer, Singapore (2015). https://doi.org/10.1007/978-981-287-936-3_5

19. Lo, S.L., Cambria, E., Chiong, R., Cornforth, D.: Multilingual sentiment analysis: from formal to informal and scarce resource languages. Artif. Intell. Rev. **48**(4), 499–527 (2017)

20. Foucart, A., Frenck-Mestre, C.: Natural language processing. In: The Cambridge Handbook of Second Language Acquisition, pp. 394–416. https://ieeexplore.ieee.org/document/6678407

Author Index

Printed in the United States
By Bookmasters